ANALYZING SYSTEMS

A [illegible] BOOK

ANALYZING SYSTEMS

JAMES A. KOWAL

AT&T Bell Laboratories, Inc.
Naperville, Illinois

North Central College
Naperville, Illinois

PRENTICE HALL, Englewood Cliffs, New Jersey 07632

Library of Congress Cataloging-in-Publication Data

KOWAL, JAMES A. (date)
Analyzing systems / James A. Kowal.
p. cm.
Bibliography
Includes index.
ISBN 0-13-033598-3
1. System analysis. I. Title.
T57.6.K69 1988
004.2'1—dc19 87-33415
CIP

Editorial/production supervision: Joan McCulley
Cover Design: Edsal Enterprises
Manufacturing buyer: Cindy Grant

Published by Prentice-Hall
A Division of Simon & Schuster
Englewood Cliffs, New Jersey 07632

This manuscript was typeset using the Microsoft's WORD™ Version 4.0 word processing software, running on an AT&T 6300 microcomputer with a Hewlett-Packard LaserJet II printer using HP® 33412AC and HP 33412AE Softfonts, under Microsoft's MS-DOS® 2.11 operating system.

Printed in the United States of America

10 9 8 7 6 5 4 3 2 1

ISBN 0-13-033598-3

Prentice-Hall International (UK) Limited, *London*
Prentice-Hall of Australia Pty. Limited, *Sydney*
Prentice-Hall of Canada Inc., *Toronto*
Prentice-Hall Hispanoamericana, S.A., *Mexico*
Prentice-Hall of India Private Limited, *New Delhi*
Prentice-Hall of Japan, Inc., *Tokyo*
Simon & Schuster Asia Pte. Ltd., *Singapore*
Editora Prentice-Hall do Brasil, Ltda., *Rio De Janeiro*

Adam and Lucille

Remember, of these parents you were born;
what can you give them for all they gave you?

Sirach 7,28

CONTENTS

PART I PHYSICAL SPECIFICATIONS

FOREWORD

Some time ago, I attended a conference on software development techniques. Many of the talks at the conference were far from lustrous in quality, and I noticed that several attendees were taking opportunities for a quiet snooze at the back of the room.

After lunch on the second day, I chose to attend yet another talk on structured analysis, the only alternative was "Dump Debugging Techniques for Large FORTRAN Programs," and I settled in for a postprandial nap. Hardly had I fallen asleep when I began to dream that I was listening to a speaker who was fresh, clear, interesting, and informative. I woke up with a start only to find that the speaker really was fresh, clear, interesting, and informative. That speaker was Jim Kowal, whose book you have in your hands right now.

But why, you ask, do we *need* another book on structured analysis, albeit one from a man who is fresh, clear, interesting and informative? After all, there are so many classics in this field, including Gane & Sarson, 1977; DeMarco, 1978; Weinberg, 1978; Flavin, 1981; McMenamin & Palmer, 1984; and Ward & Mellor, 1985.

Well, it's true that these books are classics, but they're classics that tend to emphasize some particular aspect of structured analysis. For instance, DeMarco, 1978, is about the basics of structured analysis; Flavin, 1981, is about information modeling; McMenamin & Palmer, 1984, is about the essential analysis model; and Ward & Mellor, 1985, draws many of its ideas from the so-called real-time world.

The beauty of Jim's book, on the other hand, is that it addresses *all* of the areas that a modern systems analyst needs to understand in order to specify a system. It also shows how these areas fit together. So read the structured analysis classics, by all means — especially if you wish to specialize in a particular area. But I believe that you'll gain a much broader view of structured analysis if you first read Jim's book before you tackle the others.

Meilir Page-Jones,
Wayland Systems, Seattle

PREFACE

In preparing a college curriculum in the field of computer science, the question often raised is: What are the most important things we could teach young people to best prepare them to be valuable, contributing professionals in business and industry four years from now? As part of my responsibilities at AT&T, I frequently interview and hire computer science majors as software professionals. I asked myself: What could we have taught these people four years ago that would have made them valuable to their employers today? Operating systems have been changing rapidly over the years from single-user systems to multiprogramming environments, virtual operating systems, virtual machines, UNIX®, etc. Traditional programming techniques and languages such as PL/1 and COBOL are losing popularity. Today, employers in business-applications development are looking for people with knowledge in fourth-generation nonprocedural languages and do not consider COBOL or PL/1 valuable language skills. Similarly, every 5 to 7 years, the industry has provided hardware with an order of magnitude improvement in capability, speed, and capacity. Management information systems were previously built using hierarchical data-base systems such as IBM's IMS. Today, such systems are being implemented using relational data-base management systems because of their ability to service unanticipated queries. So then, what should we teach that would make a person a valuable professional in the 1990s and beyond?

In the 1990s, as is true today, the greatest demand will be for people who are *problem solvers*. We need to teach young people how to solve problems found in business and manufacturing environments.

Traditionally, the systems analyst fulfills the prominent role of a problem solver in an organization. The systems analyst is responsible for determining what must be accomplished and how to carry on the business of the enterprise. When existing operations do not meet the expectations of management, an analyst is called on to define the source of the problem and to offer alternative solutions. To be a valuable asset to an organization, the systems analyst must define the problem and provide alternative solutions.

The first step in good problem solving is to be able to define the problem, hence structured analysis. With the methods of structured analysis, we can approach problems by immediately working to define what currently exists, how it works, and why there's a problem with the current way of doing things.

Data flow diagrams, data dictionaries, mini-specifications, E-R diagrams, input/output processing charts, event diagrams, event tables, state-transition diagrams and tables, and so forth are tools to begin the problem-solving activity. There are tools that deal with processes as well as data, content as well as structure, relationships among processes, and relationships among data elements. The tools provide ways of separating the *how it is done* from the *what is being done.* These tools also provide a means of modeling alternative solutions very economically. Models are most useful when they can be verified by the end user prior to implementation. User verification and acceptance are critical to the success of a system because it is the users who must carry out the new ideas if the problems plaguing the current system are to be overcome.

To understand the problem that the analyst faces, it is helpful to do a physical analysis. During the physical analysis, the analyst constructs a model that depicts the physical flow of data and the processes that transform the data into information. The model should be easy to understand in order to be verified by those working within the system. The model needs to be complete and detailed enough to identify problem areas and/or inherent limitations in the current way of doing things.

Processes within a system serve basically two types of functions: those that **administer to the system (infrastructure)** and are usually related to a specific type of technology, and those that **perform the basic data transformations to fulfill the needs of the business (essence).**

To understand which functions administer to the system and which carry out the necessary business functions, the analyst performs a logical analysis of the system, which contains three steps:

1. **Identify the processes and data** that serve the *system* from those processes and data that serve the *owner(s)* of the system. Processes serve the owners of a system when they carry out the basic functions and objectives of the system.
2. **Reorganize all data elements** by logically associating them to the main *objects* the data describes.
3. **Reorganize all processes** by logically associating them to the *events* to which the system responds.

Upon completion of the above three steps, the logical analysis of the system is concluded and a logical abstraction of the system can be created. A logical, stimulus/response view of a system is provided by an event diagram, which consists of all the essential processes and data organized around the events to which they respond.

By using a structured approach to system analysis, we can identify the parts of the existing system that do not meet the needs of the users. The analyst can then propose alternatives and model the necessary changes. The scope of change can be in three areas: the technology or administrative processes of the system, the basic business functions of the system, or in both. Using the event model, allows the scope and impact of any proposed change(s) to be easily assessed.

The purpose of this text is to bring together the tools and methods used successfully by analysts in modeling the physical, logical, and data aspects of a system. We can then propose changes to the model and review the proposed changes with the user to evaluate their effectiveness.

The following tools and methods are presented in the text:

- **Context diagram:** The highest-level diagram of a system that shows the boundaries of the system, all of the input and output data flows, and the origin and destination of the data.
- **Data flow diagram:** A graphic illustration that shows the data flow among the processes and data storage areas.
- **Data dictionary:** A directory that contains the definitions of the data elements and data storage areas.
- **Mini-specification:** The statement of policy that governs the transformations of the input data into the output data.
- **Pseudocode/structured English:** The detailed statements of logic that define how the input data is transformed into the output data.
- **Action diagram:** A method of illustrating the structure of the logic specified in a mini-spec.
- **Decision table:** A means of indicating the appropriate actions to be taken when various conditions exist.
- **Decision tree:** A graphic illustration of the appropriate actions to be taken when various criteria are met.
- **State-transition diagram:** A chart that illustrates the resulting change of state during the processing of a finite-state process.
- **State-transition table:** A table that lists the resulting change of state during the processing of a finite-state process.
- **Time-dependency chart:** A chart that illustrates the relationships among processes with respect to time.
- **Input/output processing chart:** A cross-reference of all the inputs, outputs, and data stores of the system and the processes that are associated with them.
- **Essential data flow diagram:** A data flow diagram that contains only the essential activities, data flows, and memories of the system.
- **Normalization and object analysis:** Processes that quantify and organize data logically according to the inherent structure of this data.
- **Entity-relationship (E-R) diagrams:** A graphic illustration of the relationships among the data used by the system.

- **Event diagrams:** A graphic illustration of the relationships among the data and the independent actions inside or outside the system that prompt the system to respond.
- **Event tables:** A list of all the input and output data and the events that cause activity within the system.

Finally, in Chapter 12, a basic system architecture is proposed that permits the analyst to take full advantage of both the process- and data-oriented methods of structured analysis. The advantages of the process-oriented methods are realized when successive decomposition and traditional structured specification tools are applied to capturing input data and transforming them into stored information. The advantages of the data-oriented methods are realized when the methods are applied to stored information, so that the users are able to apply fourth-generation languages and satisfy their own needs without any special development effort.

This text is intended for a college-level course in systems analysis. It provides the definitions and theory that permit one to create a model of a system solution that is complete, unambiguous, nonredundant, maintainable, and simple enough so that it can be verified by a nontechnical user community. The only prerequisite is basic knowledge of computer fundamentals, applications software, and the experience of at least one high-level programming language.

Technology will continue to change, providing even greater advances in hardware and software. Methods will improve, providing a means of more quickly defining and resolving the problems and not merely solving the symptoms. My efforts here present the tools and methods that have been used successfully to solve some of the problems in information systems and other software developments. These tools and methods are valuable assets in our struggle with the most vulnerable part of any system, the frailty of human communications.

J. A. Kowal

ABOUT THE AUTHOR

James A. Kowal obtained a Bachelor of Science degree in mathematics at Northern Illinois University in 1969 and a Master of Science degree in Computer Science/Industrial Engineering at Purdue University in 1976. He holds a Certified Data Processor (CDP) certificate from the Institute of Certification of Computer Professionals which he earned in 1977. He is a member of ACM and IEEE.

Mr. Kowal began his career with AT&T (formerly the Western Electric Company) at Bell Laboratories, Naperville, Illinois, in 1969 as an Information Systems Designer in the 1A ESS™ Switching System's Standards Engineering organization. As an Information Systems Staff Member, he was responsible for developing software that provided engineering and manufacturing information to AT&T factories manufacturing switching systems. In 1973, he became a systems programmer responsible for a corporate nationwide telecommunications word processing and text message system at the Warrenville Data Center, Warrenville, Illinois. In 1977, he was promoted to Department Chief of the On-line Telecommunications and Data Base Systems organization, where he was responsible for all on-line and data-base systems services for the AT&T locations in 36 states. In 1979, he was transferred to AT&T's Western Electric manufacturing plant in Lisle, Illinois, where as Department Chief of Information Systems Development he developed and supported business application software and managed data-center operations. Recently, as the Planning and Engineering Manager of the new AT&T Advanced Communications Package he directed the planning, engineering and development being done at the AT&T Network Software Center, Lisle, Illinois. Currently, he is responsible for the Centrex Applications Processor Systems Engineering group at the AT&T Bell Laboratories, Naperville, Illinois.

Mr. Kowal began teaching in 1969 at Moraine Community College and later at the College of Dupage. In 1978, he joined North Central College and taught in an MBA program in cooperation with the Illinois Institute of Technology, Chicago, Illinois. In 1980, he became a charter member of the Computer Science department at North Central College, where he is now an adjunct professor of Computer Science. He was instrumental in developing the undergraduate computer science curriculum and was involved in the development of a new program leading to a Master of Computer Science degree at North Central College, Naperville, Illinois.

Since 1970, Mr. Kowal has made his home in Naperville, Illinois, with his wife and three children. He is a musician and is actively involved in community, educational, and professional activities.

ACKNOWLEDGMENTS

Human beings do their best work by iteration and this book was no exception. I wish to thank all of the CSC 250 students and my colleagues at North Central College for their patience and tolerance in putting up with all of the early versions of the manuscript. Also, I'm grateful to the many individuals who provided suggestions, comments, and corrections to the manuscript. I'm especially grateful to my family, who gave up their husband and father for those many days, nights, and weekends so that I could complete this work.

PART I

Physical Specifications

This part presents the tools and techniques for preparing a structured specification of an existing system. The system is modeled in its entirety, including all manual and automated activities. While the model is being constructed, the users, that is, those who work within the system, must be able to verify the accuracy of the model. Thus, the notation must be complete and precise in order to accurately represent the system; yet, it must be simple enough so that nontechnical users can understand and verify the correctness of the model.

1

INTRODUCTION

. . . we will all become programmers, but programming will not be what it is today

There is an interesting parallel between the demands experienced by the data-processing industry for new information systems and the demands for telephone service that the telephone industry experienced several decades ago. In the 1920s, a survey was made to count the number of telephones in use and the number of telephone operators employed by the telephone companies. The survey was repeated in the 1930s and again in the 1940s. When the data were plotted and when some simple mathematical extrapolations were done, it was concluded that if the demand for telephones continued at the same rate, by the year 1980, the telephone companies would have to employ all the available workers in the United States as telephone operators. Well, 1980 came and went and we do not all work for the telephone company. However, if you think about it, every person using a telephone today is a telephone operator, as we defined telephone operators 40 years ago. Every person using a phone today has the ability to do the functions previously done by telephone operators. You can call anywhere in the United States and beyond by simply dialing the number directly. You can forward your telephone calls to another number. You can set up conference calls by using "three-way-calling." You can reach emergency services directly and have them know who you are and what your address is at the instant the phone is answered by dialing 911. You can even place calls and reverse the billing by dialing 1-800 numbers, or place credit-card calls without a telephone operator. Thus, every person using a telephone today is a telephone operator, as we thought about telephone operators 40 years ago.

In the 1950s, a survey attempted to quantify the number of requests for new information systems and the number of professional programmers that were employed in the country. The survey was repeated in the 1960s, and in the 1970s. It was concluded that if the current demand for new information systems continues, by the year 2000 every employable person will have to be a programmer.

Every person is not destined to become a professional programmer, at least, not as we think of programmers today. However, every person will be able to obtain information from central and/or distributed data banks by using simple inquiries, which today usually require a professional programmer. Systems are available today that show the technical feasibility of just such type of information retrieval. Query-By-Example (QBE)[Z-1] is an IBM product that exhibits just such capabilities. Persons who engineer information systems would do well by using the concepts of QBE to handle information retrieval and display.

METHODOLOGY FOR PRODUCING SOFTWARE
. . . software is more than just code

Today, software is more identifiable as a manufactured product. You can purchase software games at hardware, food, and toy stores. Furthermore, there is a better appreciation for accurate and complete documentation, in addition to good reliable code, in the *software* products we purchase. **SOFTWARE consists of the computer programs and associated documentation required to develop, operate, and maintain the programs.**[B-1]

Software products, like many other manufactured products, can be measured in terms of productivity, reliability, quality, and ease of use. The notion of engineering software as a product has gained recognition in the industry. To improve the manufacturing process of software products, *methodologies* of various kinds are being used.

A METHODOLOGY is a documented set of practices and procedures that defines the development life cycle and specifies how systems are to be developed. Companies can have a very formal methodology that contains formal publication of documents and walkthrough reviews for each deliverable. The development life cycle can also contain specific checkpoints to track the progress of the development. Companies have created development methodologies with various degrees of formality, ranging from the very formal to those that have no uniform way of proceeding and take each project as it comes.

There are some companies that modify their formal methodology with each new project. Once the methodology is agreed upon all members of the project adhere to it. Such companies are said to have evolved to an *adaptive methodology*. **An ADAPTIVE METHODOLOGY is one that chooses and/or modifies a development life cycle to fit the characteristics and needs of each particular project.**

A methodology establishes the way one does business in developing a new system or enhancing an existing system. A methodology introduces consistency among the many system development projects, provides a list of all of the activities that must be performed, provides checkpoints that enable management to monitor and control development projects, and provides a baseline set of activities that can be modified to improve the system development process. In short, a methodology is a system for building systems.

Some methodologies focus on the definition, composition, and organization of *processes*. **A PROCESS is a defined activity or task that can be accomplished completely in a finite period of time. It is used to transform input data flows into output data flows.** Developers use many types of tools and methods to structure the set of activities performed in developing a system. In this text, we will consider a process to be an activity that performs one or more tasks within a system. When a systems analyst prepares the requirements for a system, he or she identifies every process and indicates *what* must be accomplished, but avoids specifying *how* it is to be accomplished. When a systems specification is used to define an existing system, the systems analyst models processes that control the flow of data, control other processes, transform data, store data, update stored data, or provide data as well as perform a number of other tasks that produce the intended results.

There are other methodologies that focus on the definition, composition, organization, and relationships of the data and provide tools and methods to create data structures that can be easily accessed by the users of the system. Only after the data needs are understood, and the data structures and data bases are defined, do these methodologies address the processes used in a system.

DEVELOPMENT LIFE CYCLES
. . . a system to build systems

A SYSTEM is an association of interdependent devices, people, rules, and/or procedures organized to form an integral whole to achieve a common purpose.[W-4] In simple terms, a system is a set of processes and data that operate together to produce a result. Processes can refer to automated processes or human activities alike. To understand how we can meet the growing demands for new information systems, we must look at how we develop information systems today. The stages that projects go through while under development are known as the *project development life cycle*. More formally, the **PROJECT DEVELOPMENT LIFE CYCLE is a well-defined consistent set of activities that is used to translate the customer's requirements into a working system and contains checkpoints for management control.** In some companies, the project development life cycle is a very rigid formalized process that is mandated by management and is religiously followed by the technical staff. Other companies do not have any specified formal approach to the development of their systems. Still others use the *classical development life cycle*. **The CLASSICAL DEVELOPMENT LIFE CYCLE is a sequential progression from the survey of user requirements, analysis, design, code, module test, subsystem test, integration or final system test to documentation and installation** as shown in Fig. 1.1. The classical development life cycle uses minimal interaction with the users of the system. The users meet with the analyst in the beginning of the development project. During the survey and analysis stages, the users define their perception of the problem or

CLASSICAL DEVELOPMENT LIFE CYCLE

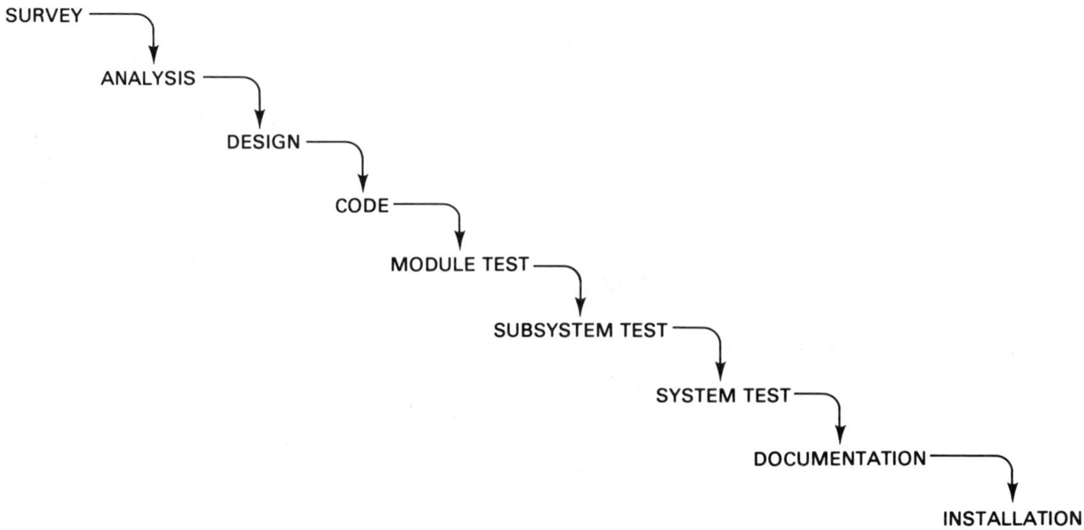

Figure 1.1

needs to the analyst. After the problem is defined, the analyst asks the users to sign the requirements document in order to "freeze" the requirements during the development stage. There is no need for any further formal interaction between the analyst and users until the new system is ready for the user to participate in acceptance testing or system demonstration prior to installation.

During project development, there are no models built, no structured techniques used, little or no opportunity exists for the users to change their minds about the requirements, and no walkthroughs or feedback loops to iterate the requirements, design, or implementation specifications. Most of the activities are performed in sequence. All documentation is usually produced near the end of the project. The classical development life cycle approach is often a *bottom-up implementation* that has been around since the 1960s and uses no modern structured techniques. **BOTTOM-UP IMPLEMENTATION produces the modules containing the lowest level of detailed logic first, tests them separately, and then integrates them to form the complete system.**

Using the simplified payroll system (SPS) example in Fig. 1.2, we will discuss how we can develop the system using the bottom-up approach. Suppose that the SPS system had a design that included the following structure for the system.

In the bottom-up approach, developers would begin by programming the following modules:

STANDARD WAGES, BONUS AMOUNTS, FEDERAL TAXES, STATE TAXES, CHARITY AMOUNTS, and the SAVING PLANS

When the tax programs are completed, development begins on the DETERMINE TAX DEDUCTIONS module of the system. Likewise, when the CHARITY AMOUNTS and SAVING PLANS modules are done, the work would begin on the DETERMINE MISCELLANEOUS DEDUCTIONS module of the system. When these routines are completed, the developers would break into subsystem teams and implement the following subsystems:

GROSS AMT SUBSYSTEM, DEDUCTION SUBSYSTEM, and OUTPUT SUBSYSTEM

SIMPLIFIED PAYROLL SYSTEM

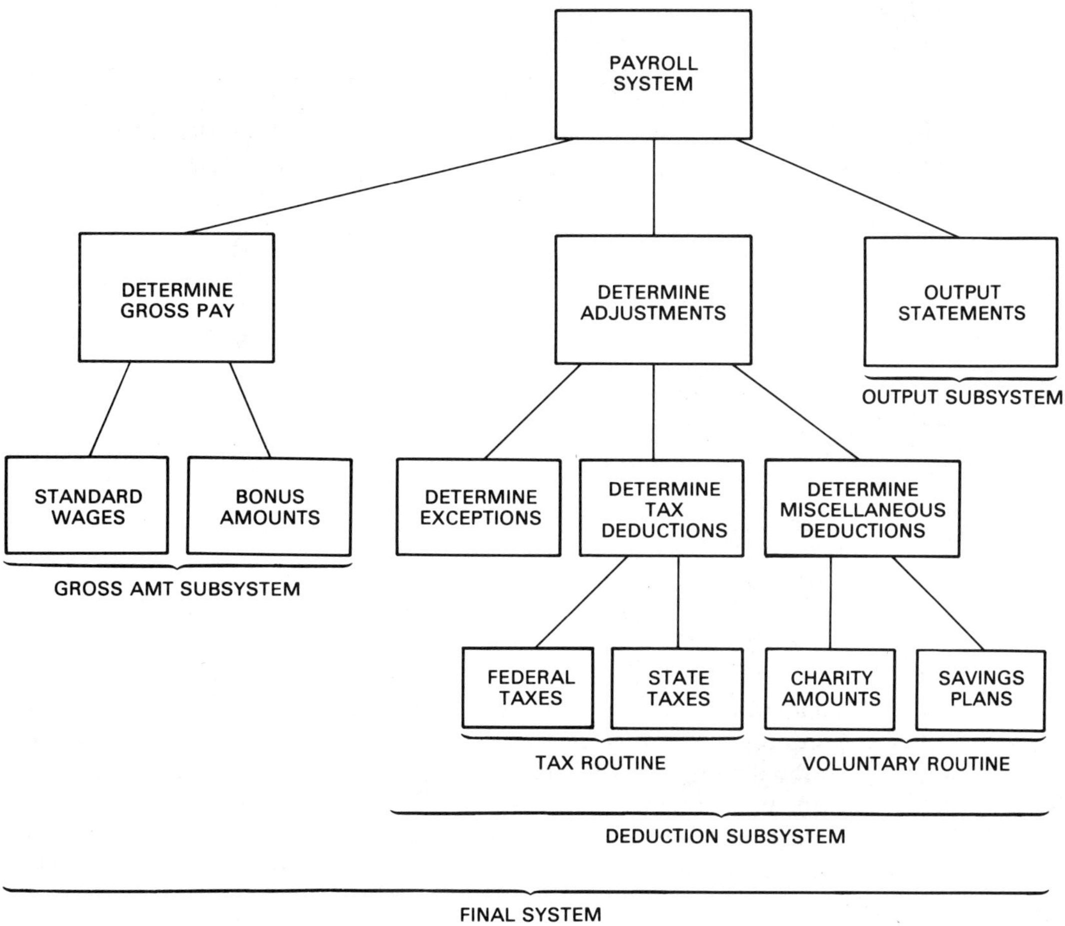

Figure 1.2

Notice that at this point in the development, there can be no demonstration of the system or any of its functions. Even if one could demonstrate a small portion of the system, there is no assurance that the subsystem would perform correctly with the rest of the system. Only after all of the subsystems are completed is there an attempt to link the subsystems and move data through the entire system structure. If the data does not flow through the entire structure successfully, a great deal of effort is required to debug the system and determine where the problem(s) exists.

Measurements indicating progress are only available in terms of percentages of modules completed or tested, amounts of code written compared to some estimate, or percentage of the estimated time expended (schedule).

The bottom-up method is contrasted with the *top-down* method. **In TOP-DOWN IMPLEMENTATION, higher-level modules are produced first. The higher-level modules call upon modules that are planned to perform tasks at a lower level of detail.** Applying the top-down approach to the SPS system example of Fig. 1.2 causes the development to proceed differently.

When programming begins, all of the lower-level modules are "stubbed off," that is, **a STUB MODULE is a program that contains simple code that responds with some fixed constant values each time the module is invoked.** The developers of the SPS system would quickly write all of the following midlevel programs:

MODULE NAMES	TASK PERFORMED
DETERMINE GROSS PAY	Return a $2,000 monthly salary for J. Smith every time the module is invoked.
DETERMINE ADJUSTMENTS	Apply a $420 deduction to all employees.
OUTPUT STATEMENTS	Develop a simplified program that lists the gross amount and deductions and prints a statement.

These modules are high-level stubs that return the same values each time they are invoked. The highest-level coordinating module (PAYROLL SYSTEM) is written to invoke the lower-level subsystems. Within a very few days, the development group can implement a simple version of the SPS system.

Admittedly, this version of the SPS system leaves a great deal to be desired in that it produces pay statements all of which are for J. Smith for $1,580 a month. However, the developers can demonstrate the overall design structure of the system. If there are problems with the structure of the system or interfaces between the modules, they can be fixed very easily at this stage. The only things the system lacks are the details, which in fact will consume the remaining time of the development schedule.

Notice that the system developers concentrate on implementing the highest-level modules and interfaces of the system structure first. The detailed modules that perform the actual data transformations that provide the intended results are implemented later. In this manner, the developers implement and test the overall system structure and critical interfaces first and produce fake output results. The next step would be to implement the lower-level modules in the same fashion using a few lines of code that usually do not make any decisions and do not perform any input or output to files.

In the initial testing of the system, no input/output takes place with any files or data bases. The programming is kept as simple as possible, even to the extent of producing very boring and monotonous output results. Stub modules would be developed to do the following processes:

MODULE NAMES	TASK PERFORMED
STANDARD WAGES	Return a constant value of $2,000 a month.
BONUS AMOUNTS	Return a constant value of $0 (no bonus).
DETERMINE EXCEPTIONS	Return a 1 exemption if single, and 3 if married.
FEDERAL TAXES	Apply a $300 monthly withholding.
STATE TAXES	Apply a 3% withholding.
CHARITY AMOUNTS	Donate $10 to the "Crusade of Mercy."
SAVINGS PLANS	Deduct a total of $0 for the savings plans.

As the programming is done, the stubs are replaced one by one in a top-down manner. Each time more detailed modules are added to the system, the system is tested and the results come closer to matching the desired output.

The developers test the same system design structure over and over again each time modules that contain more and more detailed logic are used to replace stub modules. As development proceeds, the results begin to take on a more realistic appearance. The payroll statement begins to list more and more deductions and the net amounts begin to vary.

In contrast with the bottom-up approach, the top-down implementation approach can always demonstrate a working system. If the inclusion of more details causes the system to fail, the trouble is easy to locate. Progress is seen by not only the development team but also the users. As the output results come closer and closer to the final requirements, all of the personnel involved in the system gain confidence that **real** progress is being made. Moreover, if one module could not be completed on time, the management and the owners of the system have two options. They could decide to implement the system "as it currently exists," and replace the stub module when the delayed module is available, or wait until all features of the system are completed.

STRUCTURED TECHNIQUES
. . . a move to improve software products

During the latter half of the 1960s, there developed a general interest in structured techniques that began at the NATO software engineering conference in 1968. The concept of structured programming was introduced in an article by E. Dijkstra,[D-3] and in another by C. Bohm and G. Jacopini[B-3] in 1966. In the Bohm and Jacopini article, the authors proved there were only three basic constructs needed to write the logic in any program:

1. Sequential flow of control (series of instructions)
2. Repetitive flow of control (loops)
3. Conditional flow of control (decisions with alternative processing)

If programmers limited their uses of programming logic to the three constructs outlined by Bohm and Jacopini, programs would be easier to understand and maintain. Early in the 1970s Ed Yourdon and Larry Constantine[Y-4] began teaching their ideas on structured design. In 1974, W. Stevens, G. Myers, and L. Constantine[S-4] proposed that system designers organize program functions in a hierarchical structure from which one can quantify two important design characteristics. The first is the amount of communications needed between the modules (coupling), and the second is how well the functions are factored into sets of closely related tasks and activities (cohesion). Later, D.T. Ross and K.E. Schoman, Jr.[R-1] published their article in 1977 and Tom DeMarco[D-2] published his book that provided the method of structured analysis in 1978. Structured analysis consists of modeling existing systems (automated or manual), as well as new proposed systems, by using successive decomposition. By following a well-defined documentation method, both the analyst and user would come up with a complete, accurate, unambiguous, and nonredundant set of documentation that specifies the requirements for the new system. Structured specifications use graphical aids, as well as a brief, well-organized English text known as "structured English" (or pseudocode) to specify the processing policies within the system. Toward the end of the 1970s, the focus was placed on the final aspects of system development when Glenford Myers[M-10] published his work. If done properly, structured testing is perhaps the most challenging job for the senior-level professional involved in the systems development life cycle.[1]

---

[1] The authors mentioned were not the sole persons who contributed to the development of structured methods. Their articles and books, however, are among the most widely acclaimed. Many of the articles can be found in two books edited by Edward Yourdon, (see Refs. Y-2 and Y-3).

At the end of the 1970s, there evolved a *structured development life cycle.* **The STRUCTURED DEVELOPMENT LIFE CYCLE consists of a sequential progression from survey, structured analysis, structured design, structured programming, structured testing, and installation,**[D-2] as shown in Fig. 1.3.

The structured development life cycle is characterized by the use of structured techniques, including structured walkthroughs. Several activities are done in parallel, producing documentation at each stage as an aid in the next stage of development. Formal walkthroughs are held often to detect errors at the earliest possible time during the development. The users' involvement is very high. They participate in many of the walkthroughs, where they confirm the requirements and gain an understanding of the final product. If it appears that the system will not fulfill the needs of the users, changes can be made.

Although the structured development life cycle schedules many activities in parallel, development time is still considerably long because most of the software is produced from scratch.

STRUCTURED DEVELOPMENT LIFE CYCLE

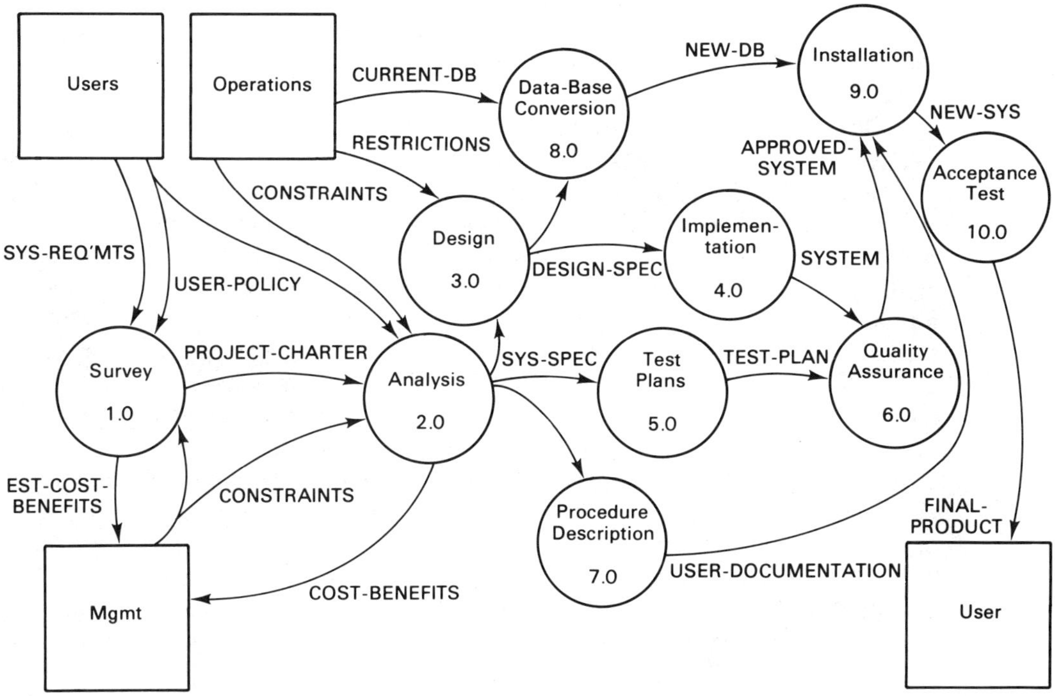

Figure 1.3

SOFTWARE ENGINEERING
. . . the time has come to engineer systems

During the 1970s, we experienced a relatively high inflation rate. The cost of developing data-processing systems rose rapidly. As more systems were developed, a greater portion of the total data-processing budget was being used for maintenance activities. As improvements in hardware were being made, the costs of program storage and computer processing were reduced. However, in the area of software development, a labor-intensive activity, the costs continued to rise. Contributing to the high cost of systems development was the fact that every information systems project consisted of new *research and development.* All component parts of the systems were created from scratch except for some sorting and computational subroutines. In the attempt to gain control of the manufacturing of software products, the idea of *software engineering* emerged. **SOFTWARE ENGINEERING is the practical application of scientific knowledge in the design and construction of computer programs and the associated documentation required to develop, operate, and maintain the programs.**[B-1]

Software engineering emphasizes programmer productivity, software quality and reliability, and overall cost effectiveness of producing software products. Software engineering represents a discipline that not only measures the cost and quality of the product, but all of the efforts involved in the *software engineering development life cycle.*[B-1]

The SOFTWARE ENGINEERING DEVELOPMENT LIFE CYCLE is a sequence of steps with certain identifiable deliverables that signal the end of one step and the beginning of another. The cycle is characterized by a process-oriented approach that focuses on the system being developed, emphasizing reusable code, peer reviews, checkpoints, and metrics for management control of productivity, quality, and cost (see Fig. 1.4). During each step, the deliverable produced is formally reviewed. If clarification or further work is needed, or missing components are identified, the activities in a previous step are repeated. The work continues until all conditions are met and the deliverable is approved by a review team. As many modules as possible are designed for general use. Some development organizations keep a library of reusable modules in order to minimize the development of new code. Relational data bases with fourth-generation languages to produce reports reduce the need of having to develop new software programs for each report the system produces.

The software engineering development life cycle also uses many of the structured techniques and particularly the formal walkthrough reviews. Reviews held at the end of each stage examine and approve the following deliverables: PROPOSAL, REQUIREMENTS, TEST PLAN, DESIGN SPECIFICATION, CODE, TEST PROCEDURES, final system SOFTWARE, and the USER'S GUIDE, as well as other user DOCUMENTATION (technical specifications, installation guides, training guides, etc.).

If problems are found in the conceptual, analysis, design, or implementation stages, the deliverable is sent back to the previous stage. In the system test, user test, or production stages, problems are recorded more formally and are diagnosed by the development group. Changes may need to be analyzed with respect to the design of the system. If a problem is noted in the final two stages, user test or production,

SOFTWARE ENGINEERING LIFE CYCLE

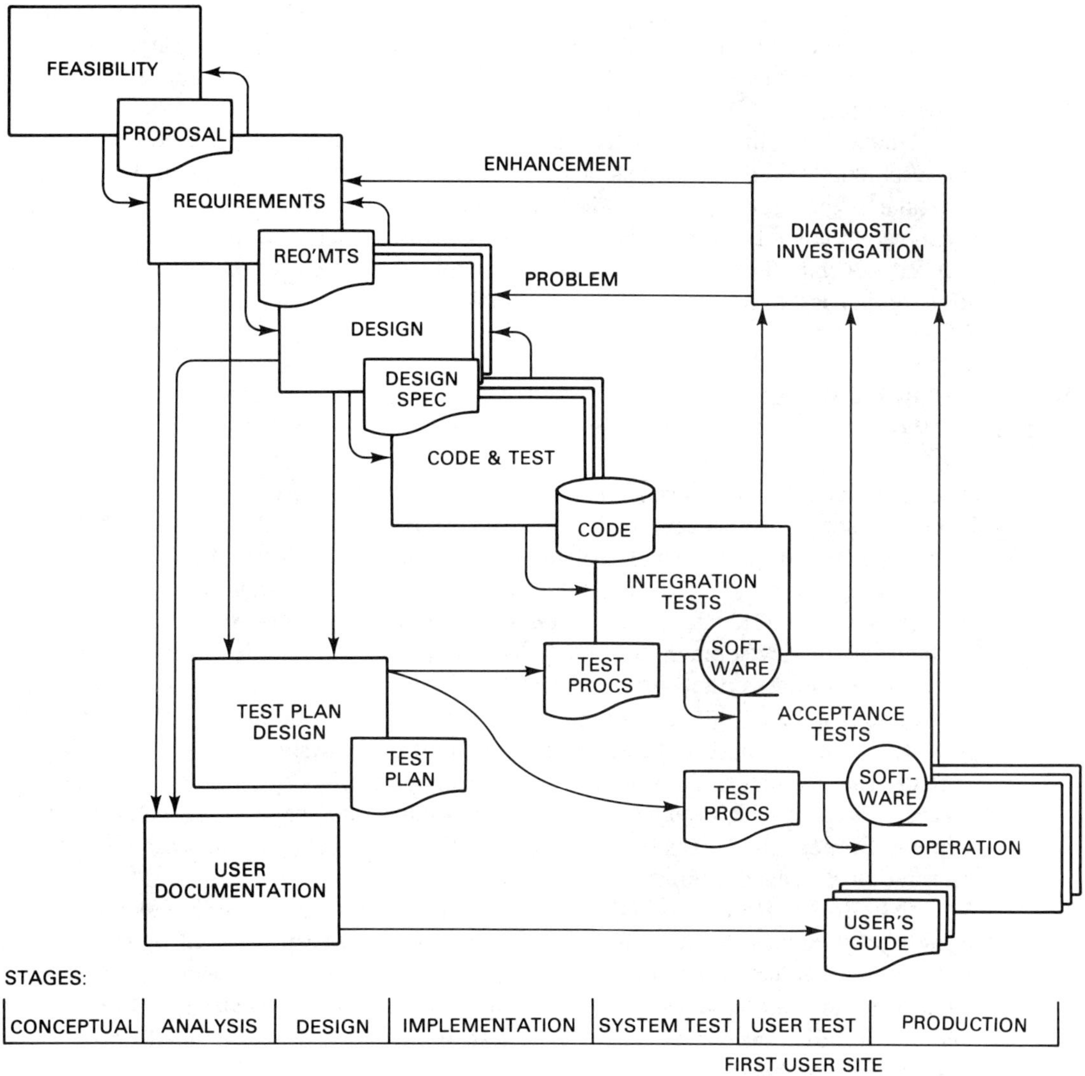

Figure 1.4

the problem is analyzed by the development team. One needs to determine if the system does not perform according to the requirements, or if a user is asking for a modification that was not specified in the original requirements. If the fix or change brings the system to operate according to the original requirements, there are no changes needed in the test plan. If the fix or change expands the capabilities of the system, or changes the user interface, the work is considered an enhancement whereby a new test plan may be necessary.

There is a great emphasis on software quality and the productivity of the system developers within the software engineering life cycle. Therefore, developers are encouraged to reuse existing code for greater productivity and higher quality. Existing code has already been tested and debugged thereby introducing few new errors into the new system.

Under the title of "software engineering" much is published on the measurements taken of various aspects of the software development life cycle, particularly the developer's productivity and the quality of software produced. Because one cannot control what one cannot measure, many software metrics have been introduced. The economics involved in software development, as well as many software metrics, have been included in a book by Barry Boehm.[B-2]

INFORMATION ENGINEERING
. . . data is more stable than the processes that use it

In an attempt to achieve even greater productivity and quality, other approaches to developing software have emerged. Based on the premise that the types of data used in an enterprise do not change very much, an approach published by James Martin and Clive Finkelstein[M-6A] introduced the concept of *information engineering* to the U.S. and Canada in 1981. Information engineering originated in Australia and New Zealand. It is a development approach that recognizes that data is much more stable than the procedures that process them. The fundamental principle underlying information engineering is that data exist and can be described independently of how they are used. Because data are at the heart of every system, the approach is to structure the data, store them in a convenient data bank, and provide an easy means to access the data. With the advances in nonprocedural languages, users can generate their own reports and display information without the need for programmers and information systems developers.

INFORMATION ENGINEERING is an integrated set of methodologies that organizes the data of an enterprise so that they can be easily accessed by end users (see Fig. 1.5). Information engineering entails (1) strategic requirements planning, (2) information analysis, (3) data modeling, (4) procedure formation, (5) data-use analysis, (6) distribution analysis, (7) physical data-base design, and (8) program specification synthesis.[M-6A]

Strategic requirements planning The objectives of the enterprise are studied to determine the kind of information necessary to make the business decisions and control the operations of the enterprise. Senior management's participation is required here to establish strategic direction and objectives of the organization.

Information analysis: The data to fulfill the requirements of the previous step are identified, as well as the data relationships to support the objectives of the enterprise.

Data modeling: A high-level, logical data-base model containing all of the data to meet the needs of the enterprise is constructed. The data-base model is independent of any particular data-base technology.

Procedure formation: The events that change or use the data are identified. Event diagrams are created and structured English can be used to describe the procedures used to transform the data.

Data-use analysis: From the event diagrams, a detailed map is produced that outlines how and where the data are obtained and used in each of the various processes. Quantities of data and frequency of access are also considered when producing the data structures during this step.

Distribution analysis: Performance requirements are considered and a data-distribution strategy is planned based on the use and volume of the data.

Physical data-base design: The data model is converted to a physical data-base structure using existing or planned facilities.

Program specifications: Processing requirements of each program are prepared to specify how each event, identified earlier in a previous step, is to be serviced.

Initial work begins with meetings of the top executives and management of the enterprise. The purpose of the meetings is to determine the type of information that is necessary to support the decision-making process in the company. Information needs vary throughout the organization and basically fall into the three categories of work: executive management, innovative management, and operations. Once the final products of the information system are understood, the analyst identifies the data elements and the relationships needed to provide the information. There are two phases of this analysis: data-discovery phase and data-consolidation phase. Given the combined set of data elements, relationships, and subject areas, an initial data-base model is constructed. This data model can serve as the framework for the corporate data base. In the data-modeling stage, a logical data-base model is produced that is intended to be the foundation for all future data processing. The data analyst (or engineer) combines the data from all functional areas of the company, normalizes the data, and attempts to produce a stable data-base structure.

In the fourth stage, the emphasis shifts from the data considerations to the procedures that create, use, or update the data. Studies are made of the business events and the types of access to the data that are necessary. The analyst sets forth procedures that are based on business needs and availability of data. Many traditional procedures, which may have been used *for forty years*, are subject to change under this analysis.

Up to this point, all of the analysis of data, their structure, and their access are done without considering technology. The physical capabilities of available technology enter the information engineering life cycle during the data-use analysis stage. During this stage, the practical accessibilities, capacities and capabilities of

INFORMATION ENGINEERING DEVELOPMENT LIFE CYCLE

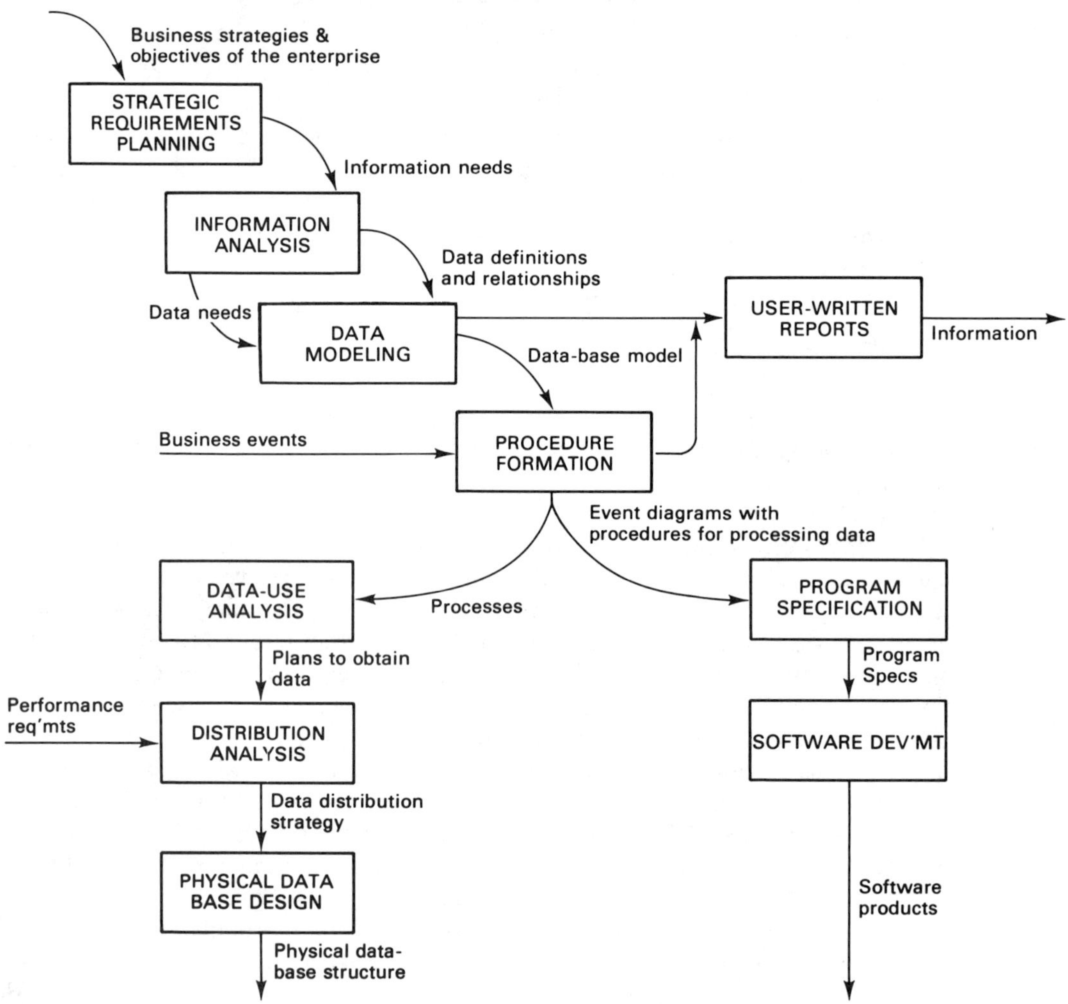

Figure 1.5

data-processing technology are factored into the system. Organization and placement of data are further examined during the distribution analysis stage. The advantages and disadvantages of small computers versus large main-frame processors, and consolidated data bases versus distributed data base are considered. Performance requirements enter the analysis and a plan of how the data base will be organized and implemented emerges.

With the above-stated analysis complete, work can begin on the physical data-base designs. Data-base integrity, operation, and performance need to be designed into the physical data base. Physical record layouts need to be defined together with linking/indexing strategies to implement the necessary data relationships.

From the business events identified earlier, the analyst defines the programs that will implement the tasks of creating, using, and updating the corporate data base. The analyst specifies detailed logic in structured English or pseudocode. Report generators using nonprocedural languages, which is covered in Chapter 12, are applied at this stage to minimize the amount of conventional programming needed. Information center, decision support systems, application prototyping, and user-written data-retrieval processes can be applied to avoid costly software developments.

As shown in Fig. 1.5, **the INFORMATION ENGINEERING DEVELOPMENT LIFE CYCLE is a sequence of steps originating with strategic planning that includes the top executives of the enterprise, and culminates in a set of user-driven and data-processing-driven methods to obtain the desired information to meet the needs of the business. The cycle is characterized by a data-oriented approach that focuses on the data needs of the enterprise across all functional areas, and emphasizes the use of data models, data normalization, and data structures that support user-written data-retrieval processes.**

To support the methodologies of information engineering, we need a notation to model data and data relationships. To support the methods that model data relationships and data structures, Chapter 7 presents the notation, developed by Peter Chen[C-3] known as entity-relationship (E-R) diagrams. A more complete discussion on information engineering is found in the James Martin books published by Prentice Hall.

SYSTEMS ANALYSIS

. . . graphical illustrations of processes and data structures are aids in understanding problems and expressing solutions

The methods and tools found in this book are aids in defining a problem, quantifying the processes and data involved, and *modeling* alternative solutions. **A MODEL is a mathematical (or logical) representation of a process, device, or concept.**[S-2] Actually, a model is an inexpensive replica that serves as a logical equivalent of the system under study. The model permits us to evaluate the characteristics of a proposed system or change prior to construction. Models permit an analyst to do what we call *systems analysis*. **SYSTEMS ANALYSIS is the process of examination,**

identification, and evaluation of the components and interrelationships involved in systems development efforts; the definition of problems, objectives, requirements, priorities, and constraints of systems, plus identification of cost, benefit, and schedule estimates for potential solutions.[W-4]

First of all, the methods, described in Part I will enable us to specify the existing system by means of graphics and detailed structured specifications. The notation, originally published by Tom DeMarco,[D-2] is kept purposely simple to illustrate the current processes, data, and relationships within a system in a manner that is verifiable by the *users*. Likewise, alternative solutions must be communicated effectively so that the users and their management are able to make intelligent decisions regarding the system.

A USER is a person, persons, or an organizational entity expected to derive benefits from the development of a system; a client of the system's development effort.[W-4] There are three types of users of systems:

1. **SYSTEM USER: The person(s) that interacts with the system.**
2. **RESPONSIBLE USER: The professional responsible for specifying the operating procedures of the system from a system user's perspective.**
3. **OWNER: The manager or person funding the development and continued operation of the system.**

The structured specification of a system, is a documented view of the system's requirements. The model should be verifiable by all three types of users. Each user needs a different level of understanding of the system. The owner usually only requires a high-level understanding of the system. The owner's emphasis is usually on the system boundary. A partitioning that breaks the system into major components and interfaces is usually sufficient to explain the system to the owner.

The responsible user needs to understand all of the details of the system down to the algorithms used to convert inputs into outputs. It is the responsible user that specifies the policy that transforms the inputs into the outputs.

The system user's interest lies only in operating a particular process, or set of processes, within the system. It is the system user that must work with the system on a regular basis, and its success is in his or her hands. Communications between the analyst and each of the three type of users is crucially important to the success of developing and implementing a system solution.

It was the importance of the communications between the system professionals and the users that have prompted many to select the data flow diagram notation published in Tom DeMarco's[D-2] book. The notation is simple and not intimidating to nontechnical persons. It uses a free format and a very limited number of symbols and notational designations. On the other hand, consider the diagram in Fig. 1.6. The discipline, formality, numerous symbols, and notational designations cause this graphical model of the data flow in a system to resemble an electronic schematic. I have found by experience that nontechnical users do not relate to system models

DATA FLOW DIAGRAM[G-1]

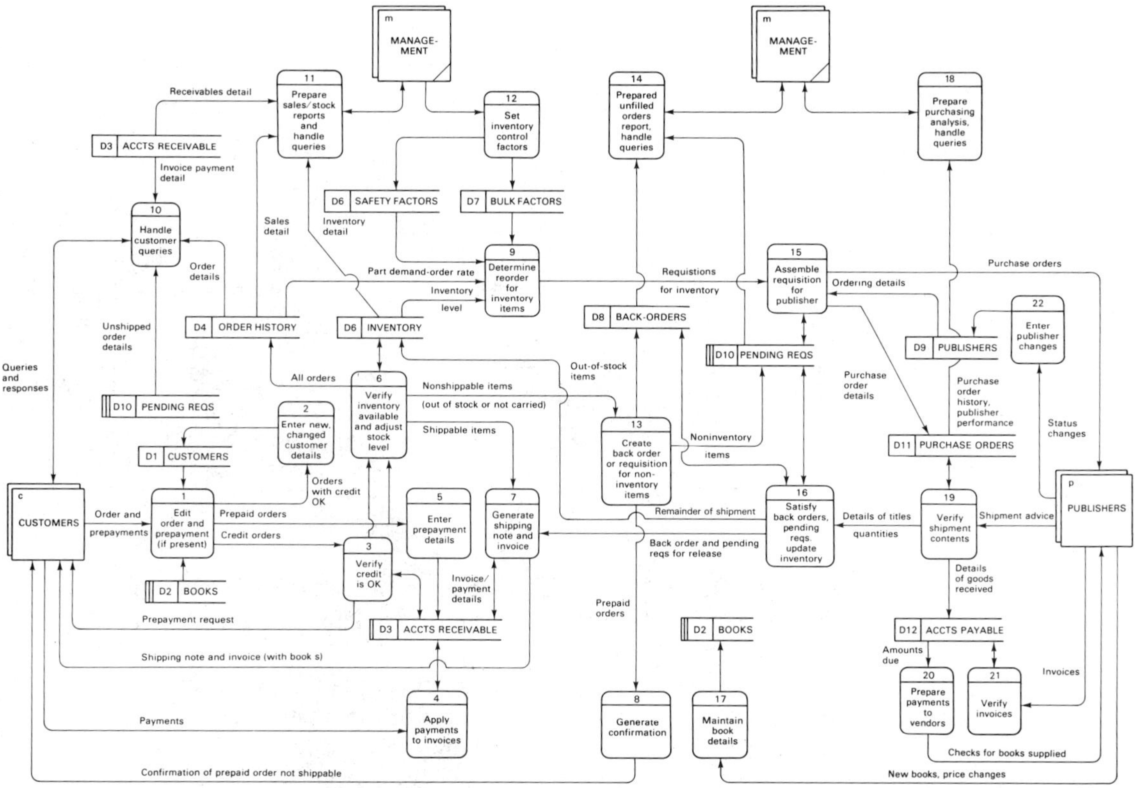

Figure 1.6

created with this type of notation and, as a result, communications between the users and the development team suffer. Fig. 1.7 illustrates the communications that take place during the system development.

Analysts have successfully communicated with technical personnel, namely, designers, programmers, and data-processing management (interfaces 4, 5, and 6 in Fig. 1.7), with such poor tools as flowcharts. Likewise, users and users' management communicate successfully with their own jargon and tools that are unique to their particular field, or profession (interface 1 in Fig. 1.7). However, the communications between analysts and users (interfaces 2 and 3 in Fig. 1.7) are the most susceptible to breakdown. Any notation used to illustrate system behavior must first of all not intimidate the user community. Furthermore, the users must participate in the modeling process and to do this effectively, they must use a mutually understandable notation. The model must also be complete and accurate and not general or vague. To best serve the iterative process of human problem solving, the model should be nonredundant so people can make changes easily. The notation published by

COMMUNICATION INTERFACES DURING SYSTEM DEVELOPMENT

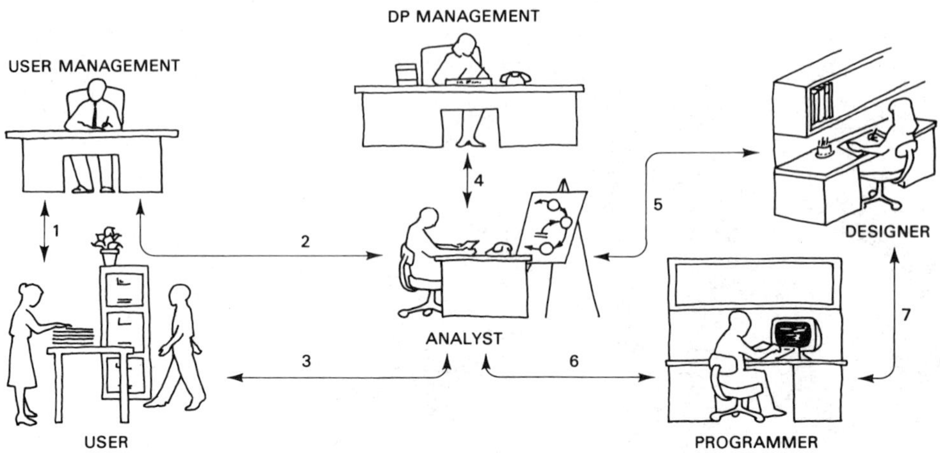

Figure 1.7

DeMarco provides a simple, nonintimidating, and easy-to-follow graphical representation of processes and data that is understandable by a great variety of groups, particularly nontechnical users.

The notation presented in Part I is so simple that the initial modeling efforts can and should be done **at the users' desks.** The model of the existing system is actually **the users' model.** It is the representation of how the processes and data are now organized. The model then serves as a reference point from which the analyst can illustrate new alternatives and proposed changes. Revisions and/or alternative suggestions made by the users are modeled in the same fashion. The users and their management can then commit to the *proposed new system*. You see, to model a system solution successfully, there must be a well-orchestrated effort between all three types of users and the systems analyst. The *analyst* provides the technical experience and knowledge and can recommend alternative processing methods. The *responsible user* knows the needs of the business and what must be done. The *system user* provides the practical perspective of performing the tasks and activities needed to produce the desired results. The *owners* provide constraints and certain limitations that must be considered. Together, they can specify the most appropriate system that satisfies the users' requirements and constraints in the most cost-effective manner.

The general steps involved in the analysis of a system are

1. Construct a physical model of the system.
   Construct a model of the current processes, data, and the associated relationships.

2. Verify the physical model.
   Verify the model with the users *for completeness and accuracy.*
3. Construct the logical abstraction of the physical model.
   The analyst reconstructs the model by separating the technology from the fundamental processes and data.
4. Organize the data around the basic *objects* of the system.
   Organize the data about the objects that the data elements describe.
5. Organize the processes around the *events* of the system.
   Organize the system's fundamental processes and data according to the events to which the system responds.
6. Construct the physical model of the new proposed system.
   Construct the new system using available technology to meet the users' needs and management's constraints.
7. Verify the new system and obtain the user's commitment.
   Verify the new system and obtain the approval of the users and/or their management.

Part II takes the physical model constructed with the methods from Part I and extracts the activities related to technology. The remaining elements form a clear picture of the essential data transformations that are the fundamental reasons why the system exists.

Data elements are the raw materials involved in the system environment. Once the data is properly organized and structured, place the data in an easily accessible data bank. By using user-oriented report-generation software that retrieves the information, users are able to obtain screen displays and formatted reports of the information they need. Users can then satisfy their changing needs and/or feed information into the decision-making process of the enterprise.

PROCESS-ORIENTED AND DATA-ORIENTED METHODS
. . . methods of the world unite

In Chapter 12, I will discuss an overall basic system architecture that combines the process-oriented and data-oriented methods presented throughout this text.

For now, let's just say that one can use the process-oriented techniques to model and engineer the input and processing portions of an information system. The data-oriented techniques can be applied to the data bank and output subsystems. **A DATA BANK refers to any collection of data, whether in the form of files, data-bases, or an information retrieval system.**[M-4] By integrating these methods, we can bring the power of software engineering to bear on modeling the users' procedures and processes and use the power of information engineering to bring the results to the users in a cost-effective manner.

The movement to use engineering principles in constructing data-processing systems has resulted in an approach to systems development whereby each new systems development project is not viewed as a new research and development undertaking. Greater emphasis is placed on the organization and management of the data, the productivity of the developers, the quality of the software, and the reliability of the systems. The engineering approach focuses on the entire system life cycle and strives to reduce the maintenance efforts by using better requirement definitions, designs, testing methods, and development techniques.

In managing the development of software as we would manage the engineering of a new product, there is a greater interest in reusable, component parts, quality control and assurance, and experimentation (prototyping) to improve the usability of the product.

The following analogy[S-3] can be drawn between the construction of a building and how software should be constructed:

WHEN A BUILDING IS BUILT	HOW SOFTWARE SHOULD BE BUILT
There is creativity in the architect's design	There is creativity in analysis and design
Models are built	Models are prepared
The later changes are made, the more they cost	The later changes are made, the more they cost
Several iterations are performed to find the final design	Several iterations are performed during analysis and design stages
Standards codes exist by which the design is measured to ensure desirable features	Standards exist to measure quality of the design
Design looks at several levels of detail	Design consists of various levels of detail
Design consists of several pieces	The model consists of several pieces
There are minimal interfaces between the pieces	There are minimal interfaces between the pieces
There is little creativity in construction	There is little creativity during the programming stage
Specialists handle the different aspects	Experts on staff handle specialized areas
Progress can be measured	Project management tracks and measures progress
Off-the-shelf components are used	Reusing existing software increases productivity
Tools and materials change	Hardware, software, and tools change

Figure 1.8

As with any product that is carefully engineered, the process to produce the product should foster creativity in the early stages (analysis and design), but constrain creativity during construction (coding). Tools to model the proposed system are necessary and should be easy to use and inexpensive. There must be standards to measure the proposed design. Human understanding and comprehension is best attained by partitioning the system. Thus, modeling methods should permit a nonredundant means of partitioning in such a way that levels of detail are shown and changes are easily made. When partitioning is done, intricate subsystems (i.e., data-base management systems) and interfaces (i.e., data communications) can be addressed by a specialist. The methodology should limit the amount of creativity in the construction phases and should permit the use of standard component parts and subsystems. Because of the rapid changes in tools and materials, maintenance efforts are reduced if the technologically related processes are separated from the processes that perform the primary objectives of the system. Finally, a means of overall measurement of progress is necessary not only to monitor the progress of the development, but to provide a means of assessing the consequences of changes in the methodology used to develop systems.

SUMMARY

. . . "you call yourselves 'engineers'? An engineer does not reinvent all of his components for each new job . . ." - Tomoo Matsubara, Hitachi

Among the productivity improvements that the information systems developers have to embrace is the concept of **reusable parts**. We no longer can afford to reinvent all of the functions and processes used in information systems. Custom software development is becoming too costly. There was an attitude in the 1960s that when a systems analyst proposed a change to the operating procedures in a company to minimize the software development costs, the users' reply was, "*Do we work for the computer or does the computer work for us?*" As a result, some analysts merely implement an automated version of the *same system* that was used to process the data manually. To successfully implement significant change, in the manner and means of processing data, clear and complete system specifications, developed and verified by the users, is absolutely essential.

As the software industry has been seeking ways to improve the development process, many methodologies have been developed. There are those that advocate methods that use the functional approach (software engineering). There are those that say one should analyze and structure data (information engineering). While this controversy has been going on, hardware technology has been providing machines with enormously low cost for memory, and fantastic processing speeds. Each generation of computers alters the approach to developing systems. The following chapters will present some of the methods and tools that have proven to be useful in solving problems for both the functional approach and the data approach. The intent is to provide methods and tools that separate the physical aspects of the problem from the logical, the technology from the essential business functions. If one separates the information-processing problem into data input, transformations, update, and retrieval, much of what we have learned from past experiences can be used to construct cost-effective, quality software products, and deliver them on schedule.

QUESTIONS

1. What is the project development life cycle?
2. What is the bottom-up method used in systems development?
3. What is the top-down method used in systems development?
4. What are stubs?
5. How does the structured development life cycle differ from the classical development life cycle?
6. What is the objective of using the engineering approach to develop software?
7. What are the steps that comprise information engineering?
8. What is the basic difference between software engineering and information engineering?
9. What is the purpose of using models in systems development and how is the cost of producing then justified?
10. What is the value of having reusable software components?
11. Who are the three types of users of a system?
12. What is a methodology?
13. Why do we have methodologies in systems development?
14. Discuss the type of methodology that would be most beneficial in each of the following cases:
    a. A software organization that develops software systems that perform process-control functions in a highly automated manufacturing environment
    b. An information systems organization that develops information systems for accounting and financial reporting purposes
    c. A contract programming enterprise that designs, develops, and implements a wide variety of software development systems.

PROBLEMS

1. Give an example of how one would use stubs in a top-down development.
2. How would the implementation proceed differently if one used bottom-up development in Problem 1?

2

DATA FLOW DIAGRAMS

We begin creating a structured specification by identifying the most basic elements involved in a system. First, there is the *source* of data to the system. The source lies outside the boundaries of the system. Similarly, we consider the destination or *sink* for the data produced by the system. The sink also resides outside the boundaries of the system. Data flows connect the sources and sinks to the system and designate the data elements that cross the boundaries into and out of the system. We can draw a high-level view of the system illustrating the sources and sinks in a *context diagram*. **A CONTEXT DIAGRAM is the highest-level data flow diagram that portrays all the net inputs and outputs of a system, but shows no decomposition.**[D-2] The context diagram is said to be at level zero, and places the system into perspective among the sources and destinations of the system's output.

CONTEXT DIAGRAM

Level Zero

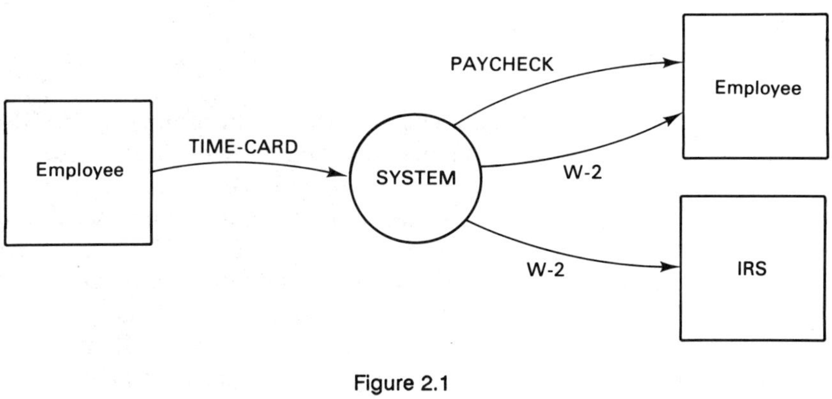

Figure 2.1

Figure 2.1 is an example of a context diagram that shows the system receiving the input data flow TIME-CARD from the source Employee, and provides output data flows PAYCHECK and W-2 to the sinks (i.e., destinations) Employee and the IRS. Because Employee and the IRS are outside the boundaries of the system, we use squares to distinguish them from the system, which is illustrated as a circle or bubble.

The purpose of this diagram is to delineate the domain of the system. The context diagram defines the source(s) and sink(s) of all data coming into and leaving the system. It also shows the data received and produced by the system. This diagram places the system into perspective by showing all the possible interfaces with the system.

Things to avoid in the *early* stages of system modeling are

- **Do not** attempt to specify the *type of technology* used to transmit the data to or from the system. Early in the analysis stage, it does not matter whether the data comes into the system by telecommunication terminals, unit record cards, telephone conversations, written documents, filled-in forms, etc.
- **Do not** attempt to specify *who* or *what* within the system receives data from a source or sends data to a sink.
- **Do not** concern yourself with the *format* or *quantity* of the data coming into or leaving the system.
- **Do not** attempt to show the *flow of control* that would prompt a response or activity from the system. (Flow of control will be included in real-time systems only; see Chapter 10.)
- **Do not** attempt to quantify the *performance factors* of the system, such as response time, capacity, etc.

In general, the analyst defers as much detail as possible in preparing the structured specification. Much of the detail of a system relates to the particular type of technology used in the system. Technology-related details are usually a design issue. During the analysis stage, only specify the system's *requirements.* Technological details are postponed until they are absolutely needed and are often added in the design stage.

Some systems carry on a *dialogue* with a user of the system. **DIALOGUE DATA FLOW is an interactive data flow that inputs and outputs data to a source/sink.** The time interval for the source/sink to respond to the dialogue data flow is not a factor. The source/sink's ability to respond in a timely manner during a dialogue is a function of the technology used. If a specific response time is a performance requirement of the system, it is specified later in the mini-specs (see Chapter 4). An interactive dialogue situation is modeled in the context diagram illustrated in Fig. 2.2.

After we have considered all of the aspects that are outside the boundaries of the system, we need to consider the **processes** internal to the system that transform

CONTEXT DIAGRAM

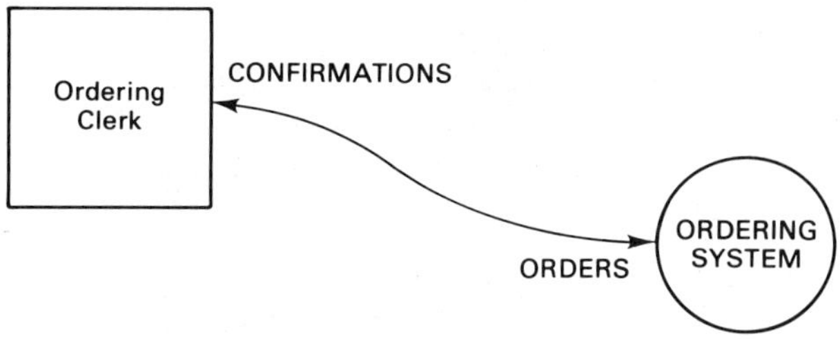

Figure 2.2

the input data flows into the output data flows. The processes can be performed by a person or machine. At this point in the analysis, the details of how the processes work are not of concern. What is of concern at the early stage of analysis are the data, where they come from, and where they go, and the processes needed to transform the data.

From the context diagram, the system is decomposed into several processes (usually, 7 ± 2). Each process becomes a major functional part of the system. Each functional part is then further decomposed into its component parts on separate data flow diagrams. The decomposition of a process continues successively until a process cannot be decomposed any further, or the policy governing the transformation of input data into output data can be written on a single sheet of paper. The written statement of policy is known as a mini-specification (mini-spec) and consists of detail logic that transforms input data into output data, or simply provides output data (see Chapter 4). The process that cannot be further decomposed is known as a *functional primitive*. **A FUNCTIONAL PRIMITIVE is a process within a system consisting of a related set of tasks that performs a basic function and cannot be further decomposed.** A mini-spec is provided for each functional primitive to state the policy that governs the transformation of input data into output data.

The third item of a system model is the **data** itself. Within a system, there are two forms of data, data at rest and data in motion. We use data flows to illustrate the data in motion. **DATA FLOWS are pipelines along which information of known composition passes to other components of the system.**[D-2]

There are two types of data flows: *elementary elements* and *group elements*. **An ELEMENTARY DATA ELEMENT is one that for the purposes of the system does not need to be further decomposed.** For example, a customer notification system can treat CITY-STATE-ZIP as a single data element. However, in a Postal Service System, CITY-STATE-ZIP may need to be a group data element consisting of two elementary data elements: CITY-STATE and ZIP. Using the data dictionary notation that will be presented in Chapter 3, we define an elementary data element with an explanatory comment:

ZIP = * 9-digit zip code used by the U.S. Postal Service.*

A GROUP DATA ELEMENT is one that is a combination of some related elementary or other group data elements. Examples are CUSTOMER-ORDER, STUDENT-TRANSCRIPT, SALES-INVOICE, etc., where CUSTOMER-ORDER consists of CUSTOMER-NAME, DATE-OF-ORDER, CUSTOMER-ADDRESS, etc. In data dictionary notation, a group data element is a series of elementary and/or group data elements:

```
CUSTOMER = CUSTOMER-NAME + CUSTOMER-ADDRESS + ACCT-NUMBER
           + CUSTOMER-PHONE + DATE-LAST-ORDER
```

Notice that the naming conventions dictate that the data names are descriptive. If more than one word is used, insert a hyphen between the words. The reason for hyphenating data element names is that a hyphenated name becomes one single entity that can be stored in a vendor-supplied software package known as a data dictionary. Software developers use that package during the design and programming stages. However, with many new automated tools becoming available in the microcomputer environment, one can benefit from the advantages of using such automated packages earlier, even in the analysis stage. In fact, there are many computer aided software engineering (CASE) products available that are designed to aid in the entire development life cycle, from system concept through implementation.

There are those who object to analysts entering data elements and data-store definitions into software packages that really benefit the designers and programmers. As a manager of software development, I have found that there are overriding benefits in using such tools during the development process, and I prefer to begin deriving those benefits as early as possible. To the software development manager, it does not matter whether the cost of entering the data is incurred during the analysis stage or during the design stage. The overall cost is approximately the same. Furthermore, there is an added benefit because the communication between the analyst and the designer is improved; and data items, no matter how insignificant, are unlikely to fall into the proverbial "crack." In addition, many vendors of CASE tools provide graphic and text facilities for creating data flow diagrams, a data dictionary, and mini-specifications, as well as other diagrams used during the analysis and design stages.

A data flow between components of a system indicates the pathway over which much data can flow. Usually, only one group data element, composed of many elementary data elements, passes between two processes in the system or between the system and a source or sink. There are only two occasions when two data flows are shown originating from the same component (source, sink, or process) and terminating at the exact same destination: first, if the data are totally unrelated, usually produced by distinctive and independent lower-level tasks or activities within the originating process. Or second, when there is a significant time difference associated with each of the two data flows, as seen in Fig. 2.1. (The system sends the payroll checks to the employee weekly, but sends the W-2 form to the employee annually.)

Data stores represent data at rest. Formally, a **DATA STORE is a time-delayed repository of data.** Not only can data stores be files and data bases of information, but they can also be memory storage, buffers, in-trays, drawers, stacks or anywhere data resides to allow for some passage of time to occur. Names of the data-stores should be descriptive of the data held and not the method or technique used to hold the data. If the data store name consists of more than one word, it is suggested that the words be connected with a hyphen, such as CUSTOMER-MASTER, ACCOUNTS-PAYABLE, EMERGENCY-NOTICES, etc. There is no need to number data stores. The definitions of data stores reside in a single, alphabetically arranged data dictionary together with data definitions and common processing routines.

The systems analyst studies the sources, processes, and destinations of the data as well as the data themselves. The analyst together with the users create a model that illustrates the flow of data through the current system. The notation used for creating the model is purposely simple in order to be meaningful to nontechnical users. The users' role in the process is to *provide information* as to the procedures used in the system, and to *verify that the model accurately represents the data and processes used in the existing system*. The analyst's role is to model the existing environment in order to have a frame of reference to propose changes or to propose a totally new means of processing data.

In preparing models of real-time systems, it is necessary to not only represent all aspects of data processing, but to model the control aspects of the system as well. Some systems contain processes that exert the power to activate and deactivate another process or regulate the flow of data or physical substances in a system. **The pipeline along which a command signal is passed to control another process in the system is known as a CONTROL FLOW.** Real-time systems and process-to-process relationships will be covered in Chapter 10.

The structured specification is a model of the system and must be complete, accurate, unambiguous, nonredundant. It must be easy to update, verifiable by the user community, and should use graphics whenever possible. The data flow diagram is one component of the structured specification that fulfills these requirements. With the advent of personal computers, there are many vendors that provide CASE tools for microcomputers to model systems using data flow diagrams and E-R diagrams. However, we will concentrate on the manual construction of the data flow diagrams because the first drafts of the diagrams should be done **at the users' desks.**

The initial task of the analyst is to create a context diagram to define the system's boundaries. The analyst then creates a high-level data flow diagram that illustrates the processes that perform the major functions of the system and the necessary data flows and data stores. After identifying the major activities, the analyst takes each process in turn and treats it as if it were a miniature system. This procedure continues as the analyst defines lower and lower levels of details using data flow diagrams until it is no longer necessary to decompose the processes.

NOTATION
. . . a picture is worth a thousand words

Before I describe how all of this is put together, let's present some data flow diagram notation. There are only five necessary symbols used in a data flow diagram:

1. A **SOURCE**, or **SINK** (or DESTINATION), of data is represented by a square. Each square represents a class of clients that originates input to the system or unique formats of output from the system.

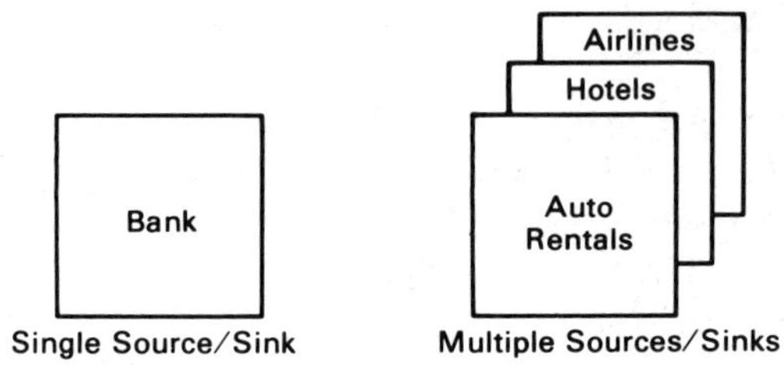

2. A **PROCESS** that transforms input data into output data is represented by a circle (also called a bubble). Each circle represents a single process. Multiple circles represent more than one process (or person) that perform the *exact same task* simultaneously. In real-time systems, when a process controls other processes but accepts and processes data flows in order to do so, it is shown as a solid bubble. Only in those rare cases when a process only performs a control function, dotted bubbles are used.

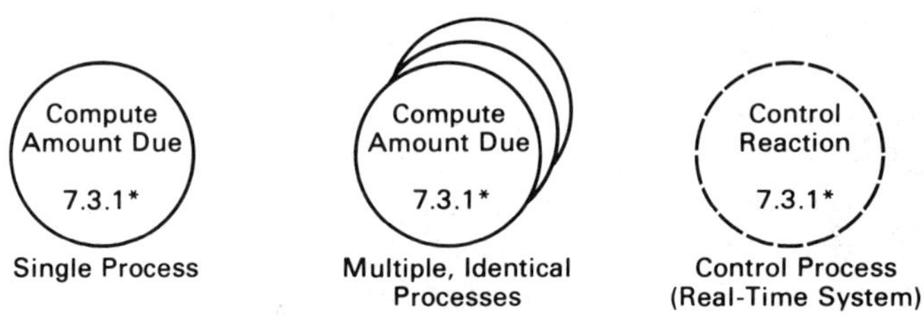

3. A **DATA STORE** and **CONTROL STORE** are represented by two parallel lines. A data store is data at rest within the system. Multiple data stores are those that contain the exact same kind of data for several identical processes that process data simultaneously. Control stores are used in real-time systems and are equivalent to *semaphores*, as defined by Dijkstra.[D-3] The parallel line symbol for a data store is shown only once on a data flow diagram. All other references to it are shown as data flows with the data store name.

Single Data Store | Multiple Data Stores (Containing Same Type of Data) | Control Semaphore (Real-Time System)

4. A **DATA FLOW**, which contains known data elements that are passed along to other components in the system at discrete moments in time or continuously over some interval of time, is represented by an arc with one or more arrowheads.

Time-Discrete Data Flow | Time-Continuous Data Flow

5. A **CONTROL FLOW** in real-time systems, which is a control signal that is sent from a controlling or supervisory process to a target process either at discrete moments in time or as a continuous flow to activate or deactivate the target process, is represented by a dash-line arc with an arrowhead.

Time-Discrete Control Flow (Real-Time System) | Time-Continuous Control Flow (Real-Time System)

Data flows and control flows can be either *discrete* or *continuous*. **A DISCRETE DATA FLOW is one that represents the movement of individual data elements at specific moments in time. It is assumed that at all other times, the data flow has no value or is undefined.** Examples are mail, phone messages, spoken instructions, reports, etc.

A CONTINUOUS DATA FLOW is one that represents a constant movement of data, whereby the flow of data exists at every instant of time over a specific time interval. A receiving process that receives a continuous data flow is usually found in real-time systems. The process receives and processes data continuously. Examples are temperature, meter, speedometer readings, etc.

A DISCRETE CONTROL FLOW is one that signals the target process at discrete moments in time. It is assumed that at all other times, the control flow is nonexistent. The signal is highly structured and is interpreted as an ON/OFF, START/STOP, ACTIVATE/DEACTIVATE, or ENABLE/DISABLE command. Examples are on/off switch, gear selector, etc.

A CONTINUOUS CONTROL FLOW is a command signal that is received continuously at every instant of time over a specific time interval and permits the

continual operation of a process. The signal can consist of relatively unstructured data, nevertheless, it has the power to permit the target process to continue to operate. Examples are electricity, pressure, wind, heat, magnetic force, etc.

Data can be passed while a process activates a target process. In such cases, the data name is shown on the control flow. Also, some control flows can only turn a target process "on" or "off." In such cases, it is helpful to label the control flow ON, OFF or ON/OFF. ON, OFF and ON/OFF are *self-defining terms* and are not defined anywhere in the structured specification. A process that has an input data flow and no control flow is assumed to always be "on," ready to process any incoming data flow when it arrives.

Continuous data and control flows are illustrated using a double-headed arrow. There are also times, such as during a dialogue, that data (or control) is transmitted back and forth between two processes. Such a data (or control) flow is a *bidirectional data (or control) flow*. **A BIDIRECTIONAL DATA (or CONTROL) FLOW is a pipeline that sends and receives data (or control instructions).** A bidirectional data flow (or control flow) is used to represent the communications channel over which a dialogue of data (or control instructions) takes place between two processes (or between a source/sink and a process) in the system.

Using these five basic graphic symbols, we can produce a model of a system that illustrates the sources, input data, processes, temporary and permanent storage, command signals, and the output results and their destinations. The following specific examples illustrate the basic components found in systems and the appropriate graphical representations illustrating the situations used in data flow diagrams.

DATA INTERFACES EXTERNAL TO THE SYSTEM

INPUT DATA to the system from some source outside the system:

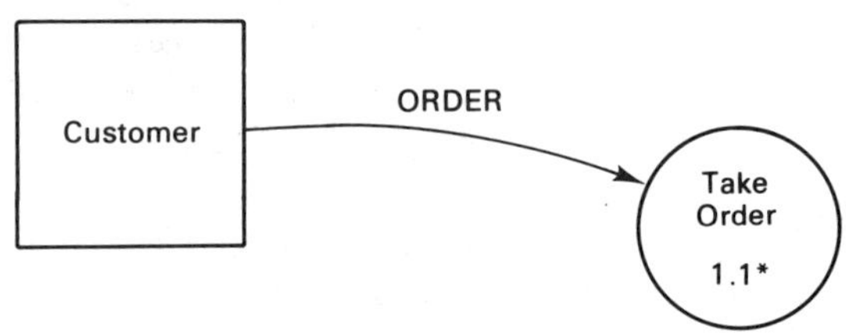

OUTPUT DATA to a sink outside the system:

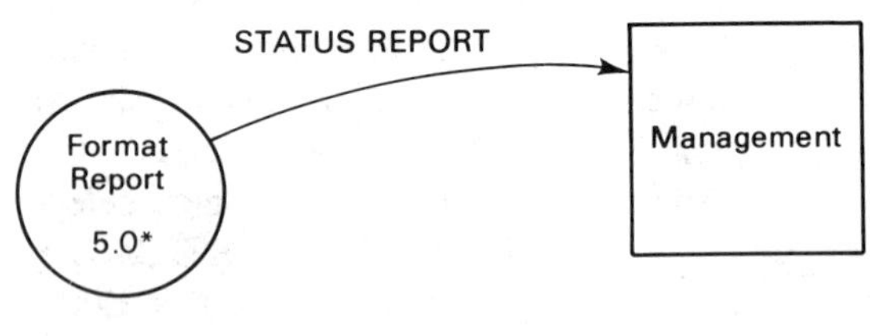

An **INTERACTIVE DIALOGUE** takes place between the system and a source/sink outside the system:

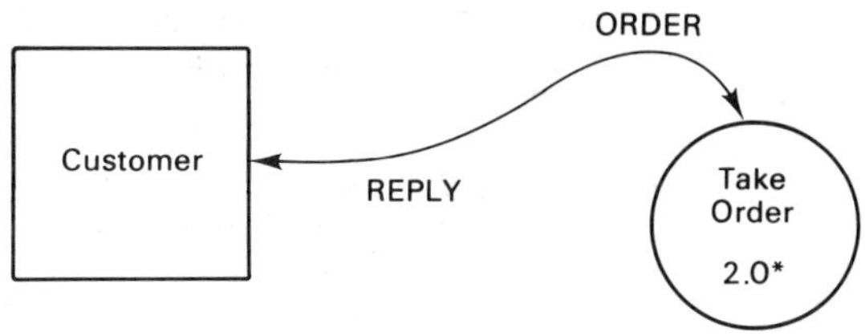

DATA INTERFACES INTERNAL TO THE SYSTEM

DISCRETE DATA FLOW to and from a process within the system:

Note: The names of the input data flow and output data flow are different.

An input data flow that is processed into another format or is certified, verified, edited, approved, or made different in any way is identified by a unique output data flow name.

CONTINUOUS DATA FLOW to and from a process within the system is shown with a double-headed arrow:

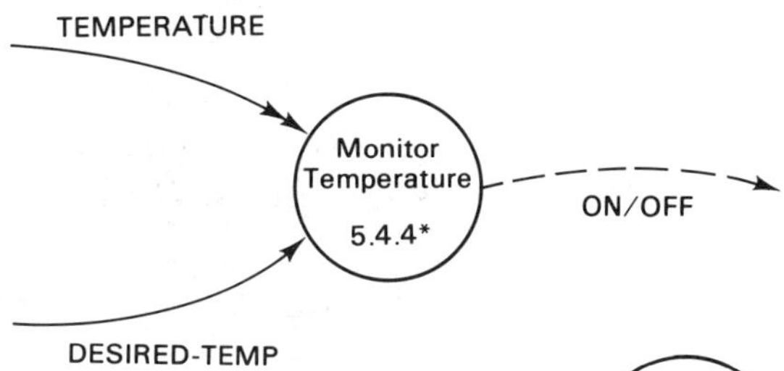

DIVERGENT DATA FLOW is a composite data flow where parts of the data flow are routed to different processes: (The data flow ORDER is a composite of PRODUCTS and SHIPPING-INFO.)

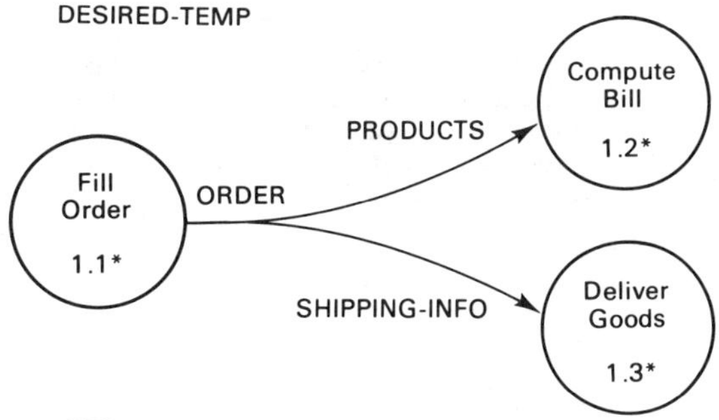

CONVERGENT DATA FLOW merges data from more than one process to form a single data flow that is routed to the same destination: (CUSTM-ORDER is a composite of CUSTM-INFO and MEAS'MTS.)

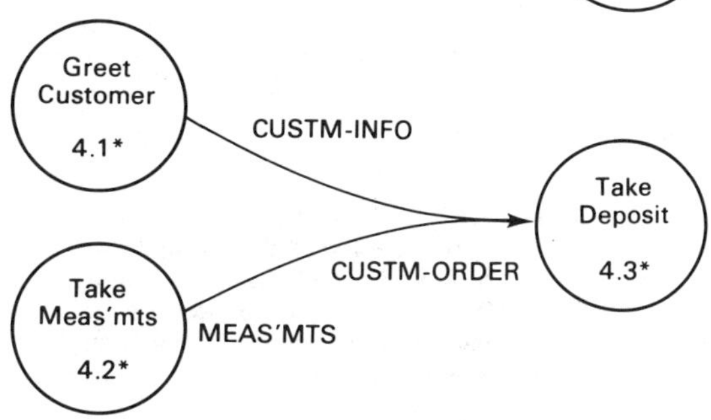

A **PHYSICAL ITEM (DATA FLOW)** is received and passed from one process to another within the system:

Note: The data flow name ICE-SKATES did not change because it is a physical item.

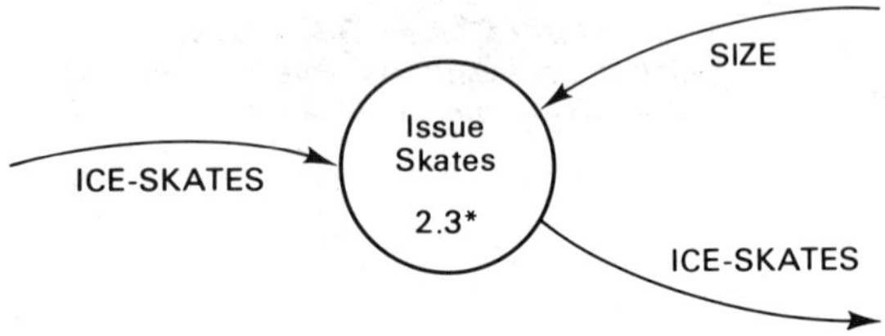

In the case where the input and output are in fact the same physical item and there is no value added to the physical item, the data flow name representing the physical item is the same for both input and output data flows.

DATA IS REJECTED by the system and no further processing is required:

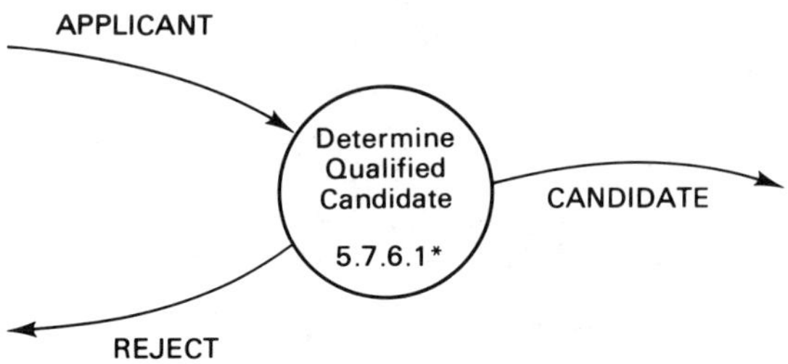

REJECT is a special data element that is not defined in the data dictionary and is used everywhere data are rejected from a functional primitive process (lowest-level process).

PROCESS INTERFACES INTERNAL TO THE SYSTEM

RECURSIVE PROCESS is one that receives initial data and then proceeds to feed itself data during subsequent executions after which some ultimate output is produced:

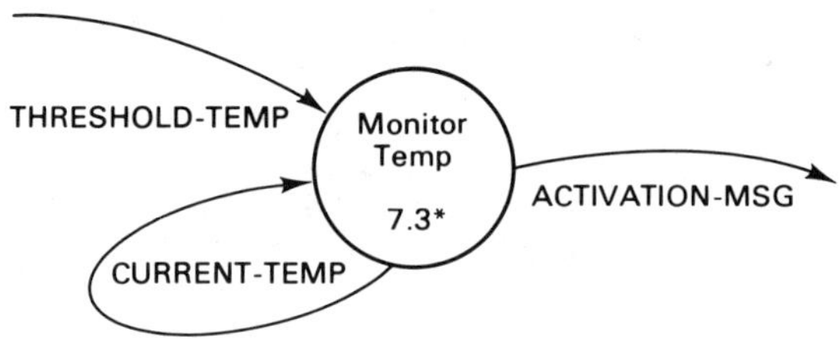

ITERATIVE PROCESS has no graphical symbol to distinguish it: *Iteration or looping of any kind is denoted in the mini-spec for a process (see Chapter 4).*

SIMULTANEOUS PROCESS is more than one occurrence of the **exact same** process active at the same time. Example: when 2 or more persons do the same job simultaneously, such as cashiers, etc.

TRANSACTION PROCESS moves data to and from only one of several similar, but distinct, alternative processes:

Note: This notation is used when there are multiple processes that have **unique** *mini-specs. In this case, the data flows do not touch the bubbles because they may apply to any one of the processes.*

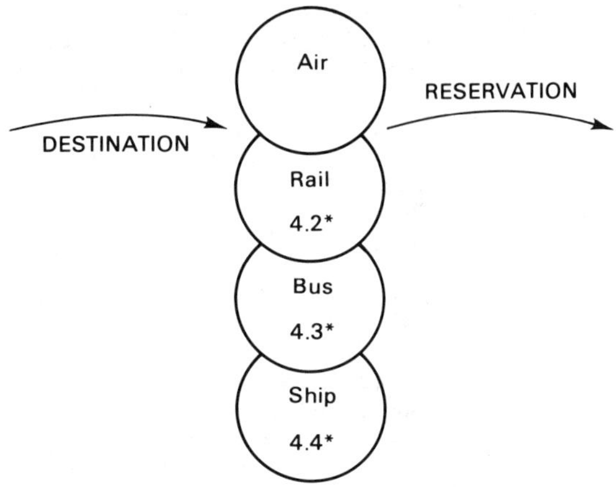

INTERFACES TO DATA STORES

A process that **WRITES DATA** to a data file or places data in a data store at individual moments in time for future processing:

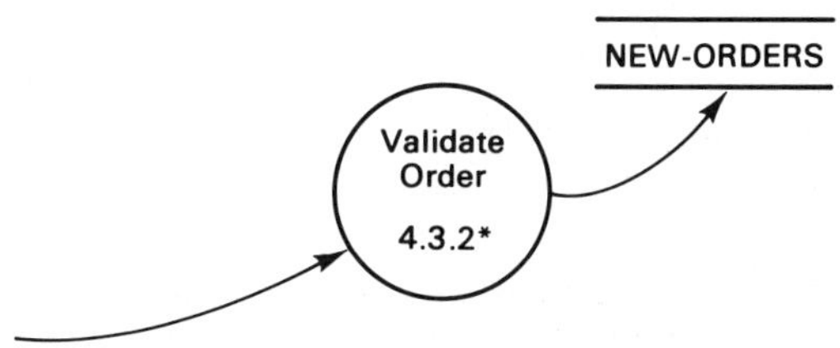

A process that **READS DATA** from a file or retrieves data from a data store:

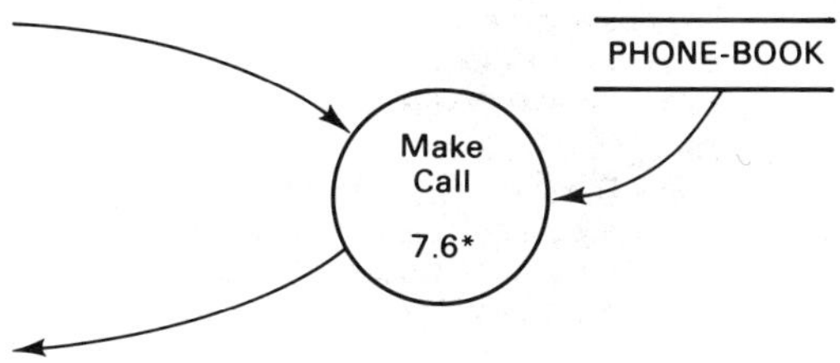

A process that **UPDATES DATA** stored in a file or designated place in the system:

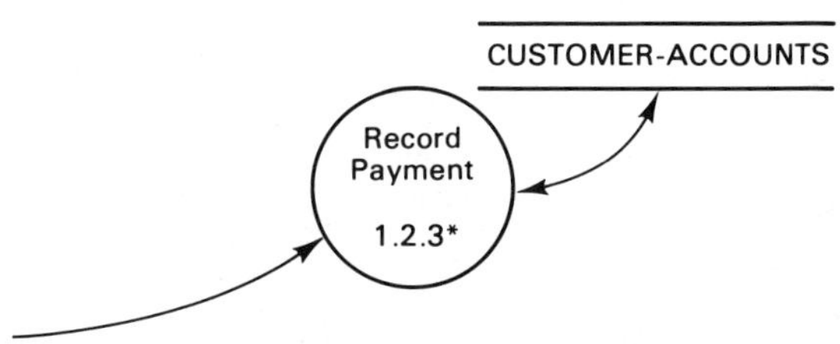

A process that **DELETES DATA** stored in a file or designated place in the system:

Note: This notation deals with the flow of data only.

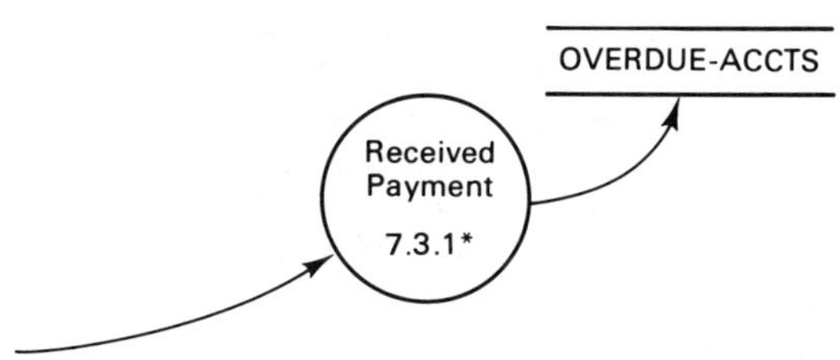

A process that **DELETES PHYSICAL ITEMS** stored in a data store or a designated place in the system:

Note: This notation deals with the movement of physical items and the arrow is shown leaving the data store.

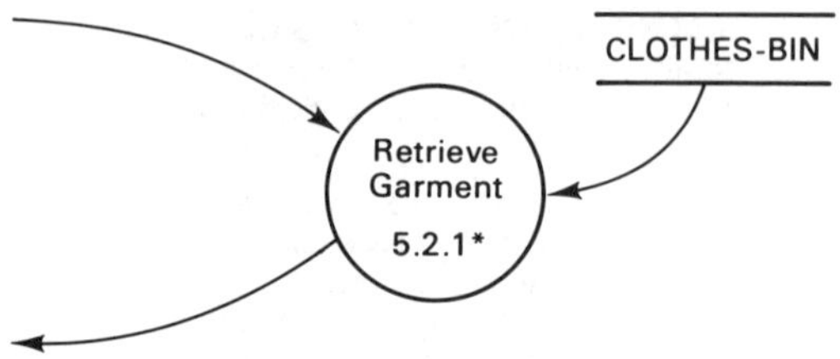

INTERFACES TO REAL-TIME PROCESSES IN THE SYSTEM

INITIATE PROCESS is an external supervisory process that initiates another process or several processes. When the initiated process completes its task, it ends and relinquishes all of the resources it uses:

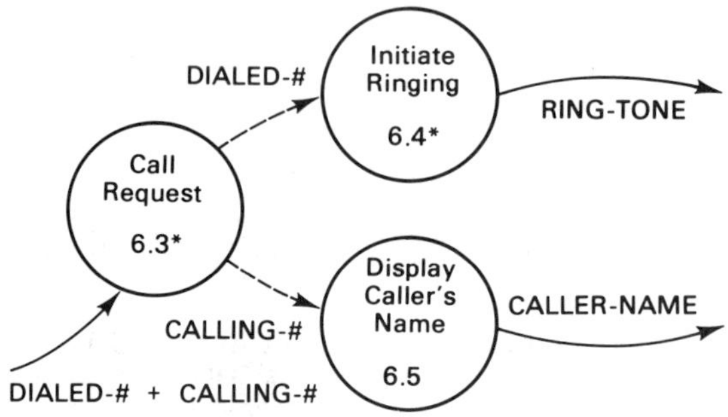

ACTIVATE/DEACTIVATE PROCESS is a supervisory process that starts or ends a target process. When the initiated process is activated, it continues to run until it is turned "OFF" by a supervisory process:

Note: The data flow ON/OFF is simply one that signals the start or end of the target process. ON/OFF does not appear in the data dictionary.

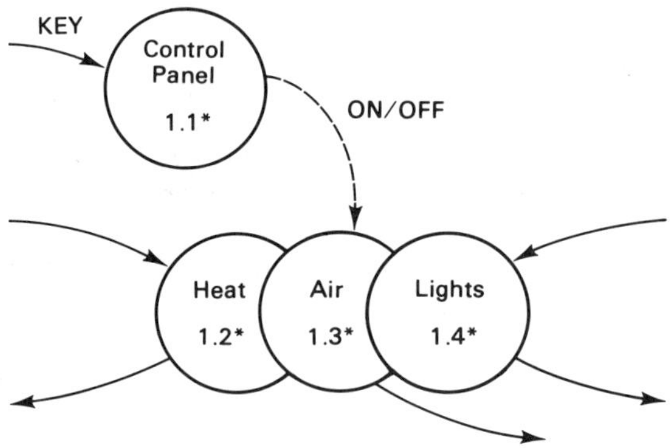

ACTIVATE/SUSPEND A PROCESS is a supervisory process that activates a target process waiting to run. When the process completes its task, it suspends its activity but retains all of the resources it uses. The process stands ready to be activated again at any time:

Note: The data store ON/OFF indicates that the target process is active or not. ON/OFF can be used as often as needed; it does not appear in the data dictionary. This type of data store is also known as a SEMAPHORE.[D-3]

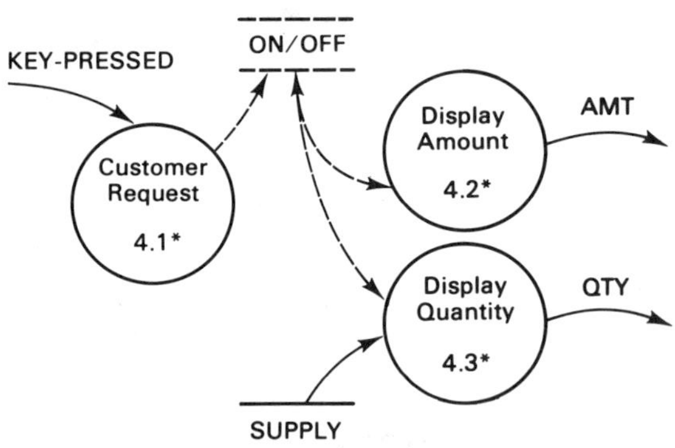

CONTROLLING THE RATE OF A DATA FLOW is a rate process that is given direction by a supervisory process on the rate at which it is to perform its task:

Note: RATE-NAME is determined by the supervisory process 2.0 and is defined in the data dictionary.*

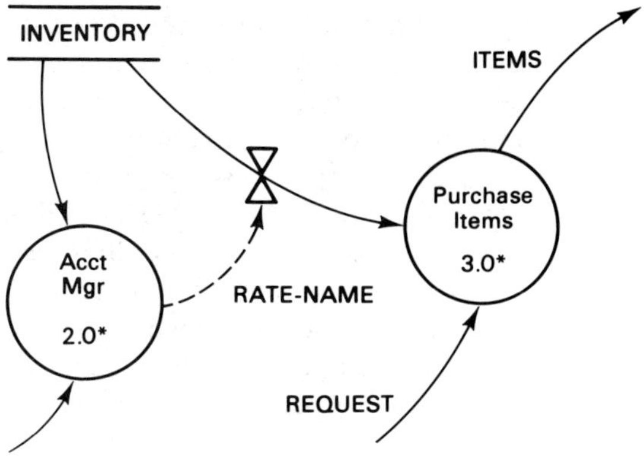

INPUT CONTINUOUS DATA FLOW from a file, and store the data or item in a data store. TAPE-SPEED is a rate variable defined in the data dictionary:

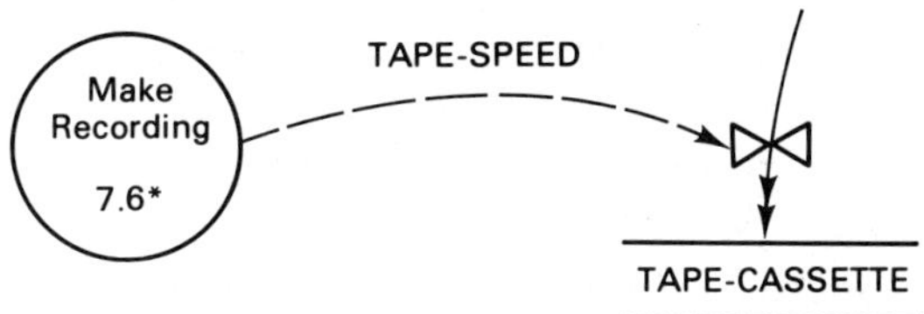

OUTPUT CONTINUOUS DATA FLOW from a data store by regulating the rate of flow. PUMP-RATE is a rate variable defined in the data dictionary:

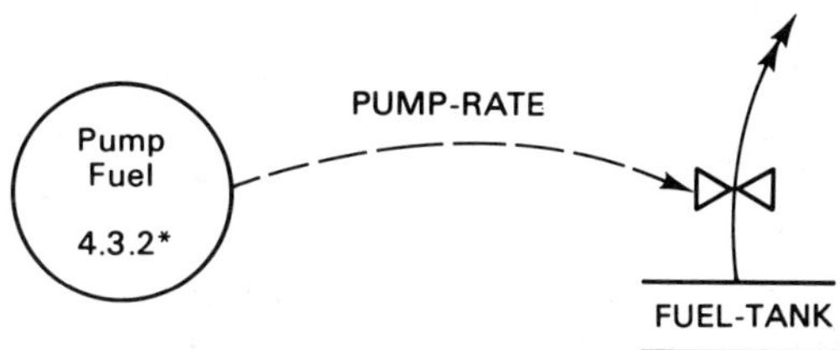

FINITE-STATE PROCESS is a process that operates differently with the same input data flow at different moments in time:

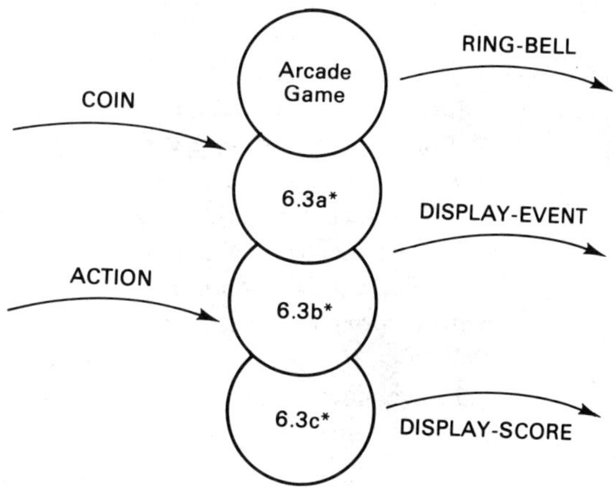

Although flow-of-control and finite-state processes are usually not necessary in most business or information systems specifications, it is important in real-time systems and will be used in Chapter 10. I include both real-time and nonreal-time examples here to be used as a complete reference. Real-time systems often have the need for one process (supervisory process) to control another process (target process). **A SUPERVISORY PROCESS is one that initiates, terminates, or otherwise controls the activities or operation of a target process. A TARGET PROCESS is a subordinate process that is activated, deactivated, or otherwise controlled by one or more independent supervisory processes.**

A target process can be under the control of more than one supervisory process. Supervisory process A can initiate the target process with an ON data flow, and control process B can terminate the target process with an OFF control flow. When there are multiple supervisory processes that are sending control flows to a target process, it is natural to assume that each input control flow is completely processed before the next one can be processed. In certain types of situations where a target process is started, stopped, and started again, the nature of the target process can be that the stop always preempts any other supervisory process. Furthermore, there can be a required "wait-time" before the process can be restarted. Such requirements are specified in the mini-spec of the target process usually by a time-dependency chart. See Chapter 10.

When two processes are linked by a direct data flow, there is a causal relationship between the two. They are required to be synchronized so that data can flow between them. If there is no requirement to synchronize the sending and receiving processes, the data flow will typically come to rest in a data store after the sending process releases it and before the receiving process accepts it.

All data flows shown coming to and from data stores are *not* labelled. When a process has access to a data store, it is assumed that it has access to all of the data elements in the data store. Likewise, when writing to a data store, the process is assumed to be able to update any or all data elements. The mini-spec provides the details as to what data elements are actually involved with a particular process. If certain data elements are required to be protected from general access, the security requirement is documented in the data dictionary. If there are changes to a data flow diagram such that additional data elements are needed from a data store, only the mini-spec has to be changed.

Only show data stores using the parallel-line symbol at the highest level where *two or more processes interact with the data store* or where a *functional primitive (lowest-level process requiring no further decomposition) interacts with the data store.* All lower-level references to the data store are simply shown as data flows labeled with the name of the data store.

Finally, what can be done to minimize overlapping data flows? Although we try to construct data flow diagrams with a minimal number of data flows crossing each other, some diagrams are plagued with congestion. To improve the readability, one or more of the following notational techniques are used to reduce congestion. Use connectors sparingly to avoid constructing data flow diagrams that become logic puzzles.

CONNECTORS

REPEAT DATA STORES near the processes that interface with them to avoid crossing data flow.

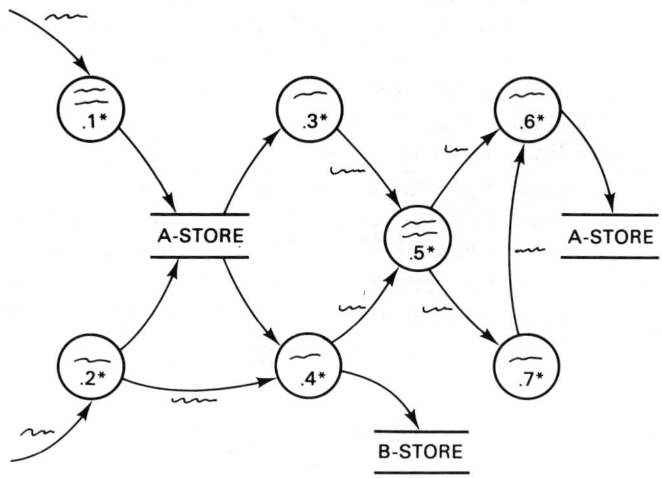

CONNECTORS are used to handle discontinuity in data flow diagrams, but only on the same diagram. No "off-page" connectors are needed.

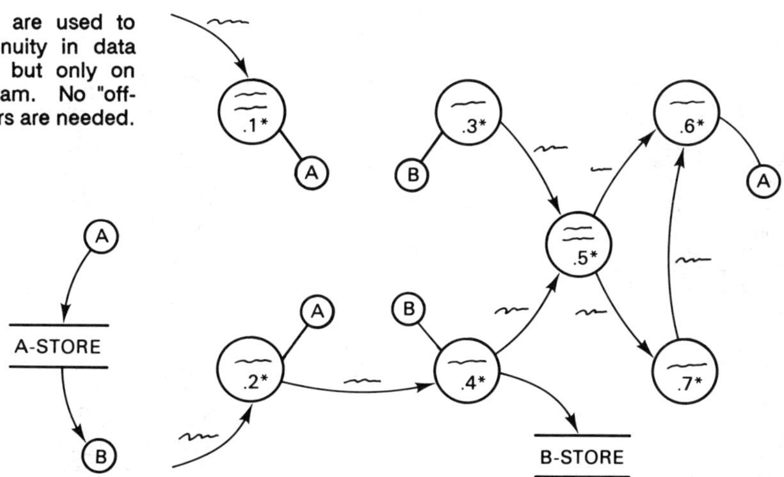

CREATING DATA FLOW DIAGRAMS
. . . documenting the flow of data

Now that we have a notation to use, let's apply it to a particular business situation. We want to create a data flow diagram model of a traditional catalog sales business.

The company takes orders for merchandise and ships the merchandise to the customer.

Let's consider the catalog sales operation shown in Fig. 2.3. Notice that a customer enters the store, chooses some merchandise from the store catalog, takes a blank ORDER-FORM from the table, fills out the form, and presents the ORDER to the order clerk. The order clerk verifies: (1) that the ORDER was properly filled out, (2) that the customer included the zip code in the ship-to address, (3) that the customer name and address, quantity, product number are listed, and (4) that the customer listed the correct price for the product. The order clerk computes the total amount due, adds all taxes, and determines the grand total. The order clerk accepts PAYMENT from the customer, places the PAYMENT in the CASH-DRAWER, dates the ORDER, approves the order by initialling the order, and places the order in the stack of APPROVED-ORDERS in the open window for the data clerk.

The stack of orders in the open window is organized by date and time written. The clerk takes the next APPROVED-ORDER off the top of the stack and enters the order into the company's computer system. The computer system performs four functions with the new order.

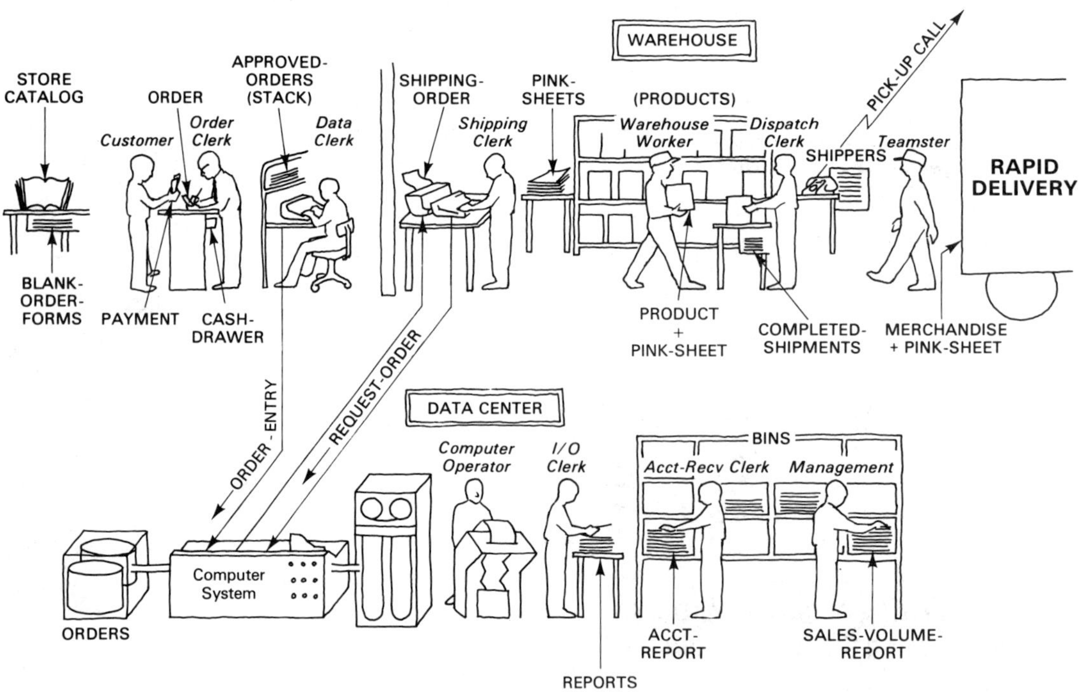

Figure 2.3

1. The system sends a SHIPPING-ORDER containing the customer's name, ship-to address, product number, and quantity to the warehouse, and a copy of the order prints out upon receipt of the shipping clerk's REQUEST-ORDER.
2. The system saves information about the order in the ORDERS data store by date of sale and customer's name.
3. The system provides information in the form of a report (ACCT-REPORT) to the accounts receivable clerk (ACCT-RECV Clerk). The ACCT-REPORT lists the following information for all sales: date of sale, order clerk's initials, total payment for merchandise, appropriate taxes, and the grand total amount collected from the customer.
4. The system produces a SALES-VOLUME-REPORT for management that indicates the current dollar volume of sales by week compared to the dollar volume of sales for the previous 12 weeks.

The shipping clerk obtains the SHIPPING-ORDER printed on a multipart pink form (PINK-SHEET). The warehouse worker picks up the PINK-SHEET and goes into the WAREHOUSE to locate and select the ordered product (PRODUCT). The warehouse worker brings the ordered product to the dispatch clerk together with the PINK-SHEET.

The dispatch clerk has four responsibilities:

1. Package the MERCHANDISE for shipment, which includes the PRODUCTs and one copy of the PINK-SHEET.
2. Telephone the delivery service that delivers to the ship-to address. A list on the wall (SHIPPERS) contains information on all of the freight delivery services (names, phone numbers, and the cities and states that they service) sorted by cities and states.
3. Give the teamster the MERCHANDISE and one copy of the PINK-SHEET.
4. Place one part of the PINK-SHEET in the COMPLETED-SHIPMENTS data store.

Figure 2.4 illustrates the context diagram for the ABC Catalog Sales Company. Now let's factor the ABC CATALOG SALES COMPANY into the major processes, as described in the statement of the system. From the illustration of the activities in the company in Fig. 2.4, we can initially assign one major process to each worker in the system. The first process is the sales clerk (1.0). The sales clerk receives the ORDER from the customer. Notice that nowhere in the structured specification of the system is there any reference to the store's catalog and the blank order forms that are available for the customer to use. In this case, both of these items are outside the boundaries of the system. They are merely a means of permitting the customer to

ABC CATALOG SALES COMPANY
CONTEXT DIAGRAM

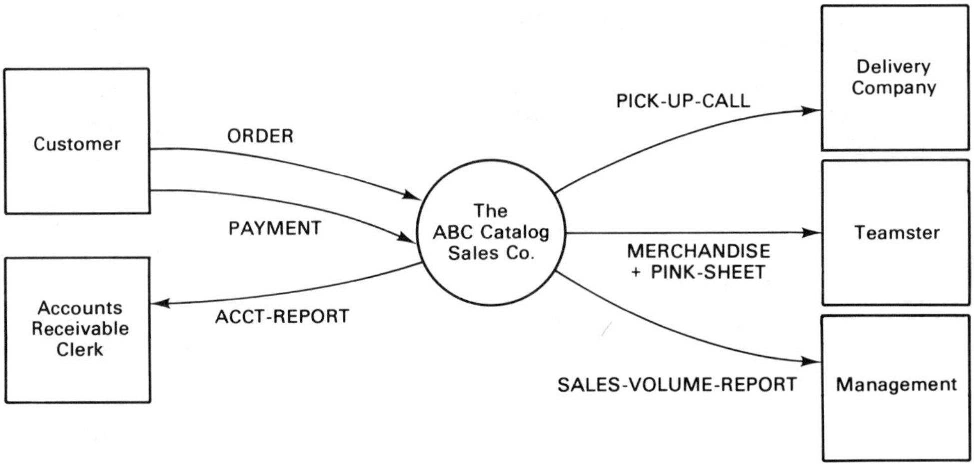

Figure 2.4

place an order, and as such they are not modeled as part of the system specifications. The user or owner of the system can choose to include such items within the boundaries of the system; however in this example we will consider them outside the system.

The customer provides PAYMENT to the order clerk, who places it in the data store CASH-DRAWER. The order clerk approves the ORDER and places it on the APPROVED-ORDERS stack for the data clerk. The approved order resides in the stack (APPROVED-ORDERS) until the data clerk has time to enter them into the system. Therefore, APPROVED-ORDERS is considered a data store.

The data clerk selects an APPROVED-ORDER and enters the data into the company's computer system. As explained earlier, the computer system provides three output data flows: ACCT-REPORT, SALES-VOLUME-REPORT, and an interactive dialogue that provides the shipping clerk a SHIPPING-ORDER upon receiving a REQUEST-ORDER. The system also maintains a data store of all orders called ORDERS. The system uses ORDERS to generate the ACCT-REPORT and SALES-VOLUME-REPORT.

The shipping clerk obtains a computer printout of the SHIPPING-ORDER on a PINK-SHEET and gives it to the warehouse worker, who must select all of the PRODUCTs from the WAREHOUSE and bring them to the dispatch clerk. Because the WAREHOUSE is a collection of products that are waiting to be selected, we model the WAREHOUSE as a data store. The products in the WAREHOUSE each have an identification number (PRODUCT-NUMBER). The WAREHOUSE is arranged by PRODUCT-NUMBER. The warehouse worker selects and brings the PRODUCTs to the dispatch clerk together with the multipart PINK-SHEET.

ABC CATALOG SALES COMPANY
DATA FLOW DIAGRAM
(Level One)

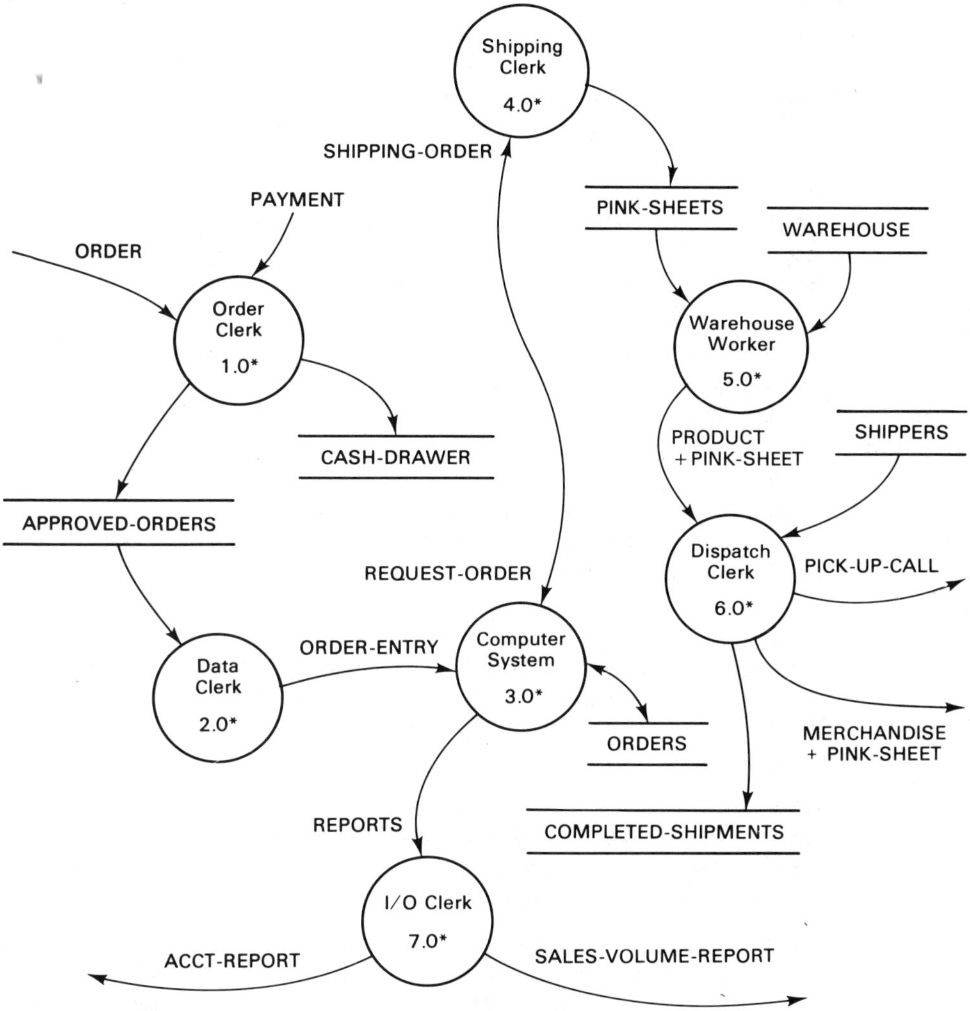

Figure 2.5

The dispatch clerk has three responsibilities as explained earlier. The dispatch clerk needs two things from the warehouse worker: the PRODUCTs and the PINK-SHEET. The dispatch clerk packages the PRODUCTs with a copy of the PINK-SHEET to form the data flow labeled MERCHANDISE. From the PINK-SHEET, the dispatch clerk learns the destination of the MERCHANDISE. The dispatch clerk selects an appropriate freight company that delivers to the destination, locates the freight company's phone number on the SHIPPERS list, and calls the freight carrier for a pickup.

When the teamster arrives to pick up the MERCHANDISE, the dispatch clerk gives one copy of the PINK-SHEET to the teamster with the MERCHANDISE. The dispatch clerk files the third copy of the PINK-SHEET in a filing drawer under COMPLETED-SHIPMENTS.

At some appointed time, the computer system prints the ACCT-REPORT and the SALES-VOLUME-REPORT. The I/O clerk in the data center takes the reports, separates them, and places each in the appropriate bin. Management picks up its SALES-VOLUME-REPORT and the accounts receivable clerk (ACCT-RECV Clerk) picks up his or her ACCT-REPORT in the data center bin area. Notice that the bin area is the final destination for the data produced by the system. The bins serve as a means of delivery and therefore we do not model them in the structured specification as a data store or sink.

In Fig. 2.5, you will notice there is no data flow shown between the sinks delivery company and teamster. However, in actuality, the person answering the PICK-UP-CALL would send information to the driver of the truck, instructing him or her where to go and what to pick up. Such communication originates and terminates beyond the boundaries of the system; therefore, do not show any communication between sources/sinks in a structured specification that does not pass through the system.

The process of creating a data flow diagram of an existing system is similar to solving a jig-saw puzzle, where:

1. Find all of the pieces that have a flat edge and form the boundary of the puzzle. Likewise, **a data flow diagram begins by defining the boundary of the system**. The source and input data received by the system and the destination (sink) and the output results provided by the system define the boundaries.
2. Gather together pieces of a particular color or pieces that have some commonality. Likewise, **the process, data flow, and data store that service a particular input data flow, or provide a specific output data flow, are associated together.**
3. Interconnect small completed sections to complete larger segments of the puzzle. Likewise, **common data flows link the small networks of processes, data flows, and data stores to form larger networks.**
4. Once you complete the higher-level data flow diagram, **consider each process as if it is a system in its own right and the process continues** but at a much lower level of detail.

The above procedure continues until all of the functional primitives are accounted for in the data flow diagrams. The detailed logic for each functional primitive is found in the mini-specs, or in the data dictionary for common processing routines.

Data Flows of Physical Items

What do we do when a process inputs and outputs a physical item such as shoes, soda, tapes, etc? First, we must represent the physical item by its equivalent data elements used by the system, for example:

SHOES = SIZE + COLOR + STYLE + MODEL-# + PRICE + MFG-NAME
+ . . . (*any other attributes used by the system that can be observed from the physical item itself*)

Second, we use the same data name for all input data flows and the output data flows that refer to the same physical item.

When the system that sells SHOES is modeled, process 2.1.3* in Fig. 2.6 must deal with the physical item SHOES. The user who deals with hundreds of pairs of SHOES each day can easily verify the model in Fig. 2.6. On the other hand, the analyst fulfills the need to quantify the item SHOES into its equivalent data elements. Later, the analyst studies both the processes and the data elements during the logical analysis stage (see Chapter 6) to determine if these components qualify as essential data and essential activities of the system. Physical items usually qualify as essential data because they ultimately move across the boundary of the system as inputs and outputs.

Dealing with physical data items does differ from the usual way systems handle data elements. The primary difference between data elements and physical data items is that the latter exists in a *one and only one copy* form. Consider the case when a process retrieves nonphysical data items from a data store or passes data on as an output data flow; only a copy of the data moves in each case. The data in fact remain

PHYSICAL ITEM DATA FLOW

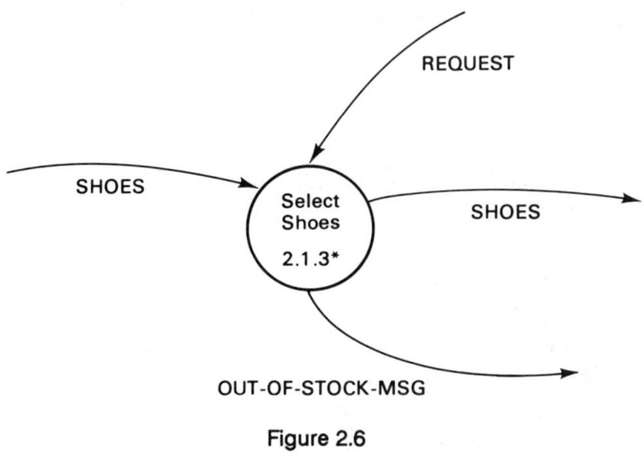

Figure 2.6

in the data store or within the sending process. When a process retrieves an actual physical item from a data store, the item is no longer in the storage area. Thus, a retrieval (or read) of a physical item from a storage area (data store) is equivalent to a *read and delete*, according to the usual manner of dealing with nonphysical data items.

In illustrating the removal of physical data items from a data store, show the data flow as leaving the data store. This is in contrast to deleting nonphysical data items from a data store, where the direction of the data flow is toward the data store as shown in Fig. 2.7.

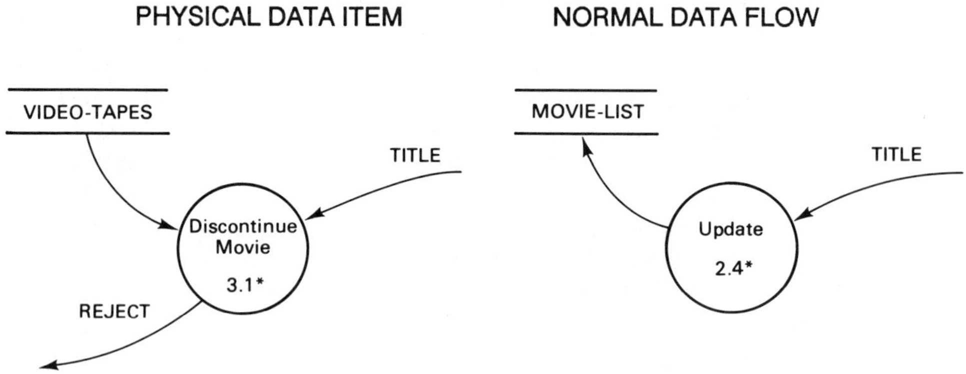

Figure 2.7a Figure 2.7b

Systems frequently deal with such physical data items as forms, tags, labels, containers, cards, ledgers, licenses, certificates, etc. as well as physical materials and products. It is important to model the existing system that uses physical items exactly as it exists, so that users can easily verify the model. Later in the analysis, the analyst will add the logical data equivalence of these physical items to the model. During the logical analysis, the analyst removes all traces of the physical implementation. In the logical model, the analyst's interest is in the data, not in the means of transporting the data.

PARTITIONING AND SUCCESSIVE DECOMPOSITION
. . . how to deal with complexity

The complexity of a system can be handled in two ways: *partitioning* a complex process into more comprehensible units, or *successively decomposing* the process into smaller more comprehensible units until each unit is easily understood and then organizing the units into a hierarchical structure, as shown in Fig. 2.8. **PARTITIONING is a two-dimensional factoring without any hierarchical or inherent structure.** With a large number of partitioned units, there are many interfaces.

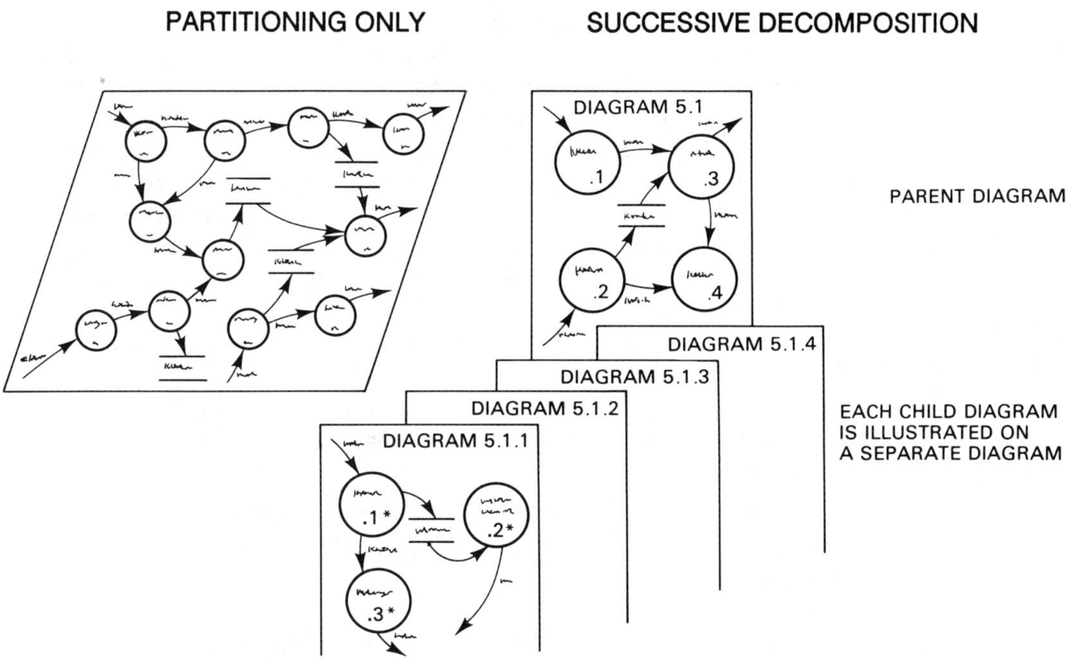

Figure 2.8

SUCCESSIVE DECOMPOSITION is a hierarchical factoring, where each "parent" node contains processes that are either functional primitives (i.e., cannot be further decomposed) or can be decomposed into a network of "child" processes, data flows, and data stores that are illustrated on the next lower-level diagram. Successive decomposition begins with partitioning. Each partitioned "parent" node is then factored into a network of "child" nodes until the desired level of detail is achieved. On each level, the factoring is done to minimize the interfaces among the components. There is no linkage between the "parent" and "child" levels of the structured specifications other than a numeric designation showing which "child" diagram is the breakdown of which "parent" node.

The following shows an example and an illustration of partitioning and successive decomposition applied to a particular situation.

Example

There are three employees working in a credit-card company. One person receives the credit invoices from the merchants. The person verifies the credit invoice for the following conditions: Is the merchant a registered client, has the credit limit been exceeded, and has the card expired (exceeded its valid time period)? The second person, who handles the payments to the merchants, performs two tasks: first, he or she prepares a list summarizing the invoices received from each merchant, and second, he or she separates the credit invoices by customer and sends them to a third person. The third person posts the charges from the credit invoices to the individual customer accounts.

The person responsible for paying the merchant computes the amount owed each merchant by summing all the charges for all customers. The person then applies a discount factor of three percent to determine the amount he or she will send to the merchant. The person then prepares and mails a check to each merchant. A record is kept indicating all of the invoices paid.

The person responsible for processing a customer had three distinct tasks to perform:

1. Receives the credit invoices and posts the charges to the customer's account.
2. Prepares a customer's monthly statement and mails it.
3. Receives payments from the customer and posts the payments received to the customer's account.

The overall system is shown in the context diagram of Fig. 2.9. Notice that the customer's initial application for credit is not included in this system.

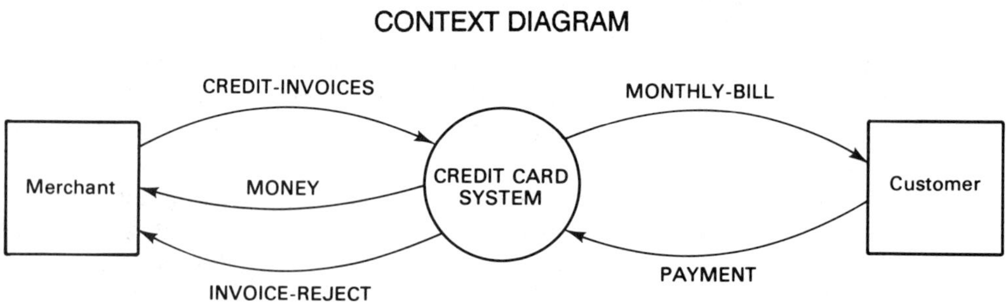

Figure 2.9

In partitioning this system, we separate all of the processes and show what input data and output data apply to each. The system is partitioned into as many processes as necessary to explain the system in understandable terms. There is no restriction as to the number of parts we can partition the system. However, there is only one level of breakdown, and all of the parts taken together must form the whole system. In this particular example, the Credit Card system is partitioned into six processes.

GRAPHIC ILLUSTRATION OF PARTITIONING

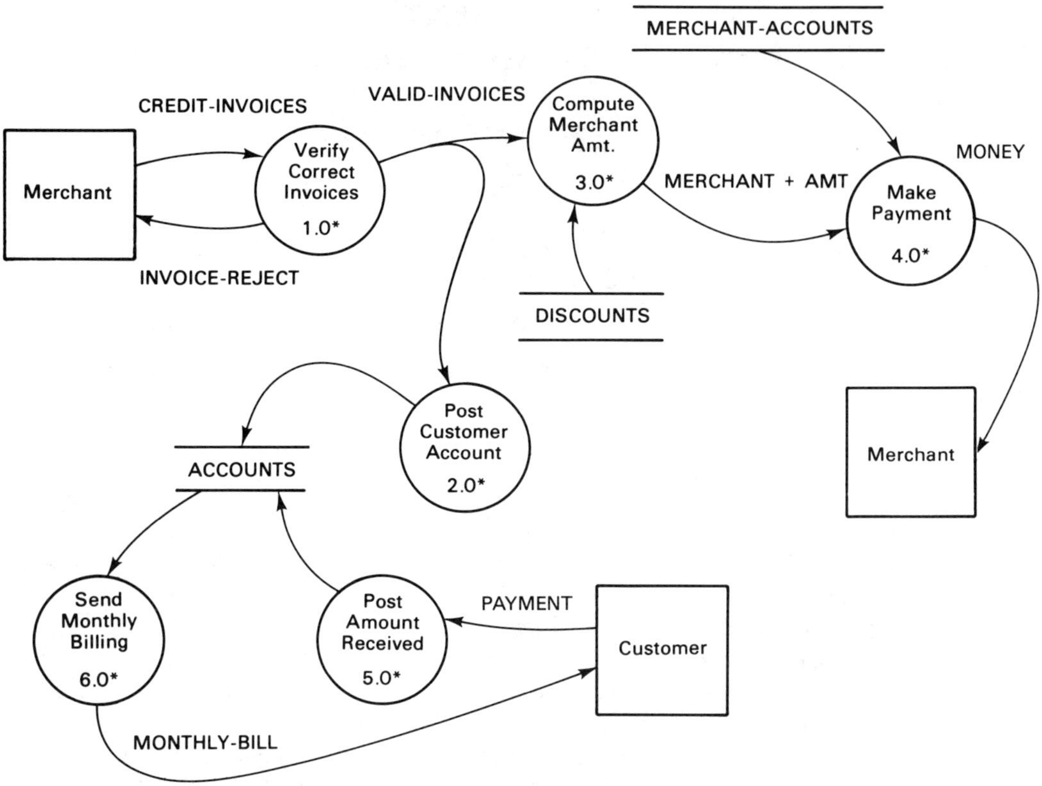

Figure 2.10

One can partition the system in Fig. 2.9 into six parts. A graphical illustration of the partitioned system is shown in Fig. 2.10. Notice that although the system was partitioned into its component parts, the graphic illustration shows all six of the components as functional primitives. Any further explanation of the component parts would be as narrative text or as an input/output table, as shown in Fig. 2.11. The table shows the six processes and the respective inputs and outputs for each process.

An alternative to partitioning a problem or system is creating a hierarchical structure of "parent" and "child" processes by successively decomposing the system into its major functions. A first-order decomposition factors the whole system into three major components (see Fig. 2.12a). Each of the three major functions are further partitioned into component parts, as seen in Figs. 2.12b and 2.12c. Our example stops at this point; however, larger, more complex systems would continue this process until those functional primitive functions that are manageable in terms of their complexity are reached. A manageable function, or functional primitive, is one in which a policy statement explaining how the inputs are transformed into the

INPUT/OUTPUT TABLE

INPUTS	PROCESS	OUTPUTS
CREDIT-INVOICES	1. Verify Correct Invoices	VALID-INVOICES INVOICE-REJECT
VALID-INVOICES	2. Post Customer Account	ACCOUNTS
VALID-INVOICES DISCOUNTS	3. Compute Merchant Amt.	MERCHANT + AMT
MERCHANT + AMT MERCHANT-ACCOUNTS	4. Make Payment	MONEY
PAYMENT	5. Post Amount Received	ACCOUNTS
ACCOUNTS	6. Send Monthly Billing	MONTHLY-BILL

Figure 2.11

outputs is usually written on one sheet of paper. The numbering convention, explained in detail later in this chapter, denotes a functional primitive by using an asterisk (*) following the process number. As an aid to the reader of a large structured specification, the asterisk used with the process number of a functional primitive informs the reader not to look for a lower-level data flow diagram. One should look for additional detail about the process in the mini-spec section.

Notice that in Fig. 2.12a, the data flow from the data store ACCOUNTS to the process Verify Correct Invoices 1.0* is purposely *not named*. To name this data flow would be redundant. If one wants to know the data elements that process 1.0* can access, the information is in the data dictionary entry for ACCOUNTS. If one wants to know specifically which data elements are in fact accessed by process 1.0*, the information is in the mini-spec for process 1.0*. This convention has the benefit that if the logic for process 1.0* required additional data from the data store ACCOUNTS, the modification needed to access the additional data is made in only *one* place, mini-spec 1.0*.

Likewise, if the content of the data store ACCOUNTS changed, whereby additional data elements were added, only one change has to be made, the change to the data dictionary definition of ACCOUNTS.

In Fig. 2.12c, the three activities are mutually exclusive. One must remember that the analyst starts by modeling the existing system. When the analyst finds unrelated tasks associated together by the users in the current system, the analyst must model the processes exactly as they exist. **The analyst must guard against modeling his or her assumptions, conclusions, or improvements that may not accurately portray the workings of the current system.** An example is adding a bidirectional data flow from MERCHANT-ACCOUNTS to and from process 2.2* in Fig. 2.12b

SUCCESSIVE DECOMPOSITION

DIAGRAM 0 (*Level 1*):

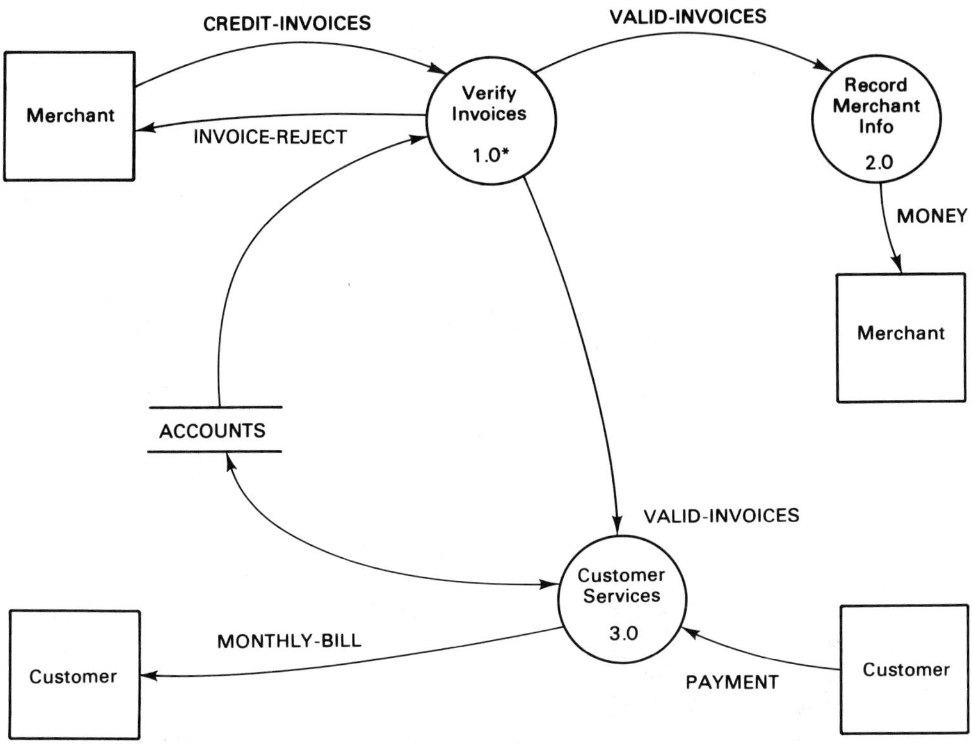

Figure 2.12a

DIAGRAM 2.0 (*Level 2*):

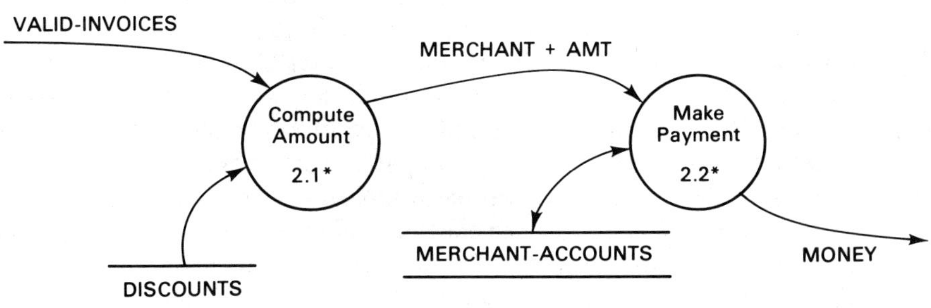

Figure 2.12b

DIAGRAM 3.0 (*Level 2*):

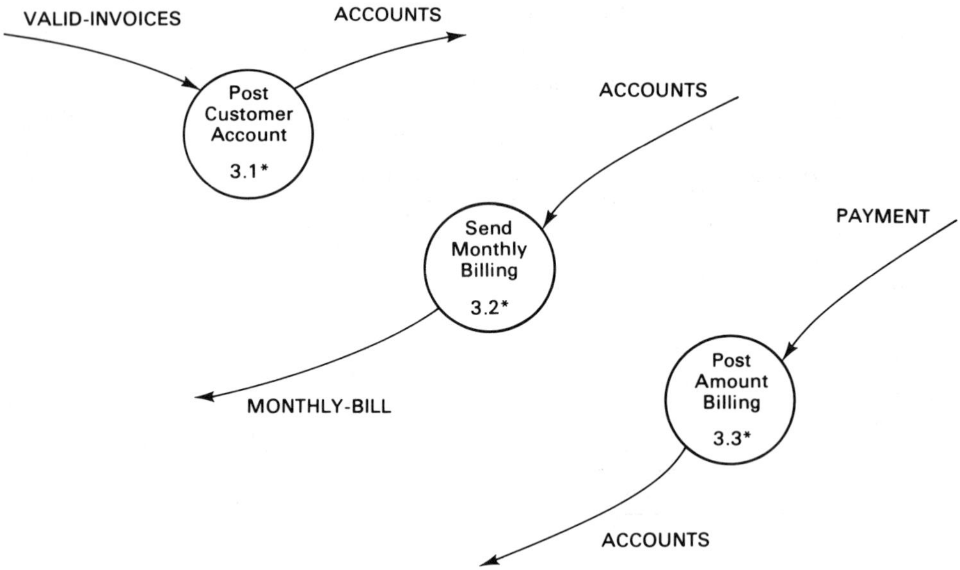

Figure 2.12c

When VALID-INVOICES is received by process 2.1, the analyst may assume an update is made in MERCHANT-ACCOUNTS. However, before using the bidirectional data flow, the analyst must be sure that process 2.2* does in fact update the amount in MERCHANT-ACCOUNTS.

A third type of decomposition exists known as *functional decomposition*. **FUNCTIONAL DECOMPOSITION is breaking a function of a system into a hierarchy of subfunctions. Each subfunction consists of a related set of activities and/or tasks that can be further factored into subfunctions until the functional primitives are reached.** The difference between successive decomposition and functional decomposition is the criteria used to factor a process into its component parts. In functional decomposition, the activities and tasks associated together to form subfunctions are highly related tasks that all contribute to the purpose of the subfunction. In successive decomposition, the association of the activities and tasks are done for economical, geographical, political, or other reasons. The criteria for selecting tasks performed by a particular organization may be to balance responsibilities in the organization, establish controls, or create a system of "checks and balances." When analyzing existing systems, we refer to the factoring of functions performed at various levels in an organization hierarchy to be a *successive decomposition*. In contrast, in new systems, particularly automated ones, it is desirable, at least originally from a logical viewpoint, to use *functional decomposition* in the statement of requirements.

LEVELING AND BALANCING
. . . building data flow diagrams

In using successive or functional decomposition, an analyst creates several levels of data flow diagrams, as illustrated in Fig. 2.13a and 2.13b.

LEVELING AND NUMBERING DATA FLOW DIAGRAMS

Level 0:

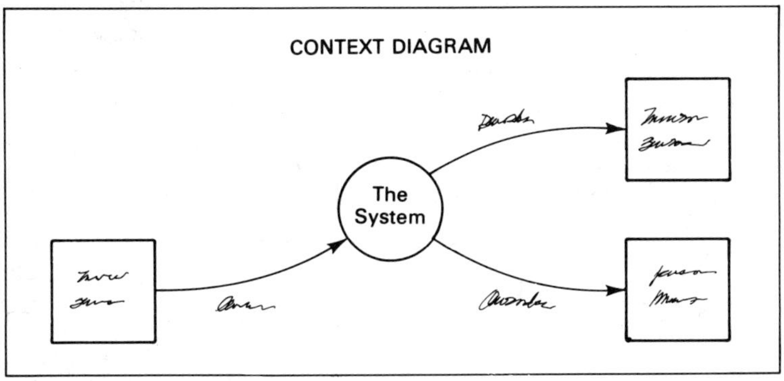

Figure 2.13a

Level 1:

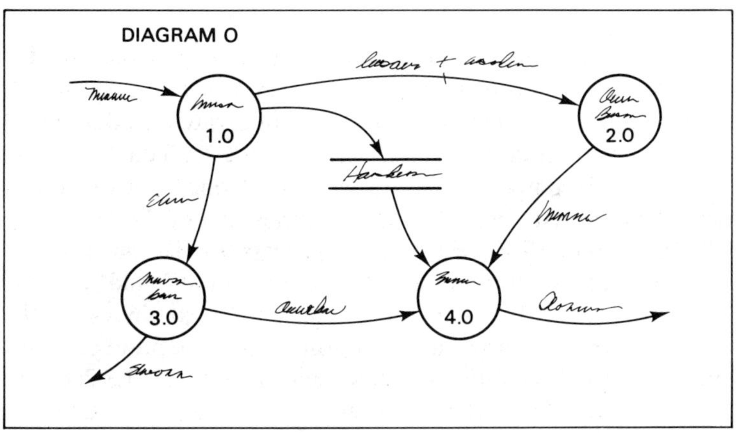

Figure 2.13b

The context diagram is defined as level 0. The first attempt of decomposing the system on the next lower level is on diagram 0, level 1. The analyst numbers each separate process on diagram 0 as 1.0, 2.0, 3.0, 4.0, etc. As the analyst decomposes each process further, the resulting diagram is on level 2 and the diagram number is the number of the "parent" process.

The level 0 context diagram is decomposed into four major processes or functional areas, as shown on level 1, diagram 0, in Fig. 2.13b. Each process on level 1, diagram 0, is further decomposed into a separate diagram that illustrates all of its processes, as shown in Fig. 2.13c.

Level 2:

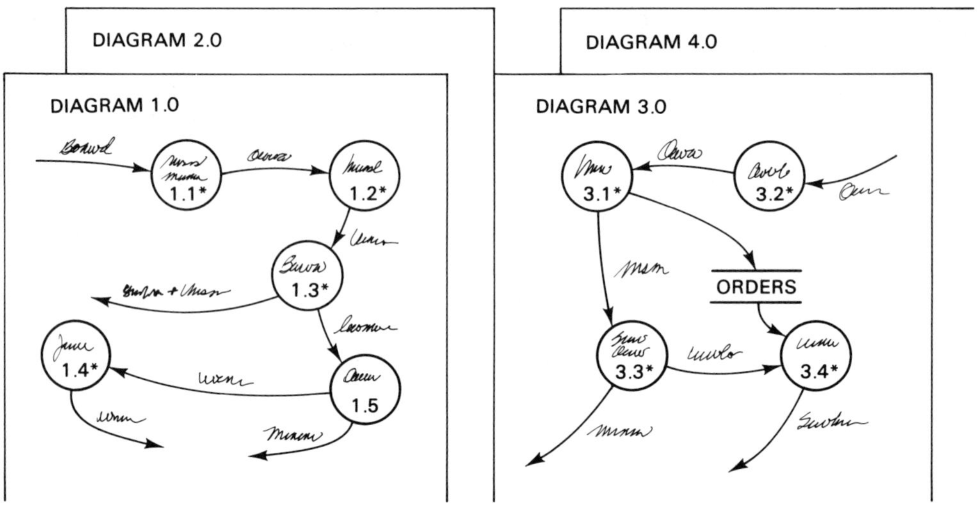

Figure 2.13c

In Fig. 2.14, there is an attempt to lay out the usual numbering convention for processes in a data flow diagram. The context diagram (*level 0*) consists of approximately 7±2 component processes illustrated on diagram 0 (*level 1*). Number each component process as 1.0, 2.0, 3.0, etc. When the analyst determines it is necessary to decompose a process on level one, a lower-level diagram of "child" processes is created. Namely, the decomposition of the "parent" process 1.0 is shown on diagram 1.0 (*level 2*). Each "child" process on diagram 1.0 adds a new suffix to the "parent" process number (1.0), which results in the "child" processes being numbered 1.1, 1.2, 1.3, etc. The numbering convention continues in this manner until all component processes are functional primitives. The numbering convention designates a functional primitive with an asterisk, such as 1.3.2.1*, 1.3.2.2*, etc.

Notice that the level of a process is the same as the number of *nonzero integers* it takes to number the process. For example, process 3.2.4.2.1* is a functional primitive at level five, found on diagram 3.2.4.2.

LEVEL NAME	DIAGRAM NAME	PROCESSES SHOWN
0	CONTEXT	*The System*
1	DIAGRAM 0	1.0, 2.0, 3.0, etc.
2	DIAGRAM 1.0	1.1, 1.2, 1.3, etc.
2	DIAGRAM 2.0	2.1, 2.2, 2.3, etc.
2	DIAGRAM 3.0	3.1, 3.2, 3.3, etc.
⋮	⋮	⋮
3	DIAGRAM 1.1	1.1.1, 1.1.2, 1.1.3, etc.
3	DIAGRAM 1.2	1.2.1, 1.2.2, 1.2.3, etc.
3	DIAGRAM 1.3	1.3.1, 1.3.2, 1.3.3, etc.
⋮	⋮	⋮
3	DIAGRAM 2.1	2.1.1, 2.1.2, 2.1.3, etc.
3	DIAGRAM 2.2	2.2.1, 2.2.2, 2.2.3, etc.
3	DIAGRAM 2.3	2.3.1, 2.3.2, 2.3.3, etc.
⋮	⋮	⋮
3	DIAGRAM 3.1	3.1.1, 3.1.2, 3.1.3, etc.
3	DIAGRAM 3.2	3.2.1, 3.2.2, 3.2.3, etc.
3	DIAGRAM 3.3	3.3.1, 3.3.2, 3.3.3, etc.
⋮	⋮	⋮
4	DIAGRAM 1.1.1	1.1.1.1, 1.1.1.2, 1.1.1.3, etc.
4	DIAGRAM 1.1.2	1.1.2.1, 1.1.2.2, 1.1.2.3, etc.
⋮	⋮	⋮

. . . continues to level n

Figure 2.14

Balancing Data Flow Diagrams

A BALANCED DATA FLOW DIAGRAM is one in which all of the inputs and outputs of the "child" diagram are exactly the same inputs and outputs of the corresponding "parent" process. This concept is mandatory in order to have a well-defined set of data flow diagrams. When one uses automated graphical aids (or CASE tools) to produce data flow diagrams, the products, in most cases, will verify that the input and output data flows of the "parent" node are all accounted for in the "child" diagram. **It takes great care and discipline to manually create balanced data flow diagrams.** No abbreviations or shorthand notations for data flows or data stores are permitted on the "child" diagram that are different from the names used on the "parent" diagram.

In each of the cases in Fig. 2.15, DIAGRAM 2.3 is the detailed data flow diagram of the "parent" process 2.3. Notice that in both Figs. 2.15a and 2.15b, process

DATA FLOW DIAGRAMS

BALANCED

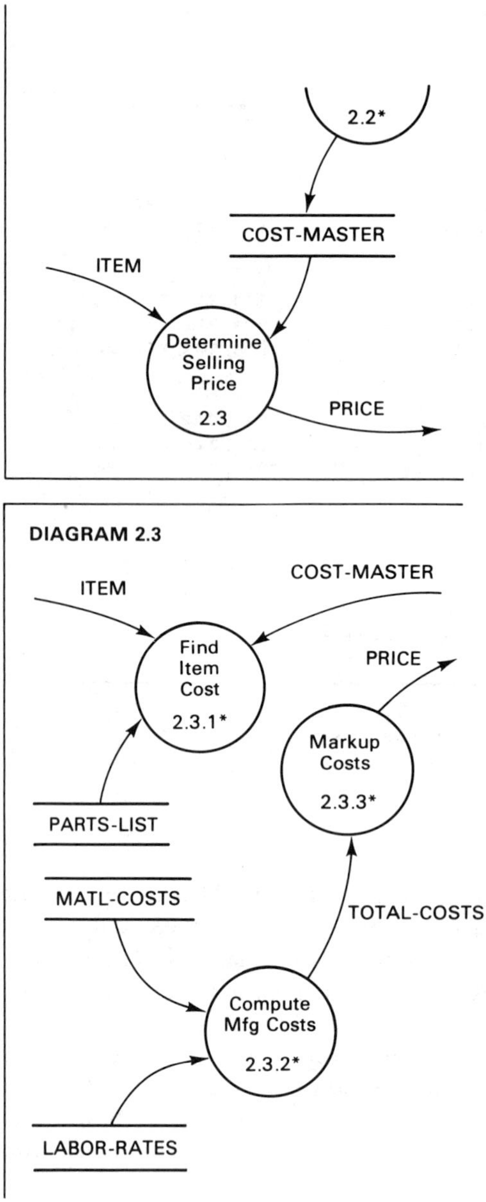

Figure 2.15a

NOT BALANCED

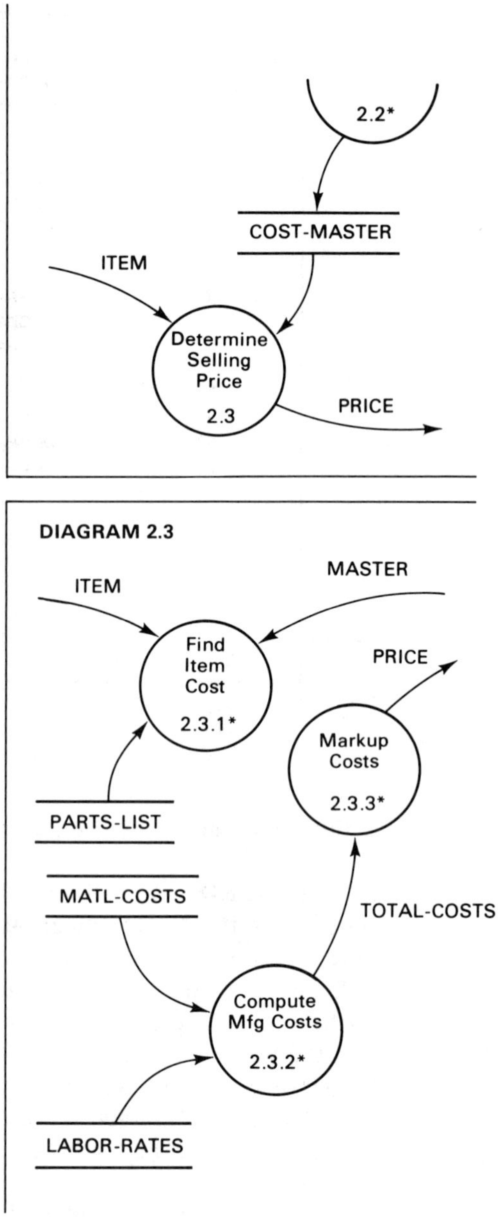

Figure 2.15b

2.3 has an input data flow, ITEM, and reads from the data store COST-MASTER. In DIAGRAM 2.3 of Fig. 2.15a, the process 2.3.1 inputs ITEM and the data from the data store COST-MASTER. The diagram in Fig. 2.15b shows that process 2.3.1 inputs ITEM and data from MASTER. MASTER is not another name for the data store COST-MASTER (*unless* MASTER *was properly defined as an alias for* COST-MASTER *in the data dictionary. In this case, we will assume that no alias exists.*) As a result, the diagram in Fig. 2.15b illustrates a common error. If such an error is made in a large, structured specification, it causes unnecessary waste of time and confusion. Care must be taken that such inconsistencies and typographical errors do not exist. Reviewers must find and eliminate them during the walkthrough reviews of the structured specification.

FACTORING PROCESSES
. . . limit on human comprehension

In creating a new level of a data flow diagram, we often ask: How many "child" processes should represent the "parent" process? There is no firm rule that applies; however, we can make use of the "magic number 7±2" as a guideline.

In 1956, George Miller[M-8] published a paper that had many intriguing examples that show that a person's ability to remember and deal with abstraction has limits. Dr. Miller's work is important in dealing with complexity and the successive decomposition approach to analysis.

One must always keep in mind that the primary purpose for using the structured specification method is to prepare system requirements. A key factor is to be able to communicate the functions and activities of a system to a user so that the user can verify the completeness and accuracy of the specification. By means of a formal review process, users and system professionals come to understand the system as a whole. From Dr. Miller's work, we know that the human mind can at best partition something into 7±2 pieces and still retain the concept of the whole. This is important when we try to define a function or activity of several processes to another person by explaining each component process one at a time.

This does not mean that if an analyst defines a "parent" process in terms of three or ten "child" processes that the structured specification becomes unintelligible or invalid as a specification. However, one should recognize the risks in explaining a process in terms of ten or more "child" processes.

COHERENCY TESTS
. . . data begets knowledge

Before we begin to break down the processes and data of a system, let's review the context diagram with all of the input and output data flows. The purpose is to be sure that the process (or system, in the case of the context diagram) has access to all of the necessary data to produce the indicated outputs. **The COHERENCY TEST is an examination of the input data flows and access to data stores that ensures that all the necessary inputs are available to produce the required outputs.** A coherency test is like a "sanity check" that verifies that the task can reasonably be accomplished. It is based on the **PRINCIPLE OF THE CONSERVATION OF DATA, which states that all of the input data to a process must be sufficient to produce the output data of the process.**[W-2]

If you find that it is impossible to produce the output data flow from the given set of input data, one of the following must be true:

1. There is a missing input.
2. The definition of a group data element is not complete.
3. The user assumes that the data store containing the missing data for the particular process is part of the process.

In case 3, the user may have neglected to tell you about the data store that is actually somewhere in the process. Frequently, such data stores are lists of data posted to the wall or *in the user's head.* The fact may be that the user has memorized the contents of a data store and now uses that knowledge in processing the data without realizing that they are using a data store. In such cases, the analyst needs to identify the source of the knowledge that the process uses and model it as a data store. Data stores of this nature are usually "read-only" data stores. In other cases, the process may actually update the data store. If the responsibility for maintaining the data store lies outside the scope of the system and other processes in the system do not use it, model the data store as a SOURCE/SINK. The internal data stores of the system should be only those that the system is responsible for and is capable of keeping sufficiently up to date.

Example (Data Flow Diagram)

> The data flow diagram is only one part of the structured specification. The graphical display of processes, data flows, and data stores contains the information on how the processes and data interrelate in the system. The details regarding the format or content of the data store areas, or the data flows, are not contained in the data flow diagram but rather in the data dictionary. The mini-specs contain the specific procedures used to transform the input data into the output data. The relationships among data stores and data elements are shown in the entity-relationship (E-R) diagrams. The organization of a structured specification requires that information about the system exists in one, and only one, place. Because of a well-defined structure, changes only have to be made in only one area of the document and thus redundancy does not exist.

The following chapters will present the data dictionary, the mini-specs, and the E-R diagrams in detail. For now, the following example concentrates on just the data flow diagram.

Early in the analysis consider each process as a *black box.* **A BLACK BOX is a component of a system where only the input data and output data are known. Nothing is known explicitly about the internal process of transforming the input data into the output data.**

Data flow diagrams consist of several levels, beginning with the context diagram. Each level is a detailed network of data and processes of its "parent" process. The data flow diagram of Fig 2.16 represents a system that issues drivers' licenses. It is an example of a two-level data flow diagram.

CONTEXT DIAGRAM (*Level 0*):

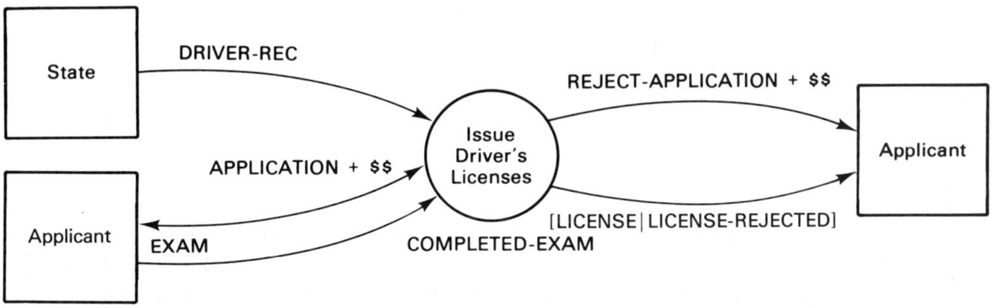

Figure 2.16

The data flow diagram in Fig. 2.17 is the first-level diagram known as diagram 0. The process Issue License 2.0 can be further decomposed into the diagram 2.0 shown in Fig. 2.18. Notice the numbering convention used among the levels. Process Issue License is numbered 2.0. All "child" processes contained in the "parent" process 2.0 are numbered as 2.1, 2.2, and 2.3. If process 2.1 were not a functional primitive but consisted of a "child" network of processes and data at level three, the processes would be numbered 2.1.1, 2.1.2, 2.1.3, etc.

Figure 2.18 is a second-level data flow diagram. This particular figure contains the detailed data flow diagram of process 2.0, and is known as diagram 2.0.

An example of how a data store is represented in two consecutive levels of data flow diagrams can be seen in Figs. 2.17 and 2.18. The data store DRIVING-HISTORY is shown between two parallel lines in Fig. 2.17 and is a normal data flow in Fig. 2.18. If DRIVING-HISTORY ever became a direct data flow, only one change (on Fig.2.17) would be

DIAGRAM 0 (*Level 1*):

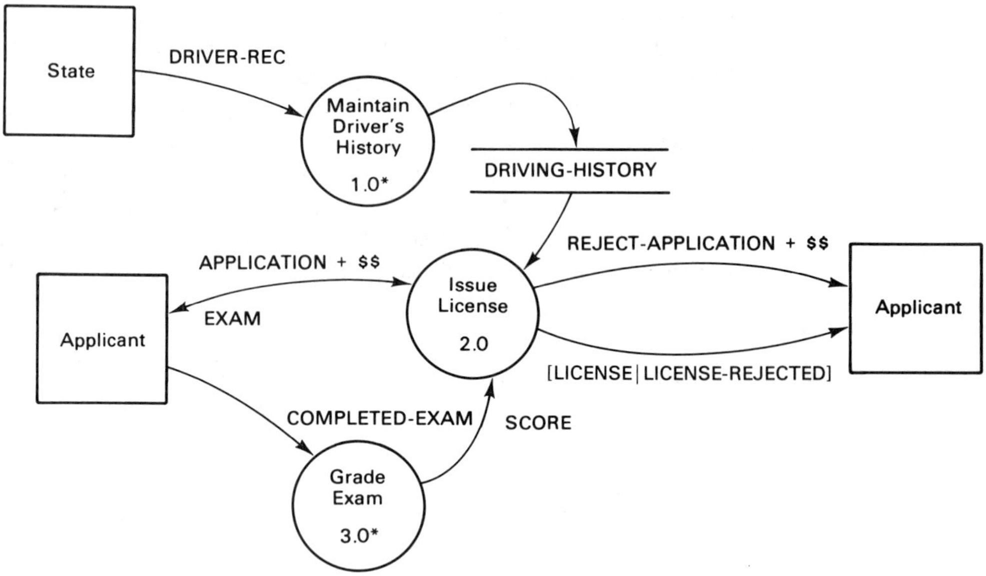

Figure 2.17

necessary, together with the corresponding change to the data dictionary.

Process 2.0 in Fig. 2.17 has three inputs: APPLICATION + $$, SCORE, and DRIVING-HISTORY, and four outputs: EXAM, REJECT-APPLICATION + $$, and LICENSE or LICENSE-REJECTED. Diagram 2.0 in Fig. 2.18 also has the same inputs and outputs; thus, diagram 2.0 is *balanced*. In diagram 2.0, there are two write-only data stores that do not appear on diagram 0 (Fig. 2.17). Because no other process other than those illustrated on diagram 2.0 access MONEY and APPLICATION-FILE, the data stores are shown on the highest level where two or more processes access the data store, or where a functional primitive accesses the data store.

One can also model the two write-only data stores as sinks. The decision on whether to model an item as a data store or as a sink may come from how the user considers the data. If the data apply to only the system being modeled, model it as a data store. If, on the other hand, the data apply to other systems, one can choose to model the write-only data store as a sink. Another criteria to examine is, "What system is ultimately responsible for the data?" If the responsibility lies within the system being modeled, then show the data as a data store. If the responsibility for maintaining the data store lies outside the system, then consider modeling the data as an output to a sink.

DIAGRAM 2.0 (*Level 2*):

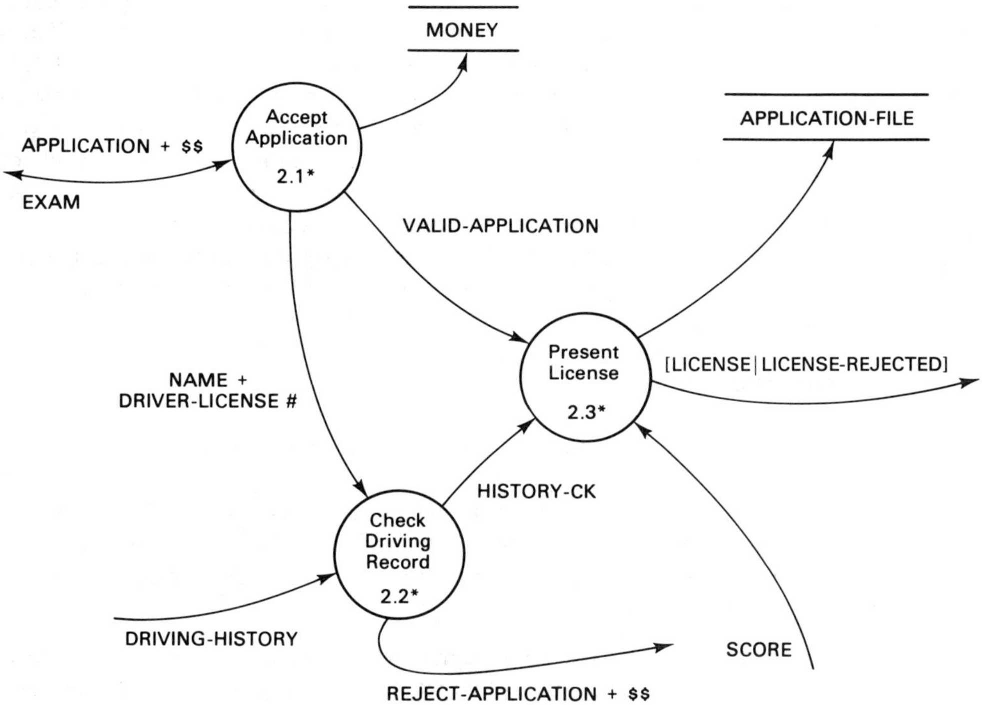

Figure 2.18

NAMING CONVENTIONS
. . . a rose by any other name . . .

The following information provides recommended standards for naming various components of a structured specification. Organizations that use structured methods may have their own established standards. These standards either evolved or are necessitated by the use of some CASE tools or other support systems. I provide this section as a guideline to consider when adopting structured specifications method.

Naming Processes

The names of processes shown on a data flow diagram should *always* contain a descriptive verb and object separated by blanks that describe the level of detail of the data transformation being performed. In addition, each process bubble **must** contain a unique number, as described earlier in this chapter, unless the process is a commonly used processing routine. There is no process number assigned to the context diagram at level zero. The processes in the first-level data flow diagram are numbered 1.0, 2.0, 3.0, etc.

Each "child" process uses the "parent's" process number as a prefix concatenated to the numbers .1, .2, .3, etc. to uniquely number each "child" process. The ability to accurately and descriptively name the processes and data flows of a "child" diagram indicates how well the "parent" process has been partitioned. All processes and data flows should be given good, honest, descriptive names that accurately represent the activities performed by the process. If you have difficulty naming "child" processes, the partitioning of the "parent" process was done poorly and did not result in well-defined related sets of activities. Similarly, one should be able to name group data elements. If one cannot find good, honest, descriptive names for a collection of data elements being passed to a "child" process, one should reexamine how the "parent" process was decomposed.

Naming Data Flows

There are four cases that one needs to consider when naming data flows:

1. Always name the data flow leaving a process differently than the data flow entering the process, except in the case of physical items. Names of data flows for physical items such as tapes, books, ledgers, etc. do not change from process to process within a system.

2. The data flow that has many data elements being passed between two processes can be shown in one of two ways: define a group data element made up of all of the data elements that are flowing from one process to another, or name the data flow as #n (n being a sequence of positive integers), where #n is an element of a table of data flows that lists all of the data elements that flow between the two processes. The table of data flows is also known as a legend table. **A LEGEND TABLE is a listing of the abbreviations used on a diagram.** Abbreviations are used for the input data flows or conditions and output data flows or actions specified on data flow diagrams and state-transition diagrams.

   Generally, we should define a group data element if the collection of data is used in more than one data flow diagram. The data-element table approach shown in Fig. 2.19 eliminates the need to look up the group data flow in the data dictionary. This approach is useful only if there is one data flow diagram that uses the particular collection of data elements.

3. When using a bidirectional data flow between processes or between sources/sinks and processes, provide a data name for both ends of the data flow.

4. Do not show any data names between data stores and processes. Sources and sinks cannot input or output directly from a data store; therefore, data flows to and from data stores are never named.

DATA FLOW DIAGRAM WITH A LEGEND TABLE

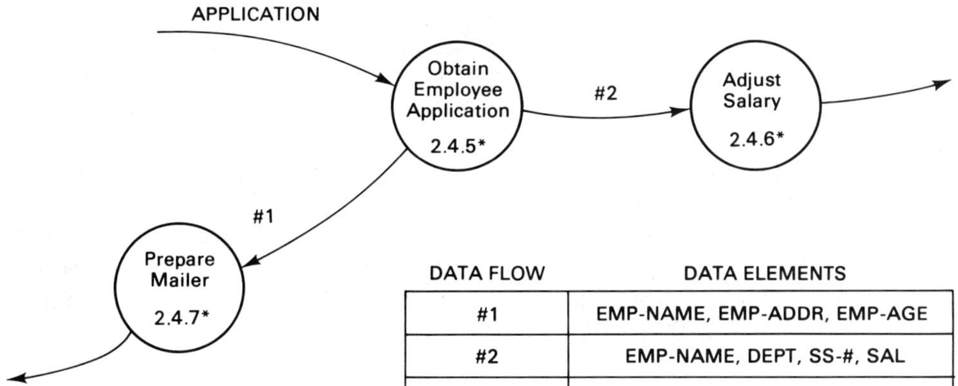

DATA FLOW	DATA ELEMENTS
#1	EMP-NAME, EMP-ADDR, EMP-AGE
#2	EMP-NAME, DEPT, SS-#, SAL
⋮	

Figure 2.19

When a process interacts with a data store, there is no need to name the data flow between the process and the data store itself. This is true even if only one data element is retrieved from a data store. The reason is to reduce redundancy. If you wish to know the data that is accessible by the process, you can simply look up the data store in the data dictionary. If the process only selects some of the data elements from the data store, document the data element selection in the mini-spec. To name the data flow between the data store and the process would therefore be redundant.

To accommodate automated data dictionaries and other CASE tools, all data names should be single identifiers. If one used two words to describe a data element, they should be connected with a hyphen to form a single identifier. Examples are YELLOW-SHEET, EMPLOYEE-NAME, and CUSTOMER-ADDR.

Naming Data Stores

Names for data stores resemble names for data flows and are sorted alphabetically together with all the data elements and common processing routines in a single data dictionary. Similar to the names for data flows, the names for data stores should be hyphenated to form a single identifier. Examples are CUSTOMER-MASTER and INVOICE-FILE. In planning system implementations there is often a need to replace a direct data flow between two processes with a data store. The data flow diagram notation and naming conventions minimize the effort necessary to make such changes. Likewise, if it is necessary to remove a data store between two processes and use a direct data flow, the effort to update the structured specification is also minimal.

Naming Common Processing Routines

The names given to common processing routines resemble the names for data flows and data stores. Common processing routines are those used within more than one "parent" process. The names of common processing routines should be an accurate description of the activities performed by the process. The logic defining the policy used to transform the inputs into the outputs of the common processing routines is put into the data dictionary and not in a mini-spec. Consequently, there is no process number associated to a common processing routine. Data flows, data stores, and common processing routines are all organized together in one alphabetical list in the data dictionary. Examples are the processes COMPUTE-FAIR-MARKET-VALUE and DETERMINE-VALID-OPTIONS.

ERRONEOUS DATA FLOW DIAGRAMS
. . . strive for accuracy and completeness

The following are examples of attempts to model situations that violate one or more of the previously mentioned guidelines on creating data flow diagrams.

ERRONEOUS DATA FLOW DIAGRAMS

DIAGRAM 0 *(Level 1)*:

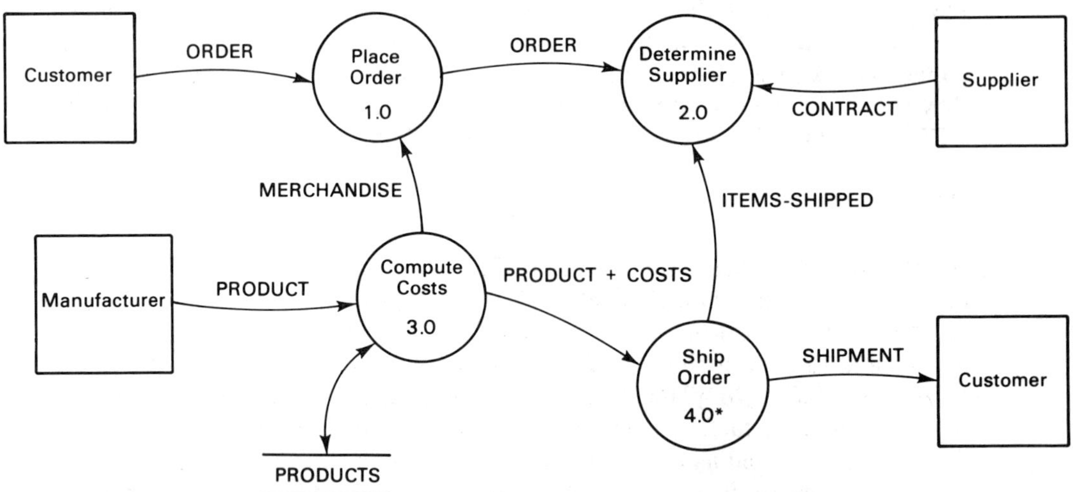

Figure 2.20a

1. Processes 2.0 and 3.0 of Fig. 2.20a both have "child" diagrams. Process 1.0 does not have a "child" diagram and thus must be a functional primitive. It is suggested to asterisk the process number of all functional primitives, such as 1.0*. The asterisk informs the reader of diagram 0 that additional details for process 1.0* are available in the mini-spec section.

2. Process 2.0 receives all input data flows and does not provide any output data flows. This is a suspicious activity.

3. The data element ORDER is a data flow processed by Place Order 1.0. The output data flow should have a unique name such as SPECIAL-ORDER, or BACK-ORDER, etc. In the case where Compute Costs 3.0 processes the data flow PRODUCT, the data element name did not change because the data flow PRODUCT is a physical item.

4. Process Obtain Availability Date is not a common processing routine elsewhere in the system, consequently it must be numbered. See Fig. 2.20b.

5. The data flow between process Items Back Ordered 2.1* and the data store BACK-ORDERS should not be named.

6. Diagram 2.0 is not balanced with "parent" process 2.0. Diagram 2.0 has an extra data flow named DELIVERY-DATE.

7. There appear to be two data names for the same data element, namely ITEM-NUMBER and ITEM#. If these two data elements are in fact the same, one name should be changed to the other. Avoid the use of aliases.

DIAGRAM 2 *(Level 2)*:

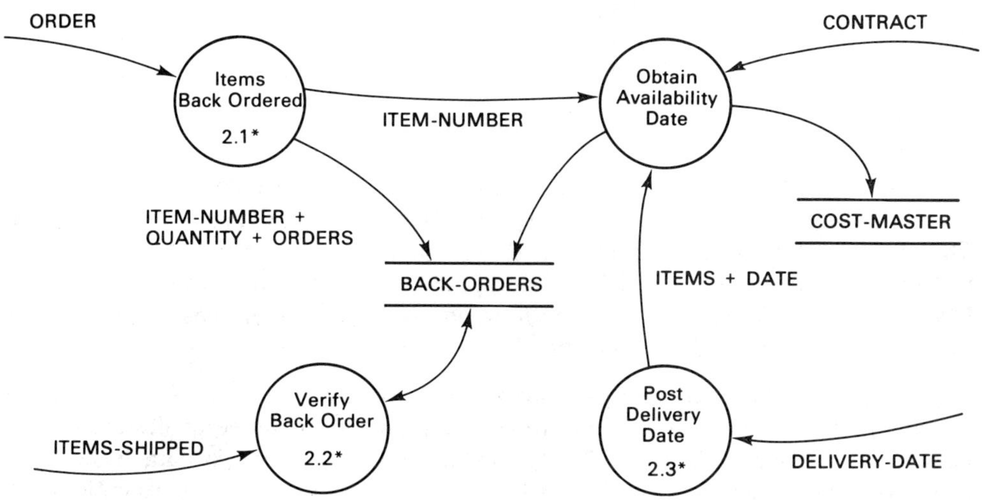

Figure 2.20b

8. The data store PRODUCTS in Fig. 2.20c only serves processes Determine Costs 3.1* and Apply Mark-Up 3.2*. This data store should not be shown at all on diagram 0 (level 1) of Fig. 2.20a. The data store is entirely contained in process 3.0 and should only appear on diagram 3.0.

9. Diagram 3.0 of Fig. 2.20c is not balanced with the process Compute Costs 3.0 in Fig. 2.20a, which has an output data flow of PRODUCT+COSTS. Process Distribute Products 3.3* outputs PRODUCT, but process Apply Mark-Up 3.2* outputs COST. COSTS is not the same as COST; therefore, the data flow diagrams are not balanced.

DIAGRAM 3.0 *(Level 2)*:

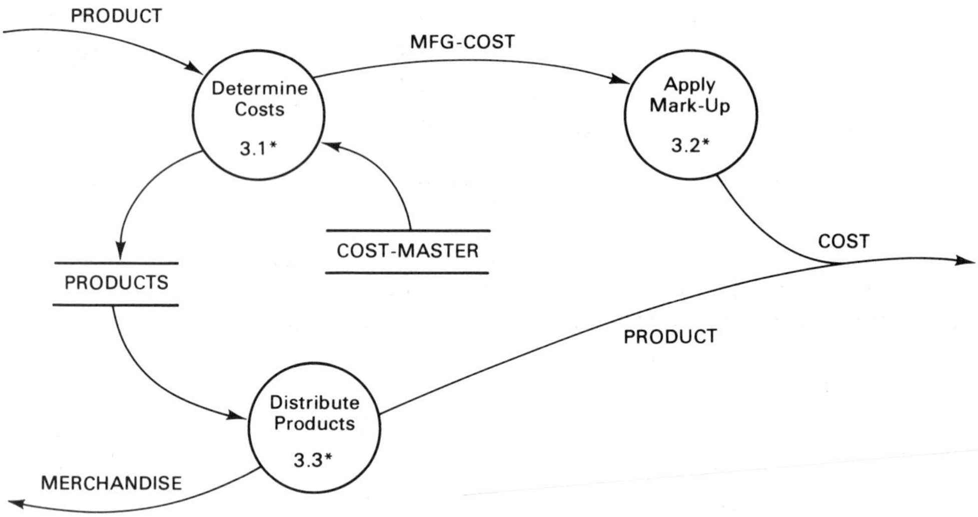

Figure 2.20c

10. Data store COST-MASTER can be a read-only data store. However, process Obtain Availability Date on diagram 2 (Fig. 2.20b) updates COST-MASTER. Therefore, COST-MASTER must be shown on diagram 0 (Fig. 2.20a) with both processes Determine Supplier 2.0 and Compute Costs 3.0 having access to it. COST-MASTER has to be shown *as a data flow* on both level-2 diagrams in order to be balanced with their "parent" processes.

11. Do not use the multiple sources/sinks notation to represent the plural of a source or sink. Each square should represent a *class* of a source or destination for data. In the example in Fig. 2.21, the system is built to receive data from many professors, and issues REPORT-CARDs to many students. In this example, each source or sink should be represented by one square.

MULTIPLE SOURCES/SINKS
(ERRONEOUS CONTEXT DIAGRAM)

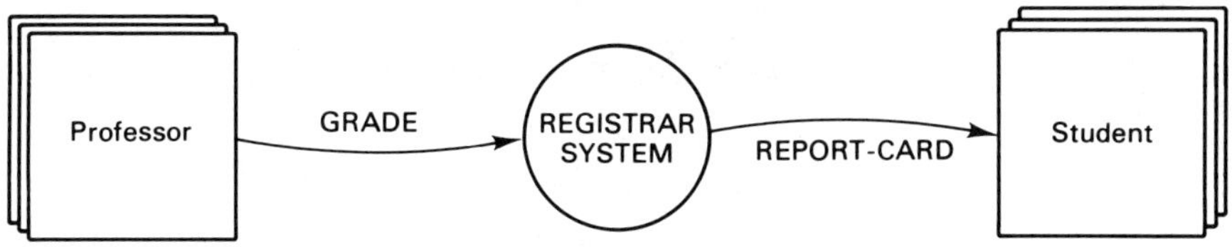

Figure 2.21

SUMMARY
. . . adherence to details

In this chapter, you have seen examples of using data flow diagrams to represent some small case studies. Remember that the structured specification is a document that must accurately, completely, unambiguously, and nonredundantly specify how data *flows* and how the system *processes* the data. If there is any human quality that is helpful in producing data flow diagrams, it is **being meticulous.** Those who follow through, carefully double checking their work and not ignoring small details, are most successful in producing fine data flow diagrams.

In choosing names for data flows, data stores, processes, sources, and sinks, cater to the vocabulary of the users. The communications channel between the analyst and the users is the most susceptible to failure. Thus, anything you can do to identify system components in the users' vocabulary should be done. Likewise, be careful of using computer jargon. Avoid technical terms such as "1200-baud modem," "full-duplex data link," "hash table," "synchronous communications analyzer," and "statistical variance subroutine" as names for components of the system.

Even though the notation presented in this chapter is fairly complete, there are times when we may wish to modify the notation to serve some specific purpose. Any additions and extensions to the notation and construction of data flow diagrams should serve the primary purpose of clearly communicating the system activities and data among the analyst, users, and management. Such extensions should be limited and withstand the following tests:

1. Is there a way to express the situation with the existing notation?
   If yes, use it instead.
2. Does the notation that you have chosen introduce any ambiguity to either a technical or nontechnical reader of the system specification?
   If yes, do not use it.
3. Does the situation introduce redundancy in the system specification?
   If yes, do not use it.

4. Does the notation or modification make the structure specification more difficult to update?
   If yes, do not use it.

Above all, when introducing, modifying, or adding to any notation, use the KISS method ("Keep It Simple Smarty!").

QUESTIONS

1. What are the basic elements of a system? Identify both those that are outside and those that are inside the boundaries of a system.
2. What is the purpose of the context diagram?
3. What is the primary purpose of a process and a data store?
4. What is the primary purpose of a structured specification?
5. What is meant by having a balanced data flow diagram?
6. What are the two forms of data shown in a structured specification?
7. How many "child" processes should be associated to a "parent" process when the "parent" process is decomposed?
8. At what level would you find activities 3.1.1, 1.0, and 2.3?
9. What would the activities be numbered that are depicted on:
   a. Diagram 1.1.2 c. Level 0
   b. Diagram 0 d. Level 2
10. How does structured analysis deal with the complexity of a system?
11. How does one graphically illustrate the following on a data flow diagram?
    a. A process that converts input data into output data
    b. A verification process that verifies and rejects data
    c. An interactive dialogue with input A and output B
    d. A process that updates a MASTER-FILE.
12. What is the principle of conservation of data and how do we verify that a system complies?
13. Give an example of a data flow diagram that would not pass a coherency test.
14. How does an iterative process differ from a recursive process?
15. Why are data flows between a data store and a process not named?

PROBLEMS

1. Modify the level-1 data flow diagram (Fig. 2.5) to reflect the following changes due to the demand to expand the ABC Catalog Sales Company to handle a larger volume of business:

   a. Additional order clerks are hired to perform the exact same job that one order clerk performs today.

   b. Additional warehouse workers are hired to handle a larger volume of products selected from the warehouse.

   c. Because of the need to have several warehouse workers, management is concerned that some may do less work than others. Therefore, the system has to include a way to objectively measure the amount of work each warehouse worker does. The agreed-upon measurement of work is the sum total of the weight of all the orders selected by each warehouse worker each week. If any worker consistently has a significantly lower total weight than the others, management is notified.

2. A small company allocates a certain amount of dollars to each department to be used to purchase things they need. The approved budget information is kept in the APPROVED-BUDGET file by department number in the purchasing system. When a department determines that it needs to buy something, the REQUESTOR fills out an order to request a purchase (REQ-ORDER). The REQ-ORDER contains the REQUESTOR's name, requestor's department number (DEPT#), quantity (QTY), name or description of the ITEM, COST and the DATE-NEEDED. Optional information that is provided by the REQUESTOR is the name and address of the supplier (SUPPLIER-NAME + SUPPLIER-ADDR). The REQ-ORDER is sent to a clerk in the purchasing system. The clerk compares the total amount of the order to the current amount remaining as specified in the APPROVED-BUDGET file for the requesting department.

   If the amount of the order exceeds the remaining amount authorized, the order is returned to the requesting department with a note saying "EXCEEDS BUDGET." If the amount of the order does not exceed the remaining budgetted amount, the order is accepted. The clerk computes the expected amount of the order and updates the APPROVED-BUDGET file to reflect the fact that additional funds are being expended. The ORDER is then passed on to the buyer and placed on the ACCEPTED-ORDERS stack.

   To process an order the buyer must verify (if provided) or obtain the name, address, and phone number (SUPPLIER-PHONE#) of the supplier from the SUPPLIERS-DATABASE file. This information is added to the order by the buyer if the information has not been provided by the

REQUESTOR. The SUPPLIERS-DATABASE also contains the amount of the DISCOUNT provided by the supplier (if any).

The buyer calls the supplier, checks to see if the ITEMs are "in-stock", verifies the COST and DISCOUNT, and obtains the expected ARRIVAL-DATE (VERIFIED-ORDER). If the ARRIVAL-DATE is on or before the DATE-NEEDED, a VERBAL-ORDER is placed with the supplier. The REQ-ORDER is marked-up indicating that the order has been placed and the expected ARRIVAL-DATE agreed upon by the supplier. The order is then passed on to the assistant buyer as ORDER-PLACED.

If the expected ARRIVAL-DATE is beyond the DATE-NEEDED, the buyer calls the REQUESTOR to obtain approval to accept the supplier's ARRIVAL-DATE. If the REQUESTOR approves the new date, the order is placed. If the REQUESTOR does not approve the new date, the buyer then looks up the ITEM being purchased in the SUPPLIERS-DATABASE file for alternative suppliers. Each supplier is checked in the data-base and ranked by the negotiated discount agreed upon between the company and the supplier. If no discount has been previously agreed to, the discount is assumed to be zero.

The buyer then calls the alternative suppliers until the expected ARRIVAL-DATE matches (or is earlier than) the DATE-NEEDED and a VERBAL-ORDER is placed. If no supplier can meet the requirements of the DATE-NEEDED, the REQ-ORDER is returned to the REQUESTOR with a note saying "cannot satisfy DATE-NEEDED." Once the verbal order is placed, the order is sent to the assistant buyer as ORDER-PLACED.

The assistant buyer makes a copy of the ORDER-PLACED and marks it as a "CONFIRMING-ORDER," and forwards it to the supplier as a written verification of the verbal order placed by the buyer. The original copy of the ORDER-PLACED is then put in a stack of OUTSTANDING-ORDERS and awaits the shipment to come into the Shipping and Receiving department.

When the merchandise arrives in the Shipping and Receiving department, the INVOICE is sent to the assistant buyer in the purchasing system. The assistant buyer matches the original copy of the ORDER-PLACED with the INVOICE and marks the ORDER-PLACED as "RECEIVED." The original copy of the ORDER-PLACED is then forwarded to the accounts payable department as REQ-ORDER + "RECEIVED." The accounts payable department is responsible for sending the payment to the supplier.

From the above narrative description of the Purchasing System, produce a CONTEXT DIAGRAM, a first-level DATA FLOW DIAGRAM, and a DATA DICTIONARY.

Figure 2.22 is an illustration of the activities that take place in the purchasing department as described in the problem statement.

PURCHASING SYSTEM

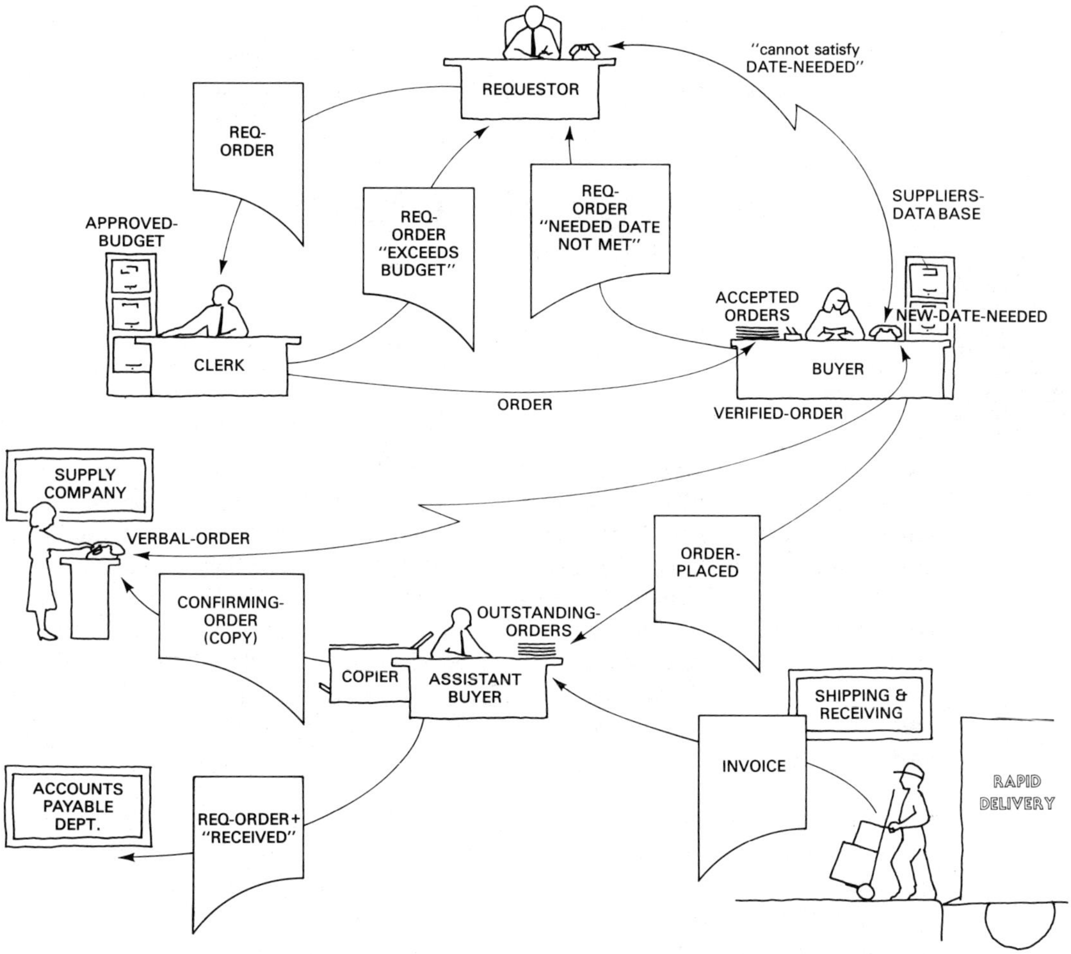

Figure 2.22

3

DATA DICTIONARY

As we create data flow diagrams, there is a need to define the exact contents of data flows, data stores, and commonly used processes. A *data dictionary* is used for this purpose. **A DATA DICTIONARY is an organized list of the composition of each data element, data flow, and data store used in the data flow diagrams, and the detailed logic specifications of commonly used processing routines and functions in the system.** Victor Weinberg[W-4] defines a data dictionary as "a repository of data about data." The data dictionary is created along with the data flow diagram during the analysis of the existing system. Entries are made anytime that data items are identified or added to the system. The data dictionary contains the following four types of entries:

- **Elementary data elements** are the lowest-level data elements that cannot be further decomposed for the purposes of the system. They are defined by comments that explain the data element or by a *data literal*. **A DATA LITERAL is the exact characters, numbers or symbols that make up the data element.** The following is an example of an elementary data element defined as a data literal:

  NO-SMOKE-MSG = "Thank you for not smoking."

- **Group data elements** are composites of elementary and/or other group data elements.
- **Data stores** are one or more occurrences of related elementary and/or group data elements.
- **Common processing routines** are procedural logics employed as commonly used routines or functions invoked by the logic of more than one functional primitive.

The data dictionary initially contains minimal specifications regarding the data used in the system. As the study of the existing system progresses, the analyst enters additional characteristics about each entry in the data dictionary. The characteristic information recorded depends on its type. Examples of some common characteristics identified during analysis are

> MINIMUM and MAXIMUM QUANTITIES, VALID VALUES, and SECURITY REQUIREMENTS.

and during the design stage are

> LENGTH or SIZE, FORMAT, INFORMATION HIDING, and SECURITY TECHNIQUES.

Some additional characteristics that apply to data stores only and are added during the design stage are

> RESPONSE TIME REQUIRED TO ACCESS THE DATA, STORAGE MEDIA, DEVICES and PROTOCOLS, ACCESS METHODS, and RESTART/RECOVERY DETAILS.

In short, just the minimum characteristics are needed during the analysis stage. The system designers add the technology-related characteristics later in the design stage. In the initial structured specification, many of the definitions of the data elements contain only a verbal comment describing the data in general terms. However, even at this early stage, I recommend using a *mechanized data dictionary* software package or a CASE support system. **The MECHANIZED DATA DICTIONARY is an vendor software product that permits entering, storing, and updating the characteristics of data.** The argument against the use of such automated tools is that they tend to require details, such as character length (size), format, etc., that do not concern the analyst. Although true, the detailed characteristics are not necessarily required upon the initial entry of the data elements into *all* data dictionary software products. Some of the newer tools permit the data elements to be placed in an environment where the details are added later by the designers and programmers. As a manager of software development, I have to pay the same cost of populating the data dictionary sooner or later in the development life cycle. If I can obtain additional benefits by populating the data dictionary sooner, it is worth it. What benefits can a mechanized data dictionary provide early in the development life cycle?

> Data descriptions are maintained in one convenient location.
>
> Data items are organized, sorted, and retrieved faster.
>
> No data element, no matter how obscure, is ever forgotten.
>
> Clerical support personnel can be employed to edit and keep the data dictionary updated as changes occur in the structured specification of a system.

The data dictionary provides the details of the data flow diagrams and makes them a rigorous specification of the system. **RIGOR is defined as the exactness implicit in a precise and complete statement of the system's requirements.**[O-1] Hence, the data flow diagrams and the data dictionary specify precisely the origin, content, movement, and destination of all data through the system. The analyst does not consider the relationships among data elements upon the initial creation of the data dictionary. Those relationships will be addressed later during the logical analysis of the system. The entity-relationship (E-R) diagram, presented in Chapter 7, will be used to show how data are related.

DATA DEFINITIONS
. . . how we specify data

The most basic notational symbol used in the data dictionary is the assign symbol "=" (equal sign). The symbol equates a data element or data store name with an appropriate definition.

CUSTOMER-INFO = CUSTOMER-NAME + *other data elements*

In 1966, Bohm and Jacopini[B-3] presented a theorem that proved that there are only three basic constructs necessary to express all program logic:

Sequence
Repetition (looping)
Selection (choice)

One could form a corollary of the Bohm-Jacopini theorem for data. That is, there are only three basic constructs necessary to represent all collections of data:

Concatenation (*sequential*) combines or groups a data item with another data item.

Repetitive (*repetition*) is the multiple occurrences of a data item .

Selection (*choice*) is a choice of one of several data elements or literal values for a data item.

An example using conventional data dictionary notation illustrates each of the three basic constructs for data as follows:

Example (Concatenation)

CUSTOMER = NAME + ADDRESS

The elementary data element NAME is placed together with the data element ADDRESS to form the group data element CUSTOMER.

Example (Repetition)

CHILDREN-NAMES = {CHILD-NAME}

The data element CHILD-NAME is repeated as many times as necessary to include the names of all the children in a family. Often it is helpful to specify the minimum and maximum occurrences of a repeating data item. This is done by preceding the repeating data item with the lower limit (minimum value) and following it with the upper limit (maximum value), such as:

$$\text{CHILDREN-NAMES} = {}_{0}\{\text{CHILD-NAME}\}^{29}$$

(Of course, I'm assuming that no one in the system has more than 29 children.)

Example (Selection)

RESIDENCE = ["HOUSE" | "APARTMENT" | "CONDO"]

The data element lists the attribute values for the elementary data element RESIDENCE as being a HOUSE, or an APARTMENT, or a CONDO. The notation above illustrates an **exclusive-or** relationship among the elements within the brackets, that is, the value of RESIDENCE is only *one* of the three options.

In the case where several options can be chosen from a list, the entry can be shown by using the repetition notation surrounding the selection items, as shown:

```
VEHICLES-OWNED = {["CHEVY" | "OLDS" | "BUICK" | "CADILLAC" | "PONTIAC"
                 | "MERCURY" | "FORD" | "LINCOLN" | "CHRYSLER" | "DODGE"
                 | "PLYMOUTH" | "AMC" | "FOREIGN"]}
```

The three basic constructs, a+b, {a}, and [a|b], are the absolute minimal set required to represent all data found in a system. However, for convenience we can use two other symbols:

Example (Optional Item)

EMPLOYEE = NAME + . . . + (AGE)

The ( ) symbols specify that the attribute AGE is optional. Optional data elements can also be denoted as a repetitive element that can occur zero or one time. However, it is more convenient to have a notation that permits us to represent data elements that are optional.

Example (Commentary)

COMMENT = * *A verbal description of the data or some explanatory remark about the data.* *

The comment notation permits the analyst to either generally describe the data item, postpone the definition of the data item, or add some additional commentary regarding a particular data item. A comment can also be added to any definition, as shown:

ADDR = * Legal residence of the client. *

TAPE-MASTER = {NAME + ADDR + . . . + CREDIT-LIMIT}
= * A new tape master is received from Headquarters each month. *

In defining data, a data dictionary entry is made up of

1. Other data elements (elementary or group) defined elsewhere in the data dictionary.

   BOOK = TITLE + AUTHOR + PUBLISHER + PUBL-ADDR + DATE-PUBLISHED

2. A data literal (the exact characters, numbers, or symbols used).

   CREDIT-CARD = ["VISA" | "MASTERCARD" | "DISCOVER" | "AMERICAN EXPRESS"]

3. A combination of items 1 and 2.

   APPLICATION = NAME + ADDR + ["MARRIED" | "SINGLE" | "WIDOWED"
   | "DIVORCED"] + (AGE) + EMPLOYER

Specifying Data Stores

Data that is held for an elapsed time is placed in a data store. Data stores are shown as repeating entities composed of elementary data elements and/or group data elements. The analyst or user selects one or more data elements to organize the collection of entries in a data store. The collection is underlined to designate it as the primary *key*.

Example (Data Store)

EMPLOYEE-FILE = {EMPLOYEE-NAME + ADDRESS + SOCIAL-SECURITY-#
+ OCCUPATION-CODE + ["MARRIED" | "DIVORCED"
| "SINGLE" | "WIDOWED"] + (SPOUSE-NAME)
+ $_0${CHILD-NAME}^{29} }

Notice that the entire concatenated list of data elements repeats. So how can one distinguish a data store from a group data item that really is a repeating set of data elements?

A data store is different from a repeating group of data elements in that it always has an *organizational structure*. The organizational structure provides a means to retrieve any single data store entry upon demand. A data store always has some type of structure or order. For example, even an "IN-BASKET" on an office desk is organized by the relative arrival time of each item.

We can distinguish data stores from repeating groups by underlining the primary key. Repeating groups do not have any organizational element. A primary key for a data store is selected from among the *candidate keys* in the data store. **A data element (or set of data elements) in a data store that has an unique value for each entry is called a CANDIDATE KEY.** A data store can have several candidate keys. However, it has only one candidate key designated as the *primary key*. **The PRIMARY KEY is the particular candidate key chosen to identify a single entry from all others in a data store.** Remember that all data store definitions must have two notational characteristics: (1) an overall repeating group designation (i.e.,{ . . . }), and (2) an underlined elementary data element (or a set of data elements) that is the primary key of the stored data.

Example (Candidate Keys)

Given the data store definition:

EMPLOYEE-FILE = {EMPLOYEE-NAME + ADDRESS + SOCIAL-SECURITY-#
+ OCCUPATION-CODE + ["MARRIED" | "DIVORCED"
| "SINGLE" | "WIDOWED"] + (SPOUSE-NAME)
+ $_0${CHILD-NAME}^{29} }

The data elements EMPLOYEE-NAME + ADDRESS or SOCIAL-SECURITY-# can both serve as an identifier for entries in the EMPLOYEE-FILE. Thus, both are candidate keys.

Example (Primary Keys)

The candidate key that is selected by the user (or analyst) to be the organizational element is the primary key for the data store. In the above example, if the user wishes to organize the data store EMPLOYEE-FILE by SOCIAL-SECURITY-#, and SOCIAL-SECURITY-# is a candidate key, then it becomes the primary key for the data store, and is underlined in the data dictionary definition.

Specifying Data Tables

A DATA TABLE is a collection of data items organized by a data element(s) known as the key that is held somewhere in the system for reference during processing. A table, you will notice, has all the characteristics of a data store, and, in fact, is shown as a data store in the data dictionary with the organizational key underlined. Because a table is basically a storage area for data, it is specified as a data store in a structured specification. The dimension of a table is sometimes difficult to determine from the data dictionary definition alone. Actually, it is always equal to the number of successive searches required to retrieve the desired data from the table.

Figure 3.1 is an example of a one-dimensional table.

STUDENTS = {NAME + ADDRESS + PHONE}

STUDENTS:

NAME	ADDRESS	PHONE
JOE BAKER	123 MAIN ST.	555-1234
BILL JONES	789 NORTH AVE.	555-4321
AMY SMITH	256 CORE LANE	555-9876
⋮	⋮	⋮

Figure 3.1

To locate the information for any particular student, only one search is necessary if the name of the student is known.

Figure 3.2 is an example of a two-dimensional table.

YELLOW-PAGES = {SERVICE-TYPE + {NAME + ADDRESS + NUMBER}}

YELLOW-PAGES:

SERVICE-TYPE	NAME	ADDRESS	PHONE
AUTO SALES	ACME AUTO	500 OGDON AVE.	555-6600
	CASEY MOTORS	1401 OGDON AVE.	555-8000
	⋮	⋮	⋮

SERVICE-TYPE	NAME	ADDRESS	PHONE
AIR CONDITIONING	ACE HEATING & AIR	100 SPRING ROAD	555-1500
	BEST HEATING CO.	149 MACEY AVE.	555-4900
	⋮	⋮	⋮

SERVICE-TYPE	NAME	ADDRESS	PHONE
ACCOUNTANTS	ADAMS BROTHERS	251 HIGHVIEW	555-2468
	ALLEN & ASSOC.	411 MILL ST.	555-1357
	BOOK'EM INC.	789 WATER ST.	555-2345
	⋮	⋮	⋮

Figure 3.2

To locate any particular phone number, we must first search for the particular "service type"; then within a "service type," we search for a particular supplier name.

A similar data store with three embedded repeating data items is

PHONE-SERVICE = {BILL-TO-NAME + BILL-TO-ADDR + {INSTALL-ADDR
+ {PHONE-NUMBER + SERVICE-TYPE + PHONE-TYPE}}}

This is a three-dimensional table with three keys. Three searches are needed to obtain the type of service (SERVICE-TYPE) or the type of phone (PHONE-TYPE) that is associated to a particular phone number, located at a particular address, and billed to a specific person.

Aliases

An ALIAS is a synonym for a previously defined data item. A data entry is considered an alias for another data entry if the two data entries are equal to the same set of elementary data elements.

I only have one thing to say about aliases: **DON'T USE THEM.** However, if you are modeling an existing system, and the users call the same data item by two or more different names, you have no choice. If there is a preferred data name, then use it as the definition. Include all other equivalent data names as aliases. The alias should only have a comment stating that the entry is an *alternate name* for the *preferred name*. Later, when the analyst specifies the new system, all aliases should be eliminated.

If you must deal with aliases in the data dictionary, I suggest that you always use the preferred name as the definition for the alias. See the following example:

```
CAPITAL-BUDGET-REQ = DEPT + SUPERVISOR + {ITEM-TO-PURCHASE + AMOUNT
                       + DATE-ITEM-REQUIRED} + DATE-REQ
                   = CAPITAL-REPORT
        ⋮
CAPITAL-REPORT     = CAPITAL-BUDGET-REQUEST
```

This practice of linking aliases may seem a bit redundant, however, keep in mind that the users must verify the structured specification. In the structured specification for a new system, the aliases should be removed, thus, eliminating whatever redundancy may be present.

DATA BASES
. . . a collection of data stores

At this point, I will present the basic types of data bases found in most business applications and briefly discuss how each type is defined in a structured specification. Later, in Chapter 7, I will discuss data bases from a more logical viewpoint after presenting normalization, objects, object data stores, and data relationships (E-R diagrams).

A DATA BASE is two or more data files (data stores) with interrelating data elements that can be accessed in more than one way.[F-1] There are basically four types of data bases used in business application systems. One can classify data bases by how the user views its structure.

Inverted-List Data Base

In an inverted-list data base, the user views the data as simply a large table of data records and some smaller table(s) of primary key elements that contains the locations of the associated records in the large table, as shown in Fig. 3.3.

INVERTED-LIST DATA-BASE STRUCTURE

CUSTOMER-ORDERS

LOC	CUSTM-NAME	CUSTM-ADDR	ORDER#	ORD-DATE	QTY	PROD#	...
1	A.C. ADAMS	123 MAIN ST.	987311	10-02-86	1	LQ-800	...
2	T.J. SMITH	10 NORTH AVE.	987426	10-02-86	20	DY-204DD	...
3	L.M. BROWN	45 PARK AVE.	987434	10-02-86	4	AMD005	...
⋮	⋮	⋮	⋮	⋮	⋮	⋮	⋮

CUSTM-NAME	ENTRY-LOC
A.C. ADAMS	1
T.J. BAKER	217
⋮	⋮

PROD#	ENTRY-LOC
AC-1008	102
AMD005	3
⋮	⋮

Figure 3.3

The inverted-list data-base structure is simple and easy to implement. In the data dictionary, the data base is defined as a simple data store with a primary key. The primary key represents the most frequently used access path. Additional access paths are possible and are mentioned in an associated comment.

The inverted-list data base is usually built to support a specific application. Retrieval time is very fast; however, updates are relatively slow. Data elements are often repeated frequently throughout the data base. Almost all accesses are preplanned.

There are basically two types of accesses that can be accommodated in an inverted-list data-base structure. The first type provides all of the data associated to a specific key value. This type of access is useful for answering the question, "What are the data elements associated to a given key?" A specific example is, "List all of the products that have been ordered by Ms. Jane Doe." The second type of access is an inversion of the first. This type of access provides all of the items associated to a specific data value. It is useful in answering the question, "What entries have a given characteristic?" An example is, "List all of the customers who have purchased the product number ATT6386." The data store definition that represents an inverted-list data base is a standard data store definition with a second definition line containing a comment that explains the types of access capabilities available.

CUSTOMER-ORDERS = {CUSTOMER-NAME + CUSTOMER-ADDR + ORDER#
+ ORDER-DATE + {QTY + PRODUCT# + DESCRIPTION
+ WEIGHT + UNIT-COST + . . . }
= * The data base can be accessed by CUSTOMER-NAME or by PRODUCT#. *

Some vendor products that implement inverted-list data bases are:

Data-Base System	Vendor
MODEL 204	Computer Corp. of America
ADABAS	Software AG
DATACOM/DB	Applied Data Research

Hierarchic Data Base

In a hierarchic data base the user views the data as organized in a tree structure as shown in Fig. 3.4..

The hierarchical data structure is more complex, and requires a data-base professional to establish and maintain the data base. A group of related data elements form a *record type*. Each record type is a node in the tree structure. A "parent" record can have many "child" records related to it, but a "child" record only

HIERARCHIC DATA-BASE STRUCTURE
(CUSTOMER-ORDERS DATA STORE)

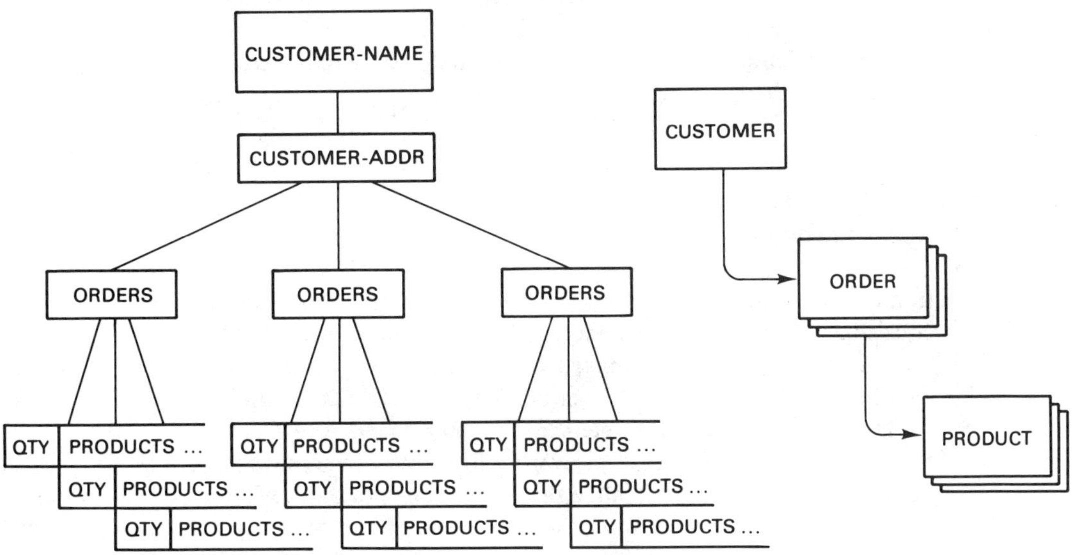

Figure 3.4

has one "parent" record. Navigation through the structure is difficult; however, because every record is linked in a "parent-child" relationship, the entire data base can be represented in a single data store entry. The primary key indicates the most frequently used access path. Secondary indexes (access paths) can be defined and are mentioned in an associated comment.

To generate reports from a hierarchical data base usually requires the services of a professional programmer. The hierarchical data base is often built to support a specific application or set of related applications. If a user decides to change the access path, a professional data-base administrator must unload the data, create a new data-base structure, and reload the data into the data base. Therefore, most accesses must be prearranged. Retrieval time can vary depending on the structure and size of the data base. Update times also vary. Because each data element is stored only once in the data base, a user updating a data element inhibits all other users from simultaneously accessing that same data element. All accesses to a particular data element reference the same occurrence of the data element in the data base. If alternate access paths have to exist, the data-base administrator can set up secondary indexes so the retrieval times are reasonable.

The data store definition representing a hierarchic data base is the standard data store definition with a second explanatory definition listing the types of accesses available as shown in Fig. 3.5.

```
CUSTOMER-ORDERS = {CUSTOMER-NAME + CUSTOMER-ADDR + ORDER#
                  + ORDER-DATE + {QTY + PRODUCT# + DESCRIPTION
                  + WEIGHT + UNIT-COST + . . . }
                = * Data base can be accessed by CUSTOMER-NAME. A
                  secondary index of PRODUCT# also exists. *
```

Some vendor products that are hierarchic data bases are

Data-Base System	Vendor
IMS	IBM
System 2000	Intel

Network Data Base

In a network data base the user views the data as a collection of sets of data records consisting of data elements linked by forward and backward links.

The network data-base structure is similar to the tree structure of a hierarchic data base with an additional level of complexity. **Each group of records (each record contains one or more data elements) is organized into two-level hierarchies called SETS (or links).** The sets can overlap to form networks whereby a given record can be part of different sets. Unlike the hierarchic data base, where each data record has exactly one "parent," a network data-base record can have more than one "parent" or possibly none. Therefore, a network data base is defined in the data dictionary as a data store, together with the definition of the sets as a comment entry.

NETWORK DATA-BASE STRUCTURE

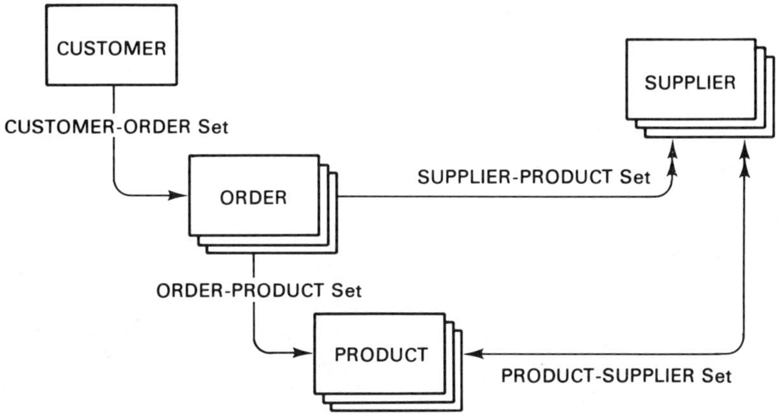

Figure 3.5

Construction and maintenance of network data bases require a data-base professional. The existence of multiple navigational paths through the network structure increases the difficulty of using a network data base. Creating programs that generate reports is mainly the chore of professional programmers. The data base is usually built to support a specific application or set of related applications. Because each group of records can be part of many different sets, the data structures can support a wide range of applications. However, each set of access paths must be prearranged. If a user decides to change or add a new access path, a data-base administrator must unload the data, build a new data-base set structure, and reload the data. Retrieval time can vary depending on the structure and size of the data base. Update times also vary. Because each data element is stored only once in the data base, a user updating a data element inhibits all other users from simultaneously accessing that same data element. All accesses to a particular data element reference the same occurrence of the data element in the data base.

The data store definition representing a network data base is the standard data store definition for each identifiable set consisting of data elements and a second explanatory definition listing the valid combinations of sets.

```
CUSTOMER-ORDERS  = * Data base of the following data stores: *
   CUSTOMER      = {CUSTOMER-NAME + CUSTOMER-ADDR + . . . }
   ORDER         = {ORDER# + ORDER-DATE + {QTY + PRODUCT# + . . . }
   PRODUCT       = {PRODUCT# + DESCRIPTION + WEIGHT + UNIT-COST + . . . }
   SUPPLIER      = {SUPPLIER-NAME + SUPPLIER-ADDR + SUPPLIER-CODE
                   + SUPPLIER-PHONE# + . . . }
                 = * The data base consists of the following SETS:
                   CUSTOMER-ORDER, ORDER-PRODUCE, PRODUCE-
                   SUPPLIER, and SUPPLIER-PRODUCT. *
```

Some vendor products that are network data bases are:

Data-Base System	Vendor
IDMS	Cullinet
DMS 1100	UNISYS
TOTAL	Cincom Systems

Relational Data Base

The user views the data as simply a set of tables of data records each containing the data elements that describe a particular object as shown in Fig. 3.6.

The relational data-base structure is simple because it is a collection of tables. However, to organize the contents of each relation properly, requires a data-base professional. Use of a relational data base structure is very easy. Programs that generate reports often have a fourth-generation query language (see Chapter 12) that makes it possible for end users to create their own reports without the aid of a professional programmer. We choose data elements to form a relation based on the inherent relationship of the data elements themselves and therefore the data base can support any application without modification. If the user decides to change the access path, the data-management system selects the necessary relations and performs the appropriate relational operations to produce the desired results. Therefore, the relational data base is able to respond to *unanticipated requests* for data. Retrieval time can vary depending on the structure and size of the data base. Update times are

RELATIONAL DATA-BASE STRUCTURE

CUSTOMERS:

CUSTOMER-NAME	CUSTOMER-ADDR	...
A.C. ADAMS	123 MAIN ST.	...
T.J. SMITH	10 NORTH AVE.	...
⋮	⋮	⋮

ORDERS:

ORDER#	ORDER-DATE	QTY	PRODUCT#	...
987311	10-02-86	1	LQ-800	...
987426	10-02-86	20	DY-204DD	...
⋮	⋮	⋮	⋮	⋮

PRODUCTS:

PRODUCT#	DESCRIPTION	WEIGHT	...
LQ-800	PRINTER	13 lb	...
DY-204DD	FLOPPY DISKS	1 lb	...
⋮	⋮	⋮	⋮

Figure 3.6

usually quick. Because data relations are shared among many applications, users accessing the same data simultaneously can lock out one another while the data is being updated. If rapid access is required, the data-base administrator can create new relations that avoid the time-consuming JOIN operation. The JOIN operation dynamically combines two relations needed to satisfy a user's request.

The data store definition used to represent a relational data base consists of a set of standard data store definitions, one for each relation (or table as the user views the data base). Each data store has its own primary key.

```
    ⋮
CUSTOMERS        = {CUSTOMER-NAME + CUSTOMER-ADDR + . . . + ORDER#}
                 = * Part of the CUSTOMER-DATABASE data base. *
    ⋮
CUSTOMER-ORDERS  = * A Relational Data Base consisting of the relations:
                     CUSTOMERS, ORDERS, & PRODUCTS. *
    ⋮
ORDERS           = {ORDER# + ORDER-DATE + QTY + PRODUCT# +. . . }
                 = * Part of the CUSTOMER-DATABASE data base. *
    ⋮
PRODUCTS         = {PRODUCT# + DESCRIPTION + WEIGHT + . . . }
                 = * Part of the CUSTOMER-DATABASE data base. *
    ⋮
```

Some vendor products that are relational data bases are

Data-Base System	Vendor
QBE, DB2, SQL/DS	IBM
INGRES	Relational Technology Inc.
ORACLE	ORACLE Corp.
Rdb/VMS	Digital Equipment Corp.
MAPPER	UNISYS
SUPRA	Cincom Systems

One defines a data base in a data dictionary by specifying each separate data store (data file) as a simple data store. The entries of a data store consist of the elementary and group data elements with the appropriate primary key. The analyst does not attempt to define the data base as it *physically* exists. Only the data and the necessary relationships among the data elements and data stores are specified. The relationships that combine data to form data bases are shown on entity-relationship (E-R) diagrams, which are discussed in Chapter 7.

COMMON PROCESSING ROUTINES

. . . reusable routines are the component parts used to build software products

When constructing a data flow diagram, we may come across a process that is used as a subordinate function in more than one place in the system. For example, the process MINIMUM-PAYMENT-REQUIRED is used in diagram 3.5.4 and again in diagram 7.8.3, as shown in Fig. 3.7.

When considering the common processing routine MINIMUM-PAYMENT-REQUIRED, several questions come up:

> Is the process documented twice?
> *No, that would be greatly redundant.*
>
> What number do we assign to the process? Number it 3.5.4.1? or 7.8.3.1? or both?
> *Neither, just give it a name.*
>
> Where in the structured specification do we place the mini-specs describing the logic for the process that is identified only by its name?
> *In the data dictionary.*

DATA FLOW DIAGRAM WITH A COMMON PROCESSING ROUTINE

FINANCIAL LOAN SYSTEM

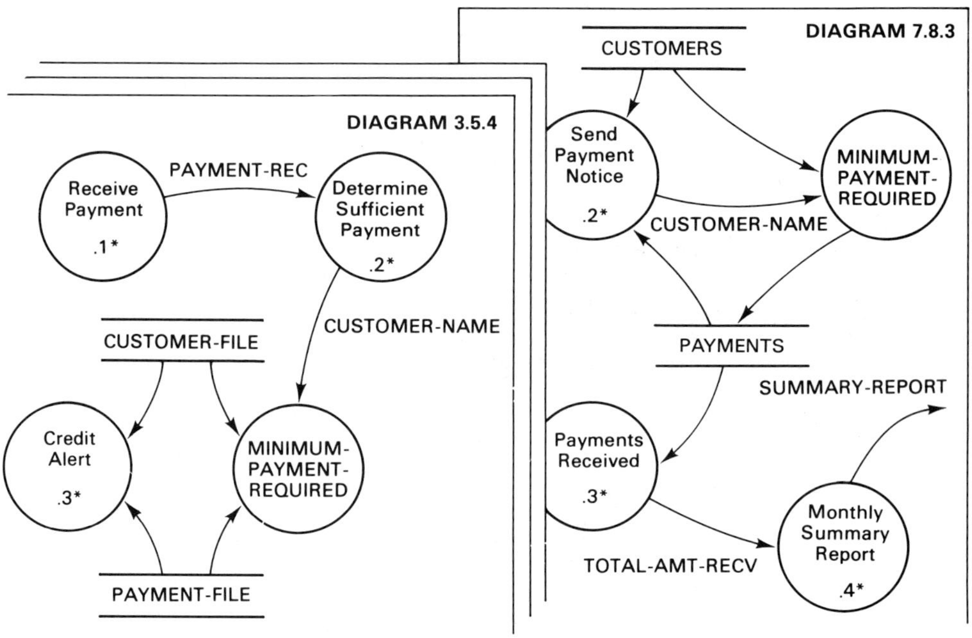

Figure 3.7

Common processing routines are used in various parts of the system. If you wish to document a common processing routine, give it a name and place the mini-spec in the data dictionary. In this way, redundancy is avoided, the detailed logic for the process is placed in one convenient place, and the process can be referenced anywhere in the structured specification. The method for documenting the details of the process itself is the same as for any mini-spec (see Chapter 4).

If you are maintaining a manual data dictionary, then add the mini-spec for the common processing routine to the data dictionary in the same manner as any other data element. Place the mini-spec among all the data elements in alphabetical order according to the name of the process.

If you are using a software package or CASE tool to maintain your data dictionary, then you can just enter a comment that refers the reader to the mini-spec for the process in the appendage section *of the data dictionary.*

Let's continue with the example of a financial loan system. Suppose there is a need to specify the policy for computing the minimum payment required (MPR) in several places throughout the system. The rule for computing the MPR is specified by a company policy and must be computed the same way throughout the system. The logic to implement the MPR is considered a common processing routine and would appear in the data dictionary as follows:

PARTIAL DATA DICTIONARY

```
    :
MIN-PAYMENT = * The minimum payment towards the outstanding loan.*

MINIMUM-PAYMENT-REQUIRED = * A Common Processing Routine *
    Input CUSTOMER-NAME.
    Search CUSTOMER-FILE for CUSTOMER-NAME.
    Input CREDIT-LIMIT from CUSTOMER-FILE.
    Search PAYMENT-FILE for CUSTOMER-NAME.
    Input AMOUNT-DUE from PAYMENT-FILE.
    If AMOUNT-DUE < CREDIT-LIMIT then MIN-PAYMENT = AMOUNT-DUE/12,
    Else, MIN-PAYMENT = (AMOUNT-DUE - CREDIT-LIMIT) + CREDIT-LIMIT/12
    Endif
    Output MIN-PAYMENT to PAYMENT-FILE for CUSTOMER-NAME.
    Stop.

MUST-PAY-NOTICE = CUSTOMER-NAME + CUSTOMER-ADDR + . . .
    :
    . . . normal data entries continue . . .
```

Figure 3.8

There is no provision to use functional decomposition to control the complexity of common processing routines. If the logic of a common processing routine is complex, we could simply partition the logic into more comprehensible units. Also, techniques such as case structures, decision trees, etc. (see Chapter 4) can be effective in controlling the complexity of common processing routines. In some cases we could define a common processing routine that calls other common processing routines which are documented in the data dictionary in the same fashion as shown in Fig. 3.8. This techniques can also be used to control complexity of common processing routines.

SUMMARY
. . . a repository of data about data

The analyst lists all of the data dictionary entries, including all common processing routines, together in one alphabetized list in the data dictionary. During analysis, the elementary data elements are shown only as comments or as data literals. Later, in the design phase, the systems designers add the detail characteristics of the data elements to the data dictionary. Group data elements are shown as a combination of elementary and/or group data elements using the standard notation. The data dictionary and the mini-specs are the two parts of the structured specification that must be capable of random queries. We locate information and read a data flow diagram in a top-down manner. The data dictionary supplies small amounts of information upon demand. Thus, the overall data dictionary is maintained as a single collection of entries all alphabetically arranged as one section in the structured specification.

The data dictionary can also be used to hold information on any item that is referenced on a random basis to add clarity to other parts of the structured specification. If a particular formula or operation is used in the mini-spec, the place to define that item in greater detail is in the data dictionary. What *does not* belong in the data dictionary is the policy that determines the values of the data elements. Such details belong in the mini-spec unless the statement of policy is a common processing routine. Data flow diagrams show the source, destination, or flow of data exclusively. Each component of the structured specification has its own responsibility. The data dictionary is that component that makes the structured specification a rigorous, well-defined, accurate model.

QUESTIONS

1. What is a data dictionary and what is its purpose?
2. During what stage of the development life cycle is the data dictionary created?
3. What are the three basic and the two optional constructs used to represent data in the data dictionary?
4. What is a table? How does one determine the dimension of a table?
5. What is an alias? Why should one avoid using aliases?
6. How and where does one document common processing routines used in the system?
7. a. How is a data store different from a repeating group?

   b. How is a data store different from a table?

8. What are the four types of data dictionary entries?
9. What are the benefits of populating a data dictionary software package early in the development cycle?
10. What is a data base? Where is it defined in a structured specification?
11. What are the four basic types of data bases?
12. Can a table of data elements be considered a data base?
13. In Fig. 2.5 from Chapter 2, APPROVED-ORDERS is a data store that resides as a stack of paper in the open window for the data clerk. What is the organizational structure of that data store?

PROBLEMS

1. Give an example of a three-dimensional table in data dictionary notation.
2. Create a data dictionary for the ABC Catalog Sales system in Chapter 2. Identify all assumptions used that were not explicitly stated in the problem.
3. Identify the alias(es) in the data dictionary for Problem 2.
4. Create a data dictionary for Problem 2 in Chapter 2

4

MINI-SPECIFICATIONS

So far in our pursuit to model an existing system, we have developed a graphical network of data flow diagrams (DFD) that indicate how processes and data relate in the system. We have also introduced the data dictionary that defines the data elements, data stores, and common processing routines used in the system. However, we have yet to specify how the input data are transformed into the output data.

We now want to concentrate on the processes themselves. Each functional primitive process on a data flow diagram transforms some input into some output, or simply provides some output. In our leveled set of data flow diagrams, we successively decomposed each activity until we reached the functional primitives. The internal policy applied in those functional primitives has to be recorded so that there is a clear understanding of the policy that governs each process in the system. **A MINI-SPECIFICATION (mini-spec) provides a detailed explanation of the internal processing policy of a functional primitive that transforms input data into output data or simply provides output data.** There are various means and tools used to document the policy of a process:

Structured English (pseudocode)

Decision trees

Decision tables

Action diagrams

Mathematical algorithms

Other means of specifying system processing logic that transforms input data into output data.

The mini-specifications (mini-specs) need only specify the processing policies necessary to transform data or to simply provide data. When stating a policy, only specify the *what*, rather than the *how*, with regards to processing data. Specify the data transformation policy from the *users'* vantage point. Page-Jones[P-1] states five rules for creating mini-specs:

- A mini-spec must exist for every functional primitive in the data flow diagram.
- A mini-spec must state the ways in which the data flows that enter the functional primitive are transformed into the data flows that leave the functional primitive.
- A mini-spec must state the policy that governs the transformation, not the method for carrying out that policy. That is, a mini-spec should say *what* is to be done, but should be inscrutable about *how* it is done.
- A mini-spec should seek to control redundancy. In particular, it should not restate those things already stated in the data flow diagram or in the data dictionary. (However, a little redundancy is permissible if it improves the readability of a mini-spec.)
- The set of constructs used to build a mini-spec should be small and simple. In addition, they should oblige us to express ourselves in a standard way.

In addition to specifying the logic of the data transformation involved in a process, there are several other types of appropriate information to include in the mini-spec:

Processor requirements necessary to perform the activities stated in the mini-spec for the process are specified.

If a computer performs the process, and there is a need to have some special processing capabilities or a particular type of environment, such as a real-time operating system, one needs to specify the requirement in the mini-spec. For processes performed manually, the mini-spec specifies the job title, occupation, or qualifications of the person(s) who performs the activities.

Security requirements are specified in the mini-spec for those processes that have access to sensitive or private data, or provide a communications path to the rest of the system.

Time requirements for performing the tasks associated with the process are included in the mini-spec. Time constraints and/or dependencies that affect the process or that the process affects are declared in the mini-spec. Examples are ramp-up time, ramp-down time, wait time (time between processes), etc.

Processing priorities relative to other activities or processes that are internal or external to the process being specified are stated in the mini-spec.

Both manual and automated processes have activities to perform that have relatively high-to-low priorities. Also, there are activities that cannot or should not be interrupted. When processing priorities are a policy decision of the users, the analyst has to state them in the mini-spec.

Performance requirements of the process in terms of both capacity and responsiveness are specified in the mini-spec.

Safety requirements (software and manual) used particularly for process control or robotic activities are specified in the mini-spec.

Software safety requirements include safeguards, process behavior following abnormal sequencing or error recognition, reset capabilities, process interruption or termination features, etc. This area of specification includes the requirements for defining the fault tolerance of the system. The degree to which the systems designers have to build fault-tolerant procedures into the system are stated in the mini-spec. One aspect of software safety is to describe the sensitivity of the process to faults in the system.

Manual safety requirements are usually specified by OSHA and the company's standard practices and procedures.

Other requirements or explanations associated to the policy of carrying out the process are specified in the mini-spec. The analyst needs to specify any requirements that aid in planning, designing, and implementing the system.

From the data flow diagram, we know exactly where the process is in the system and what data flows and data stores serve as inputs and outputs. For specific detailed information about the data itself, we have the data dictionary. Thus, we only need to specify how each process works with the available data. One numbers each mini-spec to correspond with the functional primitive on the data flow diagram. The actual format of the mini-spec is determined by the analyst. However, there are some commonly used ways of specifying general processing logic that will be described later.

Analysts prepare mini-specifications in varying degrees of detail, although the only absolute requirement is to specify the *policy* that governs the transformation of the input data into the output data. To provide more detail over and above the policy statement is to encroach upon the territory of the systems designers. There is a fine art of conveying to someone specifically *what* to do, in a well-defined manner, while leaving them to use their own ingenuity of *how* to accomplish the objective. The intent is to permit the greatest degree of freedom very early in the development life cycle of constructing a system. During the conceptual stage, we are only limited by our imagination. Little regard is given to technical feasibility. As plans move into the feasibility and analysis stage, one becomes more specific as to exactly what is the range, bounds, and objectives of the system. The analyst, although required to specify the data and processes used to accomplish the objectives, still has a great deal of latitude in planning the system. As the project moves into the design stage, the degrees of freedom become more limited. However, the systems designer does have the freedom to select the implementation plan to accomplish the objectives set forth in the structured specification. Once the design stage is completed, the implementation stage of converting the system design into software and building the data files has fewer degrees of freedom with regard to creativity.

Figure 4.1 illustrates the relationships among the major parts of a structured specification:

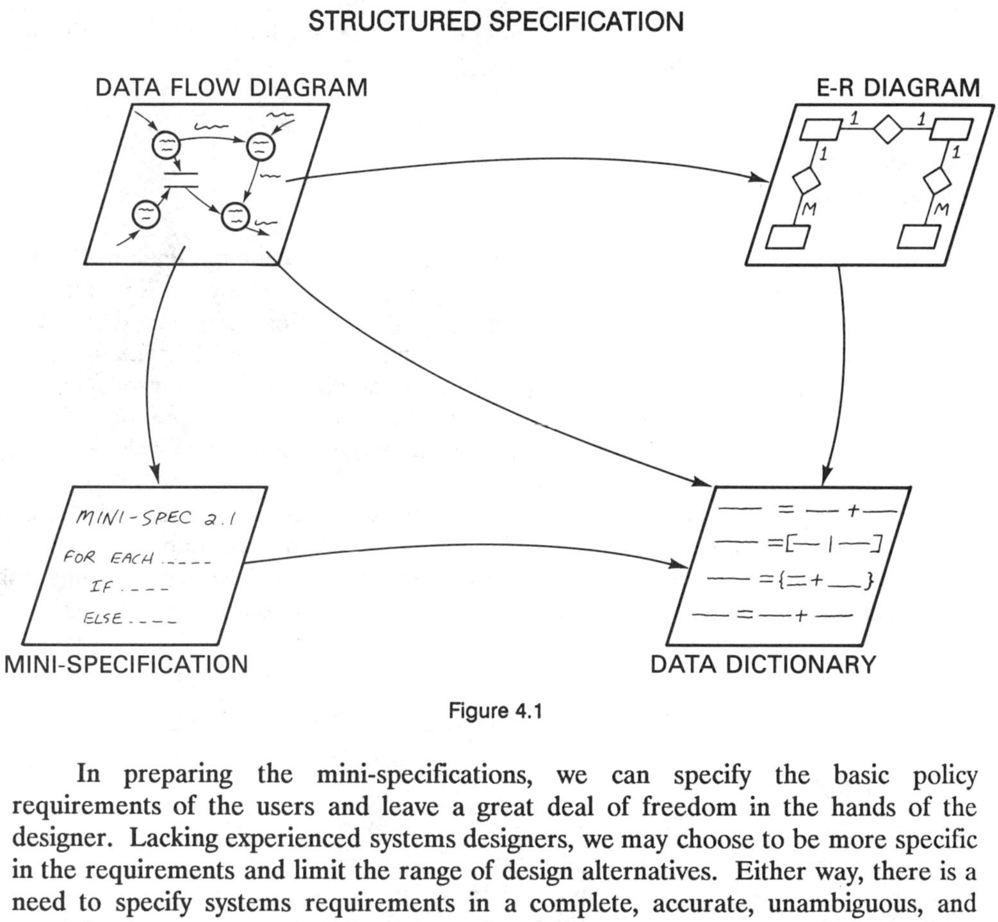

Figure 4.1

In preparing the mini-specifications, we can specify the basic policy requirements of the users and leave a great deal of freedom in the hands of the designer. Lacking experienced systems designers, we may choose to be more specific in the requirements and limit the range of design alternatives. Either way, there is a need to specify systems requirements in a complete, accurate, unambiguous, and nonredundant manner. This chapter presents various means of specifying the transformations of input data into output data as part of the system's requirements.

Structured English perhaps is the most broad and least restrictive means of specifying data transformation policy next to using natural language. **Pseudocode** has a more restrictive vocabulary and resembles high-level programming languages that may appear to give less latitude to the staff in selecting an implementation. Although decision trees and decision tables provide an excellent way of completely and unambiguously specifying logic, they can confine the creativity of the designers. And finally, action diagrams provide a means of specifying not only the *logic* by using structured English or pseudocode, but also the *structure* of the logic used in a process. One can use action diagrams with structured English to help clarify the intent of the written requirements and can leave a large degree of freedom to the designers. On the other hand, action diagrams can specify in great detail the internal logic structure of the software programs necessary to implement a process and leave almost no freedom or creativity to the implementation staff.

To prepare structured specifications, tools should be chosen and applied according to the organization's methodology for developing systems, and the available skills and talents of the development staff.

SPECIFYING LOGIC
. . . three basic building blocks

Ideally, one should specify only the policy of transforming input data into output data in preparing a mini-spec for a structured specification. However, if it is desirable or necessary to be more explicit in specifying the transformations of data, we should keep in mind the work of Bohm and Jacopini.[B-3] Their work laid the foundation for structured programming and has proven to be an effective way of producing software that is easily maintained. They showed that there have to be only three basic constructs for program logic. Figure 4.2 shows the various commonly used diagrams as variations of the three basic constructs.

Logic is easier to understand and maintain if we are limited to the basic three constructs shown in Fig. 4.2. This recommendation for implementing logic in programs does not in any way limit what software can achieve in performing tasks. It does provide a means of factoring the software solution into manageable and understandable components.

THREE BASIC CONSTRUCTS OF PROGRAM LOGIC

1. SEQUENCE:

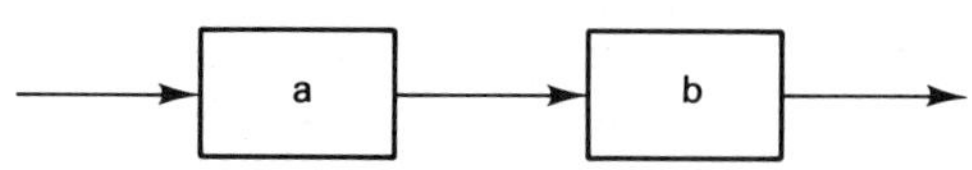

Process Series

Figure 4.2

2. REPETITION:

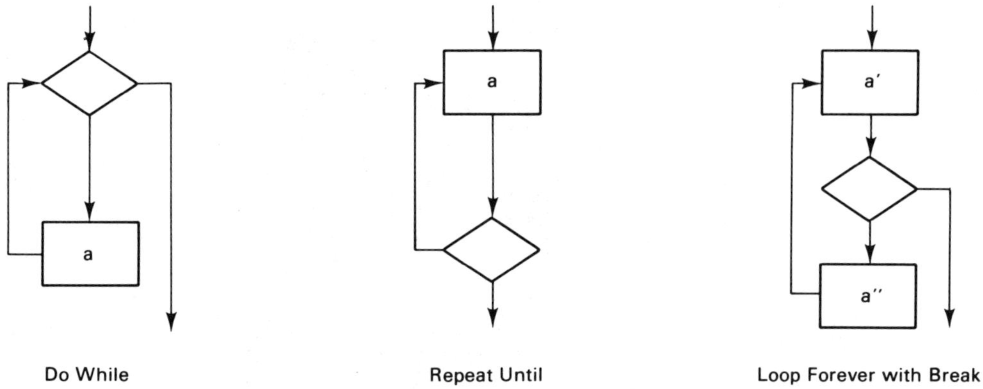

3. SELECTION:

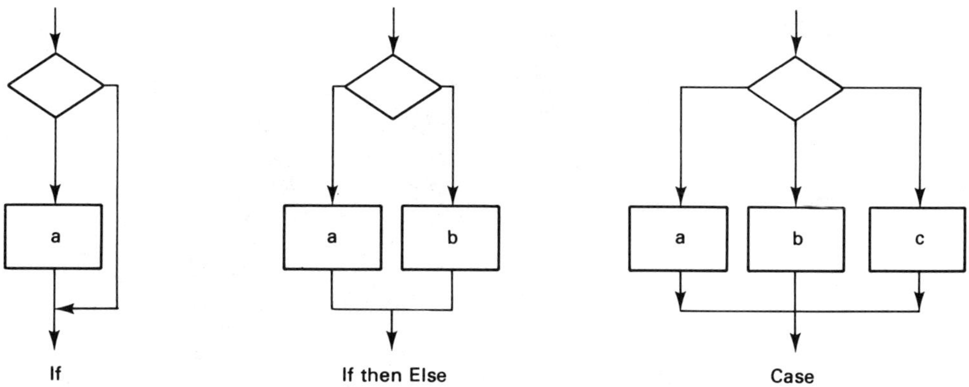

Figure 4.2 (cont.)

Consider the flow diagram in Fig. 4.3. Although this figure represents a more complex logic structure, it is a combination of only three basic construct, sequence, repetition, and selection, as shown by the dotted lines. Logic that is constructed in this fashion is easier to understand and requires less effort in making modifications. The most important thing is that there are less unexpected results when maintaining and enhancing logic that is constructed with only the three basic constructs. *Structured programming* is the term used to describe software that was created using this technique. More formally, **STRUCTURED PROGRAMMING is a programming technique for developing source code logic that is understandably and verifiably correct; specifically, a technique that employs a top-down refinement strategy to**

LOGIC-FLOW DIAGRAM

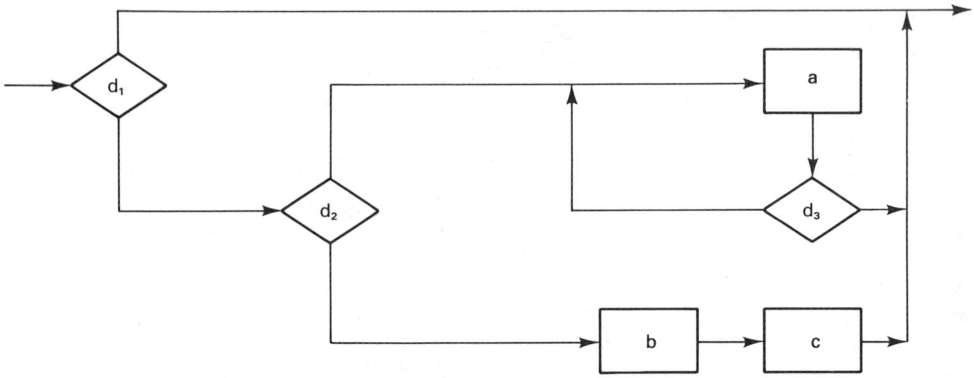

This flow diagram is a combination of an overall IF structure (SELECTION) applied to an IF . . . Repeat Until (REPETITION), THEN ELSE . . . Process Series (SEQUENCE), as shown by the dotted lines.

Figure 4.3

produce code built from a small set of logical constructs (chiefly, the sequence construct, the decision construct, and the loop construct.[P-1]

An example of some logic structures that do not conform to the recommendations and should be avoided are shown in Fig. 4.4.

CONSTRUCTS TO AVOID

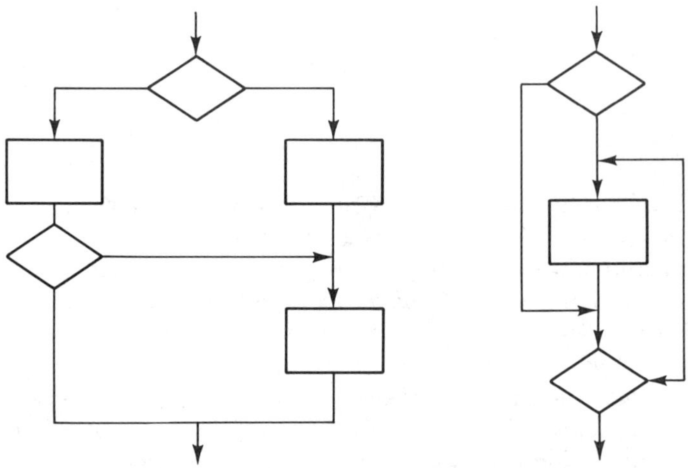

Figure 4.4

The logic that is illustrated in Fig. 4.4 leads to what is commonly called "spaghetti code." Code written without regard to structured programming is difficult to understand, costly to maintain, and should not be used in the final product.

STRUCTURED ENGLISH (PSEUDOCODE)
. . . the bridge from design to code

It is in the mini-spec that we want to specify the user's policy regarding the transformation of data by using *natural* language. However, we find that English is a very ambiguous language. There is a need to restrict the repertoire of words to avoid the inherent ambiguity in the language. Defining *structured English* is not easy, even if you traced the origin of the idea back to the paper by Caine and Gordon,[C-1] where the authors introduced programming design language (PDL), a tool for design software. In the paper, they describe a *pidgin* language as one that "uses the vocabulary of one language (i.e., English) and the overall syntax of another (i.e., a structured programming language)." The resultant context is a **"PSEUDOCODE" that has all the necessary language for expressing logic clearly and unambiguously, but lacks the rigor of a fully compileable programming language.**

Tom DeMarco[D-2] defines **STRUCTURED ENGLISH as a specification language that uses a limited vocabulary and a limited syntax. The vocabulary of structured English consists only of:**

Imperative English-language verbs

Terms defined in the data dictionary

Certain reserved words for logic formulation

After searching many texts on the subject, I was not able to derive any type of consensus on the composition of structured English or pseudocode. Structured English is a bit less exacting than pseudocode. It permits more freedom in phrasing the command logic, so that it resembles a *natural* language. The following subsection identifies the necessary component syntax needed to form a minimum set of structured English commands. There are many variations of this minimal set to specify the processing policy for functional primitives in a structured specification. The following pages list the basic components of structured English and form a basis for a pseudocode language. Included are recommendations based on experiences of using structured techniques in large-scale data processing and real-time software development.

Syntax and Variables

List all data flows, data stores, and common processing routines as uppercase hyphenated words and properly define them in a formal data dictionary. List local variables (i.e., variables defined only within a single functional primitive, such as CURRENT-TIME, CURRENT-DATE, ON/OFF, and REJECT, in uppercase hyphenated words, but do not put them into the data dictionary. The structured English commands, the process names of each functional primitive, the names of sources and sinks are in lowercase with the first letter of each word capitalized. All prepositions and verbs, such as is, are, to be, in, from, to, etc., are in lowercase and aid the specification in achieving a "natural" language flow.

Variables can be elements of a table, and when so, they can be indexed by a subscript. Typical notation is **VARIABLE (sub1,sub2, . . .)**. The subscript itself is a local variable and is in lowercase. As a local variable, the subscript is not put into the data dictionary.

Assignment Statements

Assign values to a single variable with the use of the ASSIGN statement.

VARIABLE = Value, Literal, or Expression

or VARIABLE <-- Value, Literal, or Expression

or **Set** VARIABLE **to** Value, Literal, or Expression.

or **Compute** VARIABLE = Expression.

There is frequently a need to copy several data elements from one group data element (i.e., usually a source document) to another group data element. It is convenient to be able to provide a command that accomplishes this in one operation. Corresponding values are assigned to individual data elements within group data elements where the elementary data elements have the same names by using the Copy command.

Copy GROUP-DATA-ELEMENT-1 **to** GROUP-DATA-ELEMENT-2.

Operators

Expressions can be either arithmetic or boolean. In the arithmetic case, the results are numeric and the operators are addition (+), subtraction (-), multiplication (*), division (/), or exponentiation (** or ^{superscript}). In the boolean case, the results are either **true** or **false**. In order to produce these true or false values, the logical operators **and**, **or**, and **not** are used. Also, the relational operators of less than (<), less than or equal (<=), equal (=), not equal (<> or ≠), greater than or equal (>=), and greater than (>) are available for use. We can specify computations in three ways: Assignment statement (=), **Compute** statement, and the **Calculate** statement. All three perform the identical results.

GROSS-PAY = HOURS * RATE

or **Compute** GROSS-PAY = HOURS * RATE

or **Calculate** GROSS-PAY = HOURS * RATE

Sequence of Execution

All commands are intended to be executed in a forward sequential order as written unless the logic or a command indicates otherwise. It is helpful to establish indentation rules so that computational units of logic are more easily identified. We can use the **Call** command to activate the logic of a common processing routine or function specified in the data dictionary.

Repetition Statements

There are three ways to represent a repetitive sequence of logic as shown in Figs. 4.5a to 4.5c.

DO WHILE statement:

While *condition* **Do**
  ⋮ (statements)
Endwhile.

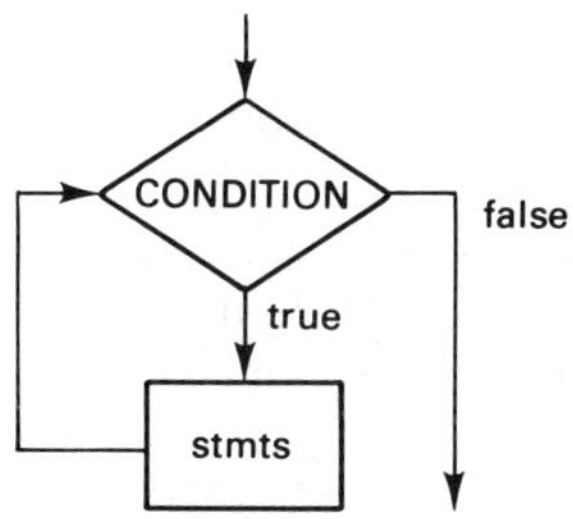

This looping construct tests for the condition prior to executing any logic in the control loop. Thus, the loop will not execute unless the condition is true.

Figure 4.5a

REPEAT UNTIL statement:

Repeat
  ⋮ (statements)
Until *condition.*

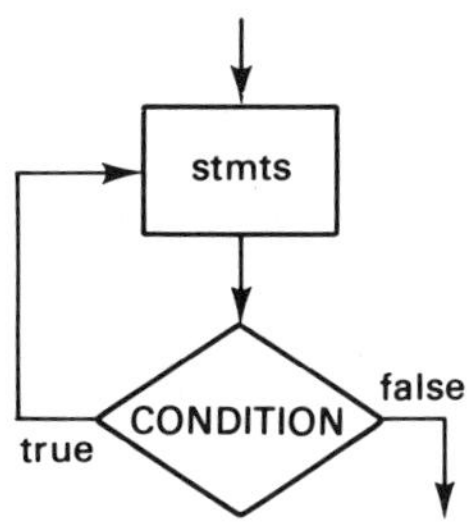

This looping construct tests for the condition after one full execution of the logic loop. Thus, at least one full pass of the control loop will always occur.

Figure 4.5b

LOOP FOREVER statement:

Loop
⋮ (statements)
If *condition* **then exit loop.**
⋮ (statements)
Forever.
⋮ (statements)

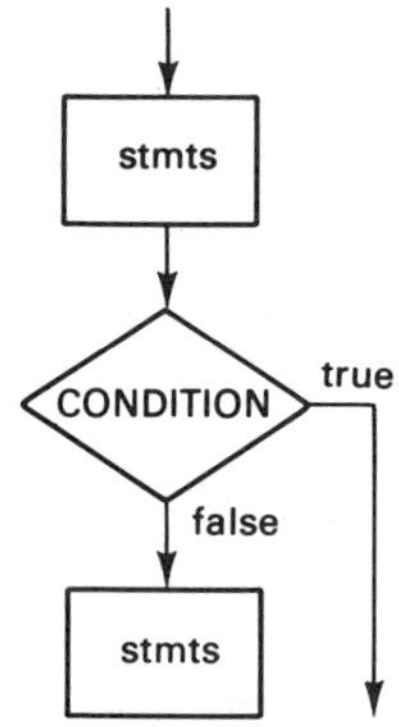

In contrast with the other looping mechanisms, this one provides complete control as to how much of the process is executed at least once, and how much is dependent on the condition being true. When the ***Exit*** *is taken, control passes to the statement following the command* ***Forever.***

Figure 4.5c

Decision Statements

The control logic to execute an alternate path of logic statements is available with the conditional statement **If . . . then . . . Else**. In Fig. 4.6, one of two logic paths can be selected based on a specified condition.

If *condition* **then** statements$_1$,
⋮ (statements)
Else statements$_2$.
⋮ (statements)
Endif.
statements$_3$.
⋮

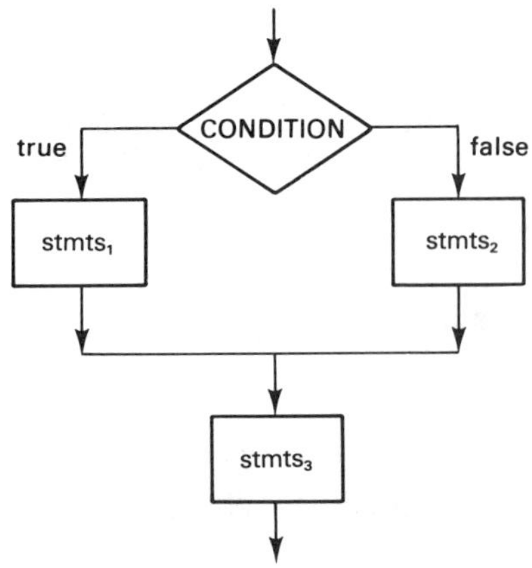

If the condition is "true," statements$_1$ execute. If the condition is "false," then statements$_2$ execute. Control always falls to statements$_3$.

Figure 4.6

Input/Output Statements

There are various acceptable ways of designating the flow of data into and out of a functional primitive process. Some analysts will use **Input** (data elements) **and Output** (data elements), or **Accept** (data elements) and **Send** (data elements). For those who began their careers as system programmers, **Get** (data elements) and **Put** (data elements) are favored. Those who grew up in COBOL prefer **Read** (data elements) and **Write** or **Print** (data elements). Whichever syntax you prefer, you have to be able to specify a basic input and output to the process.

⋮ (statements)

Input (*data elements . . .* **) from** *process, data store or source.*

⋮ (statements)

Output (*data elements . . .* **) to** *process, data store or sink.*

⋮ (statements)

In writing a mini-spec for a functional primitive process, we must input the data flows and data from data stores or sources before they are used in the logic. Likewise, we must send out all output data flows, data written to data stores, or data passed to sinks by means of an output statement. There are those who feel that specifying the **from** and **to** options in the **Input** and **Output** statements is redundant because the data flow diagram indicates the origin and destination of the input and output data for each process. The argument is, in theory, correct, although there are those who feel that the logic specification reads more smoothly with the **from** and **to** phrases specified. Thus, the use of the **from** and **to** clauses is left for the authors of the structured specifications to use if they wish.

File Manipulations

There are other pseudocode commands that can indicate certain types of file manipulations. These commands, although not absolutely necessary, do aid the reader in understanding very clearly the kind of interactions that take place with data stores within the system.

Create *data-entry** **in** *data-store-name.*

Delete *data-entry** **from** *data-store-name.*

Search *data-store-name* **for** *condition.*

Locate *data-element(s)* **in** *data-store-name* **for** *condition.*

Match *data element-1* **to** *data element-2* **in** *data-store-name.*

Merge *data-store-name-1* **into** *data-store-name-2.*

or **Merge** *data-store-name-1* **with** *data-store-name-2* **creating** *data-store-name-3*.

> ** The data-entry is usually identified by specifying the primary key value for the primary key data element(s) for the data store.*

Example

> **Create** CO-NAME **entry in** INSURANCE-COMPANIES.
>
> *where in the data dictionary:*
>
> ⋮
>
> INSURANCE-COMPANIES = {<u>CO-NAME</u> + CO-ADDR + CO-PHONE + . . . }
>
> ⋮

Case Statements

Another statement that analysts frequently use in specifying policy within mini-specs is the **Case** statement. This statement allows one to distinguish easily between several alternative sets of processing logic without using multiple **If..then..Else** statements.

⋮		⋮
Case 1 (*condition*$_1$):		**For** *condition*$_1$:
statements$_1$.		statements$_1$.
⋮		⋮
Endcase 1.		
Case 2 (*condition*$_2$):		**For** *condition*$_2$:
statements$_2$.		statements$_2$.
⋮		⋮
Endcase 2.		
Case 3 (*condition*$_3$):	or	**For** *condition*$_3$:
⋮		⋮
Endcase 3.		
⋮		⋮
Case n (*condition*$_n$):		**For** *condition*$_n$:
statements$_n$.		statements$_n$.
⋮		⋮
Endcase n.		
Else: statements$_{n+1}$.		**Else:** statements$_{n+1}$.
Endcases.		**Endfor.**
⋮		⋮

> *Here one can specify a set of logic that is used only when a specific condition applies. When a condition is met, all of the conditions in lowercase are not tested and the statements are not executed. The format of the* **Case** *statement offers the optional* **Else** *clause, which is used in the event that none of the above conditions apply.*

The use of the case structure is a convenient way to partition segments of logic that apply only to specific conditions. It may also be useful to partition segments of logic within a particular case segment. When we use a case structure within a case structure, it is known as *nested case structure.* The first-level case structure is traditionally numbered **Case 1**, **Case 2**, **Case 3**, etc. When case structures are nested it is helpful to use the traditional numbering system for data flow diagrams.

NESTED CASE STRUCTURES

```
Process Name                                         MINI-SPECIFICATION n

Input . . .
  :
Case 1  ( . . . condition for case 1 . . . ):
   Case 1.1  (first subcondition for case 1 . . . ):
      Set . . .
        :
      Stop.
   Case 1.2  (second subcondition for case 1 . . . ):
      Set . . .
        :
      Stop.
  :
Endcase 1.
Case 2  ( . . . condition for case 2 . . . ):
   Input . . .
     :
   Case 2.1  (first subcondition for case 2 . . . ):
      Case 2.1.1  (first subcondition for case 2.1 . . . ):
            :
            Stop.
      Case 2.1.2  (second subcondition for case 2.1 . . . ):
            :
            Stop.
        :
   Endcase 2.1.
   Case 2.2  (second subcondition for case 2 . . . ):
      Set . . .
        :
      Stop.
   Endcase 2.2.
     :
Endcase 2.
  :
Endcases.
```

Common Processing Routines (Call)

By calling a common processing routine defined in the data dictionary, control passes to a predefined set of logic that is available for use more than once in the system. The **Call** statement lists the parameters needed by the called process to perform its task. Likewise, the calling process receives the output data elements from the common processing routine. Define the name of the common processing routine in the data dictionary, as well as all of the data elements used in the parameter list. Common processing routines can **Call** themselves – direct recursion – or they can be part of a sequence of statements resulting in indirect recursion. (See the comments on recursion versus looping in Chapter 2.)

```
  ⋮
**Call** COMMON PROCESSING ROUTINE NAME (*parameter list*).
  ⋮
```

The parameter list contains all input and output variables passed to and from the common processing routine and the calling process. The logic in the common processing routine uses all of the variables in the parameter list as if they were variables defined in the common processing routine itself. No specific **Input** or **Output** statement is necessary to access or specify the values of these parameter variables. The common processing routine uses the parameter variables immediately as if they were local variables. When the common processing routine completes processing, it returns control to the statement immediately following the **Call** statement in the calling process. The common processing routine sends values back to the calling process by setting the specific parameter variables in the parameter list to the desired value before the common processing routine ends.

An example of how the **Call** statement would be used is shown in the mini-spec for the process 7.1.2* in Fig. 4.7.

CALL TO A COMMON PROCESSING ROUTINE

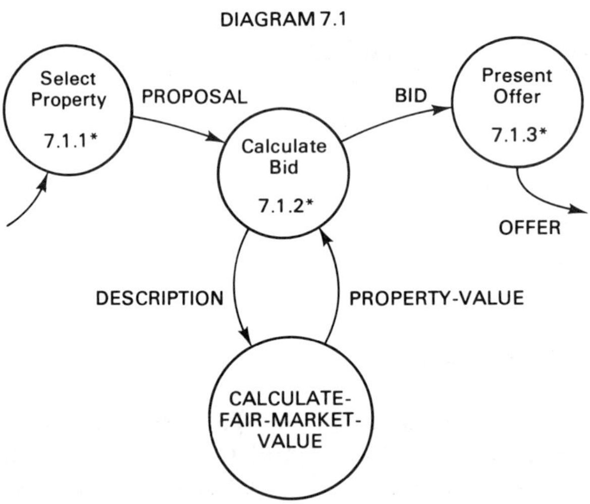

Calculate Bid MINI-SPEC 7.1.2

Input PROPOSAL from Select Property.
⋮
Call CALC-FAIR-MKT-VALUE
(DESCRIPTION, PROPERTY-VALUE).
Compare PROPERTY-VALUE to ASKING-PF
⋮
Output BID to Present Offer.
Stop.

Figure 4.7

Suspensions/Terminations/Comments

There may be a necessity to halt processing temporarily to allow a passage of time, or, in the case of a real-time process, to suspend execution until a control process reactivates the process. This operation is performed by the **Wait** command.

Temporarily Halt Execution

```
   ⋮
Output REQUEST-FOR-INFO-MSG to Customer.
Wait.
Input CUSTOMER-INFO from Customer.
   ⋮
```

Suspend a Real-Time Process

```
   ⋮
Output CUSTOMER-REC to MASTER-FILE.
Wait.
```

Sometimes there is a need to show the end or completion of the logic stream that processes the data. Also, comments can appear anywhere within the logic statement by using asterisks as used in the data dictionary.

```
   ⋮ (statements)  * This is a comment. *
Stop.
```

Use the **Stop** statement to halt execution and signify the end of the processing logic within the body of the mini-spec. When a **Stop** is executed, all resources dedicated to the process are *released* and become available to the operating system for reassignment to other processes. When a process issues a **Wait**, there is only a temporary halt in execution. All resources dedicated to the process *remain* dedicated to the process even though the process is not using them.

Other Real-Time Commands

There are several special pseudocode commands needed when specifying processes that operate in a real-time environment. In that environment, control processes initiate, activate, deactivate, suspend, and terminate other target processes. When a finite-state process is specified, it must set the value of the state variable to control which set of logic will receive control next. A complete discussion of real-time processing is presented in Chapter 10. However, to complete our discussion of pseudocode commands, the following commands are necessary to specify real-time requirements:

Initiate *new-process name.* * Control process starts new process. *

Activate *suspended-process name.* * Control process restarts a process that was suspended. *

Suspend *process name.* * Control process temporarily halts the activities of another process while allowing it to retain its resources. *

Deactivate *process name.* * Control process stops another process and causes it to release its resources. *

Set STATE-1.2 **to** b. * Finite-state process prepares to change states. *

Wait for *condition.* * Process temporarily suspends itself until the condition is satisfied. The process retains its resources. *

Wait. * Process suspends itself but retains its resources. *

Stop. * Process deactivates itself and releases its resources. *

A control process uses the command **Deactivate** to terminate a target process. It is similar to the **Stop** command in that all resources dedicated to the target process are released and made available to the operating system for reassignment to other processes. Likewise, control processes use the **Suspend** command, which is similar to the **Wait** command in that a temporary halt is obtained and all resources dedicated to the target process remain dedicated to the process even though the process is not using them. Control processes use the **Initiate** command when they begin a new process and cause the process to claim resources from the operating system. Similarly, control processes use the **Activate** command to resume a target process that has been temporarily halted and has resources assigned to it. All functional primitives terminate and release all resources by executing the **Stop** command.

You may notice that there is no command that implements an unconditional branch, such as a **GoTo**. There is no need to break out of logic unconditionally when only specifying policy. The **Call** command executes common processing routines and functions documented in the data dictionary. *Do not* attempt to pass control from one functional primitive process to another via a **Call** statement. Use the **Call** statement to activate subordinate common processing routines only.

When the policy for a process in a mini-spec performed by a human being is documented, additional verbs may be necessary. Remember to keep statements short, simple, and direct, using verbs, objects, and prepositional phrases. The subject *you* is always understood. Statements should avoid taking on the form of a narrative text.

The command statements used in pseudocode presented here are flexible so as to permit an author to represent any policy or set of activities necessary to process the data in the desired manner. However, commands such as **Check, Verify**, or **Calculate** (without specifying a condition or mathematical formula) may not be sufficient to unambiguously specify the processing requirements. Care should be taken when introducing additional pseudocode commands and formality so as not to specify too many details of *how* to do the transformation of data. There is a varying degree of detail used in a mini-spec. However, it is only necessary to specify the policy at a relatively high level. Consider the following example, shown in Figs. 4.8a to 4.8c, where one could specify, in detail, the pseudocode steps necessary to balance a personal checkbook (Fig. 4.8b), and the mathematical formula (Fig. 4.8c) that is sufficient to specify the *policy* of determining if a personal checkbook balances with the bank statement.

CHECKBOOK BALANCING

DATA FLOW DIAGRAM

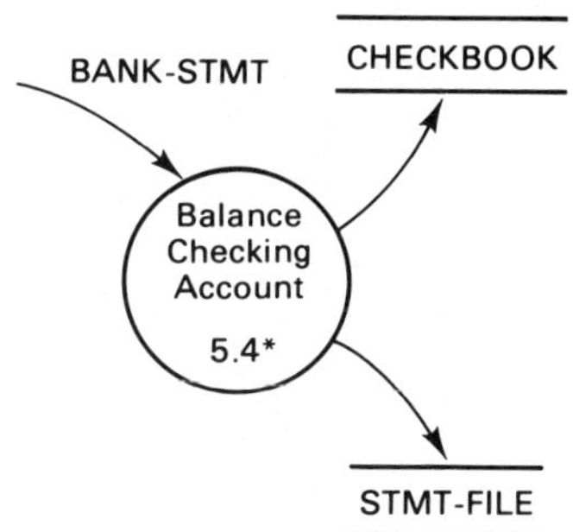

PARTIAL DATA DICTIONARY

⋮

BANK-STMT = ACCT-NO + STMT-DATE + {CANCELED-CHECK} + {DEPOSIT-SLIP}
+ BANK-BAL + (SERVICE-CHARGE)

CANCELED-CHECK = CHECK-# + CHECK-DATE + CHECK-AMT + PAYEE + ACCT-#
+ DATE-CLEARED + PAYOR-SIGNATURE + PAYEE-SIGNATURE

CHECKBOOK = {DATE-TRANS + TRANS-CLEARED + [CHECK-# + CHECK-AMT
+ PAYEE | AMT-DEPOSIT] + CURRENT-BAL}

DEPOSIT-SLIP = DEPOSIT-DATE + AMT-DEPOSIT + ACCT-NO

SERVICE-CHARGE = * $2.00 + $.10 per CANCELED-CHECK. *

STMT-FILE = {MONTH-YR + BANK-STMT + ["BALANCED" | "NOT-BALANCED"]}

⋮

Figure 4.8a

DETAILED PSEUDOCODE

Balance Checking Account MINI-SPECIFICATION 5.4

```
Input BANK-STMT from Bank.
If SERVICE-CHARGE > 0,
    Output New-entry in CHECKBOOK by:
        Set DATE-TRANS to STMT-DATE.
        Set PAYEE to "Bank Service Charge".
        Set CHECK-AMT to SERVICE-CHARGE.
        Subtract SERVICE-CHARGE from CURRENT-BAL.
Endif
For each CANCELED-CHECK on BANK-STMT:
    Match CHECK-DATE to DATE-TRANS.
    Set CHECK-CLEARED to "√".
Endfor.
For each DEPOSIT-SLIP on BANK-STMT:
    Match DEPOSIT-DATE to DATE-TRANS.
    Set CHECK-CLEARED to "√".
Endfor.
Set OUTSTANDING-CHECKS to zero.
For each DATE-TRANS where CHECK-CLEARED = "√",
    If CANCELED-CHECKS, Add CHECK-AMT to OUTSTANDING-CHECKS.
    Endif.
    If DEPOSIT-SLIP, Add AMT-DEPOSIT to OUTSTANDING-DEPOSITS.
    Endif.
Endfor.
Add OUTSTANDING-DEPOSITS to BANK-BAL.
Subtract OUTSTANDING-CHECKS from BANK-BAL.
If BANK-BAL = CURRENT-BAL,
    Create new entry in STMT-FILE:
        Set MONTH-YR to STMT-DATE.
        Set new entry to "BALANCED".
        Output BANK-STMT to STMT-FILE.
Else, Create new entry in STMT-FILE:
        Set MONTH-YR to STMT-DATE.
        Set new entry to "NOT-BALANCED".
        Output BANK-STMT to STMT-FILE.
Endif.
Stop.
```

<u>Processor Requirement:</u> . . .

<u>Security Requirements:</u> . . .

<u>Time Requirements:</u> . . .

<u>Processing Priorities:</u> . . .

<u>Performance Requirements:</u> . . .

<u>Safety Requirements:</u> . . .

<u>Other Requirements:</u> . . .

Figure 4.8b

STATEMENT OF POLICY

Balance Checking Account MINI-SPECIFICATION 5.4

Input BANK-STMT.
AMT = BANK-BAL - SERVICE-CHARGE - OUTSTANDING-CHECKS + OUTSTANDING-DEPOSITS
If AMT = CURRENT-BAL then Output STMT-DATE, "BALANCED" to STMT-FILE,
Else, Output STMT-DATE, "NOT-BALANCED" to STMT-FILE.
Endif.
Stop.

OUTSTANDING-CHECKS = Σ CHECK-AMT in CHECKBOOK -Σ CHECK-AMT in BANK-STMT

OUTSTANDING-DEPOSITS = Σ DEPOSIT-AMT in CHECKBOOK -Σ DEPOSIT-AMT in BANK-STMT

Processor Requirement: . . .
Security Requirements: . . .
Time Requirements: . . .
Processing Priorities: . . .
Performance Requirements: . . .
Safety Requirements: . . .
Other Requirements: . . .

Figure 4.8c

Some companies may choose to specify more or less detail in their mini-specs, however, it is only necessary to specify the business policy, as shown in the STATEMENT OF POLICY example of Fig. 4.8c.

In real-time applications, there are frequently *time* and *performance* requirements to consider in the processing of the data. State these requirements in the lower section of each mini-spec.

Some companies find it necessary to explicitly define the process in a greater level of detail by using pseudocode, as shown in the DETAIL PSEUDOCODE of Fig. 4.8b. As a general rule, avoid falling into the trap of specifying *how* a task or activity is to be performed during the analysis stage. A mini-spec should explain details of the input preparation, the policy governing the basic data transformation, and the requirements of the output data.

Later in this chapter I will present action diagrams. Often action diagrams can be used to specify the policy of processing the data while permitting the author to specify the structure of the logic to perform the policy. At later stages of the development, system designers can specify a more detailed level of logic within the action diagrams. During the coding stage the programmer can use the action diagrams and pseudocode, and supply the specific syntax necessary to implement the solution in a particular programming language for a specific type of computer.

The organizational style of a mini-spec varies greatly. Some of the most common types follow, however none of these examples use action diagrams.

Example (Indented Phrasing)

```
Input DAILY-SHIPMENTS from Shipping Clerk.
While for each SHIPPING-INVOICE Do:
    Set SHIPPING-WEIGHT on SHIPPING-INVOICE to WEIGHT-OF-SHIPMENT.
    If CUSTOMER-CREDIT-RATING = "Good",
        Set SHIPPING-$ on SHIPPING-INVOICE to SHIPPING-COSTS.
        Output SHIPPING-INVOICE to Transportation.
    EndIf.
    If CUSTOMER-CREDIT-RATING = "No Credit",
        Stamp the SHIPPING-INVOICE with "PREPAYMENT REQUIRED".
        Compute SHIPPING-$ = SHIPPING-COST + SPECIAL-HANDLING-FEE.
        Set SHIPPING-$ on SHIPPING-INVOICE.
        Output SHIPPING-INVOICE to Cash Sales.
    EndIf.
    If CUSTOMER-CREDIT-RATING = "COD Only",
        Compute SHIPPING-$ = SHIPPING-COST + COD-CHARGES.
        Set SHIPPING-$ on SHIPPING-INVOICE.
        Output SHIPPING-INVOICE to Transportation.
    EndIf.
EndWhile.
Stop.
```

Processor Requirement: . . .

⋮

Other Requirements: . . .

Example (Outlining)

```
1. Input DAILY-SHIPMENTS from Shipping Clerk.
2. While for each SHIPPING-INVOICE Do:
   2.1    Set SHIPPING-WEIGHT on SHIPPING-INVOICE to WEIGHT-OF-SHIPMENT.
   2.2    If CUSTOMER-CREDIT-RATING = "Good",
      2.2.1    Set SHIPPING-$ on SHIPPING-INVOICE to SHIPPING-COSTS.
      2.2.2    Output SHIPPING-INVOICE to Transportation.
   2.3    If CUSTOMER-CREDIT-RATING = "No Credit",
      2.3.1    Stamp the SHIPPING-INVOICE with "PREPAYMENT REQUIRED".
      2.3.2    Compute SHIPPING-$ = SHIPPING-COST + SPECIAL-HANDLING-FEE.
      2.3.3    Set SHIPPING-$ on SHIPPING-INVOICE.
      2.3.4    Output SHIPPING-INVOICE to Cash Sales.
   2.4    If CUSTOMER-CREDIT-RATING = "COD Only",
      2.4.1    Compute SHIPPING-$ = SHIPPING-COST + COD-CHARGES.
      2.4.2    Set SHIPPING-$ on SHIPPING-INVOICE.
      2.4.3    Output SHIPPING-INVOICE to Transportation.
3. Stop.
```

Processor Requirement: . . .

⋮

Other Requirements: . . .

Example (Case Structure)

```
Input DAILY-SHIPMENTS from Shipping Clerk.
While for each SHIPPING-INVOICE Do:
    Set SHIPPING-WEIGHT on SHIPPING-INVOICE to WEIGHT-OF-SHIPMENT.
    CASE 1 (CUSTOMER-CREDIT-RATING = "Good"):
        Set SHIPPING-$ on SHIPPING-INVOICE to SHIPPING-COSTS.
        Output SHIPPING-INVOICE to Transportation.
    Endcase 1.
    CASE 2 (CUSTOMER-CREDIT-RATING = "No Credit"):
        Stamp the SHIPPING-INVOICE with "PREPAYMENT REQUIRED".
        Compute SHIPPING-$ = SHIPPING-COST + SPECIAL-HANDLING-FEE.
        Set SHIPPING-$ on SHIPPING-INVOICE.
        Output SHIPPING-INVOICE to Cash Sales.
    Endcase 2.
    CASE 3 (CUSTOMER-CREDIT-RATING = "COD Only"):
        Compute SHIPPING-$ = SHIPPING-COST + COD-CHARGES.
        Set SHIPPING-$ on SHIPPING-INVOICE.
        Output SHIPPING-INVOICE to Transportation.
    EndCases.
EndWhile.
Stop.
```

<u>Processor Requirement:</u> . . .

⋮

<u>Other Requirements:</u> . . .

Example (Other)

```
For each SHIPPING-INVOICE in DAILY-SHIPMENTS:
    Write the SHIPPING-WEIGHT on the SHIPPING-INVOICE as the WEIGHT-OF-SHIPMENT.
    For CUSTOMER-CREDIT RATING for CUSTOMER-NAME is "Good",
        Set the SHIPPING-$ to SHIPPING-COST.
        Send the SHIPPING-INVOICE to Transportation.
    Endfor.
    For CUSTOMER-CREDIT-RATING for CUSTOMER-NAME is "No Credit",
        Set the SHIPPING-$ to SHIPPING-COST + SPECIAL-HANDLING-FEE.
        Send the SHIPPING-INVOICE to Cash Sales.
    Endfor.
    For CUSTOMER-CREDIT-RATING for CUSTOMER-NAME is "COD Only",
        Set the SHIPPING-$ to SHIPPING-COST + COD-CHARGES.
        Send the SHIPPING-INVOICE to Cash Sales.
    Endfor.
Endfor.
Stop.
```

<u>Processor Requirement:</u> . . .

⋮

<u>Other Requirements:</u> . . .

LOOPING AND RECURSION
. . . a function of who calls who

Both recursion and iteration cause repeated execution of a procedure or some set of instructions. The difference lies in who invokes the repetition of the process and how the process receives its input data.

An ITERATIVE PROCESS is one that is invoked externally and continuously repeats until some criteria is satisfied. Some logic or process outside the iterative process initially invokes the execution of the process. The repetition can be controlled by logic externally or internally. A process is controlled internally when its logic determines if a task is to start over again until it meets some specific criteria. The initiation of the second pass through the logic of the iterative process permanently ends the execution of the first pass.

Iterative processes receive their data only from external processes, data stores and/or sources. Internal variables are established by the process when it is invoked for the first time. These variables can control the number of times the process repeats. If the process requires additional data, they come from external processes, data stores, and/or sources.

A RECURSIVE PROCESS is one, that during an execution cycle, invokes itself and initiates another execution cycle. Notice that the control to initiate subsequent executions is done *within the process itself.* The initiation of the second pass or a call to the logic of a recursive process *does not end* the execution of the first pass. When the process satisfies the criteria to end the processing cycle, control is passed back to "the appropriate place" in the midst of the previous execution cycle to complete the execution of the previous pass. To maintain *where the appropriate place is*, the process places a return address on a *stack*. **A STACK is a last in, first out (LIFO) queue of data elements.** Recursive processes use a stack to save all pertinent data elements, so that when a previously interrupted cycle is resumed, all of the values are restored as of the time of the interrupt. The process takes the return address and any

DATA FLOW DIAGRAM OF ITERATIVE LOOPING & RECURSION

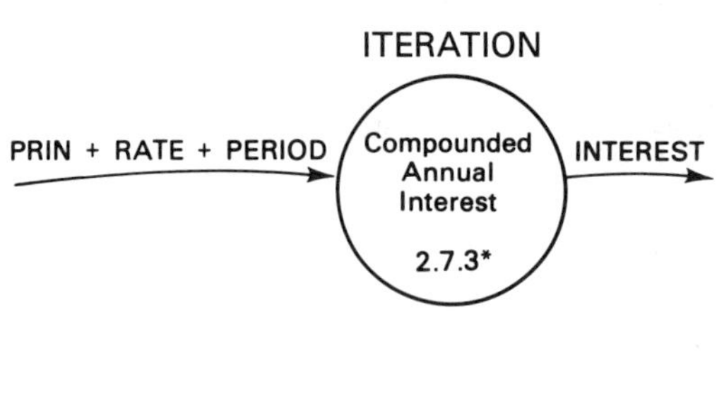

Figure 4.9a

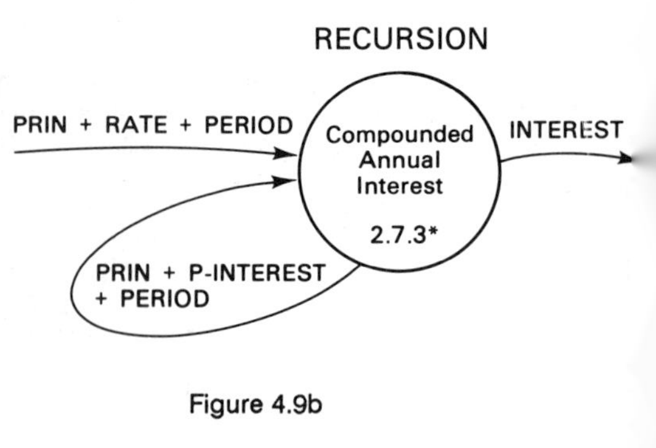

Figure 4.9b

necessary variables off the top of the stack and execution continues until the next cycle completes. At this point, the process again returns control to the point in the previously interrupted cycle. This reverse sequencing continues until the logic that originally invoked the recursive process receives control or the entire recursive process has completed its task.

Recursive processes receive their data only initially from external processes, data stores, and/or sources. The process itself provides the input or output for all additional execution cycles, including the return address used when the entire process completes. A graphical illustration that indicates whether a process repeats iteratively or recursively is shown in Figs. 4.9a and 4.9b.

Notice in Fig. 4.9a and 4.9b that the notation for an iterative process does not imply any type of repeated execution. To determine that the process does in fact repeat, we must reference the mini-spec. However, when a process suspends its own execution, invokes itself, and provides new data values for use in subsequent execution, the data flow diagram does indicate the additional data flow. When and how the process determines it should invoke itself is in the mini-spec. The data flow diagram is interested in only indicating the source(s) and destination(s) of the data flows that exist in the system.

MINI-SPECIFICATIONS

ITERATION

Compounded Annual Interest Mini-spec 2.7*

```
Input PRIN, RATE and PRD.
If PRD is "annual", Set P=1.
If PRD is "semi-annual", Set P=2.
If PRD is "quarterly", Set P=4.
If PRD is "monthly", Set P=12.
While P>0, then
    P-INTEREST = RATE/P * PRIN
    INTEREST = INTEREST + P-INTEREST
    PRIN - PRIN + P-INTEREST
    P = P - 1
Endwhile.
Output INTEREST.
Stop.
```

Computes the interest for an entire year.

Figure 4.10a

RECURSION

Compounded Annual Interest Mini-spec 2.7*

```
Input PRIN, RATE and PRD.
If PRD is "annual", Set P=1, RATE=RATE/P.
If PRD is "semi-annual", Set P=2, RATE=RATE/P.
If PRD is "quarterly", Set P=4, RATE=RATE/P.
If PRD is "monthly", Set P=12, RATE=RATE/P,
    P-INTEREST = RATE * PRIN
    PRIN = PRIN + P-INTEREST
    PRD = PRD -1
If PERIOD > 0,
    Call Compounded Annual Interest (PRIN,
    P-INTEREST, RATE),
Else, Return.
INTEREST = INTEREST + P-INTEREST
Output INTEREST.
Stop.
```

Computes the interest for 1 period in the year.

Figure 4.10b

PRIN is principle, PRD is period (length of time of the loan).

The mini-spec of Fig. 4.10a contains a loop that repeats for each period and accumulates a total year's interest. In the mini-spec of Fig. 4.10b, the process computes the interest for just one period. Each time a cycle ends, the subsequent cycle inherits the principle and the accumulated interest, and proceeds to compute the next period's interest. This action continues until the process computes an entire year's interest.

Let's consider the recursive process for PERIOD = "QUARTERLY":

```
Initially:  PRIN=$1,000.00 RATE=6.0% PERIOD="QUARTERLY" P-INTEREST=$0

   During execution, the values
   are changed. The values that
   exist just prior to the
   Call are: PRIN=$1,015.00  RATE=1.5%  PERIOD=3  P-INTEREST=$15.00

      During the second execution, the
      values are changed. The values
      that exist just prior to the
      Call are: PRIN=$1,030.23  RATE=1.5%  PERIOD=2  P-INTEREST=$15.23

         During the third execution, the
         values are changed. The values
         that exist just prior to the
         Call are: PRIN=$1,045.68  RATE=1.5%  PERIOD=1  P-INTEREST=$15.45

            During the fourth execution, the
            values are changed. The values
            that exist just prior to the
            Call are: PRIN=$1,061.37  RATE=1.5%  PERIOD=0  P-INTEREST=$15.69

            The fourth pass completes and the process now begins to unravel
            because PERIOD=0, and sets INTEREST=$15.69 and returns.

         The third pass completes and sets INTEREST=$31.14 and returns.

      The second pass completes and sets INTEREST=$46.37 and returns.

   The process completes and sets INTEREST=$61.37. The value of INTEREST
   is outputted and the entire process stops.
```

Although most recursive processes are automated and are deliberately designed to accommodate recursion, manual processes can sometimes be recursive. Think of a manual process that performs some time-consuming detailed calculations for a client. For example, consider the case of an accountant filling out a client's income tax. Page 2 of the 1040 form instructs the accountant to compute the Amount of Itemized Deductions on Form A. The accountant suspends working on page 2 of the 1040 form and begins filling out Form A.

While Form A is being filled out, there is a need to determine the client's Adjusted Gross Income. To determine the Adjusted Gross Income the accountant must complete page 1 of the 1040 form. Therefore, the accountant suspends working on Form A and begins working on page 1 of the 1040 form.

While page 1 of the 1040 form is being filled out, there is a need to know the Total Amount of Interest and Dividends, obtained by completing Form B. The accountant then suspends working on page 1 of the 1040 form and begins on Form B.

After completing Form B, the accountant resumes working on page 1 of the 1040 with the Total Amount of Interest and Dividends. Upon completing page 1 of the 1040 form, the accountant resumes working on Form A, knowing the client's Adjusted Gross Income.

By completing Form A, the accountant knows the client's Amount of Itemized Deductions, so work can resume on page 2 of the 1040 form. After completing page 2 of the 1040 form, the accountant finally determines the client's Amount of Tax Liability.

FEDERAL INCOME-TAX PREPARATION PROCESS

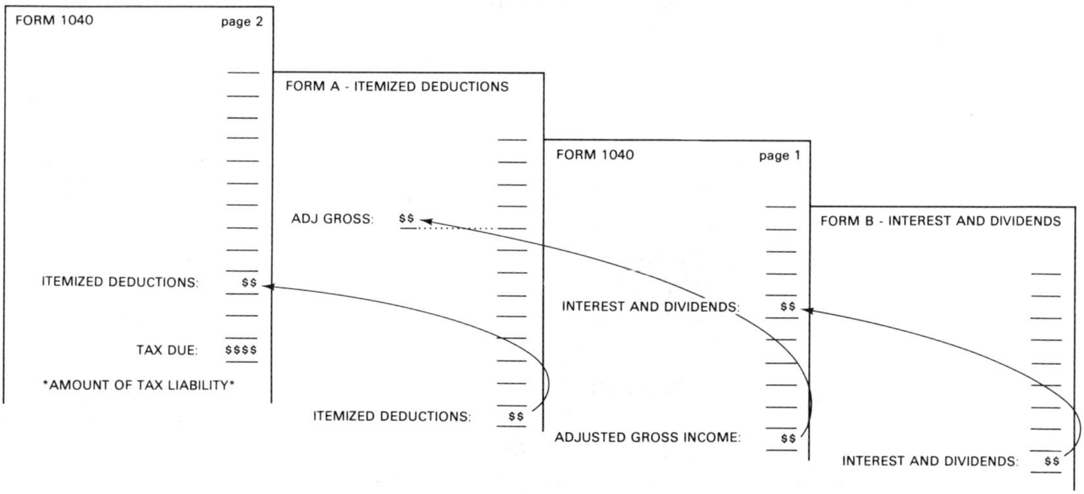

Figure 4.11

Figure 4.11 shows the accountant of our example acting as a recursive process that fills out tax forms. The data flow diagram for the process is shown in Fig. 4.12.

While one tax form is being filled out, the process needs to know a subtotal from another tax form. These intermediate subtotal amounts continue until the process computes the final INTEREST AND DIVIDENDS. Then, the process begins to unravel itself and continues completing each form until the objective of the process is reached.

The above example is a recursive process. Generally, a process has to be recursive if the result from one cycle is necessary to complete the processing of another cycle, and the processing of that cycle is necessary to complete the processing for the previous cycle.

TAX FORM DATA FLOW DIAGRAM

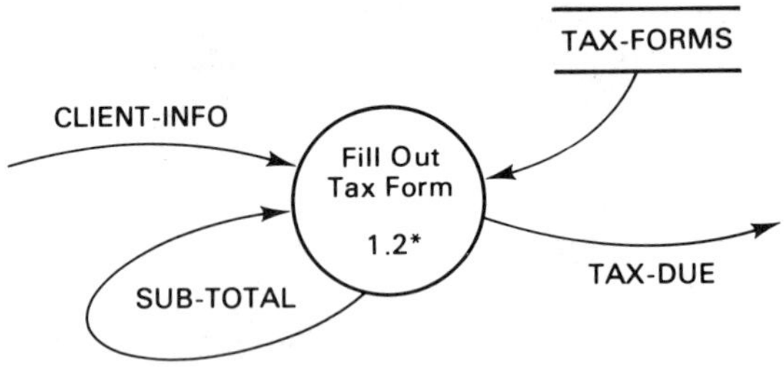

Figure 4.12

TOOLS TO SPECIFY THE PROCESSING OF DATA
. . . if the only tool you have is a hammer, you tend to treat everything as if it were a nail

Decision Trees

A DECISION TREE is a graphic tool for portraying a hierarchy of independent conditions and the activities resulting from each valid combination of conditions.[P-1] The tree consists of a hierarchy of nodes, where each node is either a condition (decision point), or an action point that specifies the action to take, or a specific value (leaf). The output of a decision point is either another decision point, or a leaf. Consider, in the example of Fig. 4.13 the possible actions that a driver of an auto can take when approaching a traffic light.

Decision trees are useful when there are statements of condition that have several outcomes, or when all statements of condition do not apply, such as, is there "TIME TO STOP?" when the "COLOR OF LIGHT?" is GREEN.

In constructing a decision tree, organize the statements of condition from the general (leftmost) to the specific (rightmost). Position the statements of condition that have the greatest effect in determining the outcome to the left of those that have a lesser impact. The leftmost node in a decision tree is the *root node*. **The ROOT NODE is the single, highest-level node in a decision tree from which the entire tree is generated.** In the example of Fig. 4.13, the "COLOR OF LIGHT?" has the most dominant effect in determining the resulting action, particularly because the condition "TIME TO STOP?" does not even apply in all cases. The "COLOR OF LIGHT?" condition is the root node of the decision tree.

To read a decision tree, begin at the left and trace a path from left to right using the values of each statement of condition to determine the path. The paths of a decision tree are usually mutually exclusive.

Tree structures are useful in showing the structure of data, systems, programs, as well as logic in a mini-spec. Decision trees are not only useful in computer science

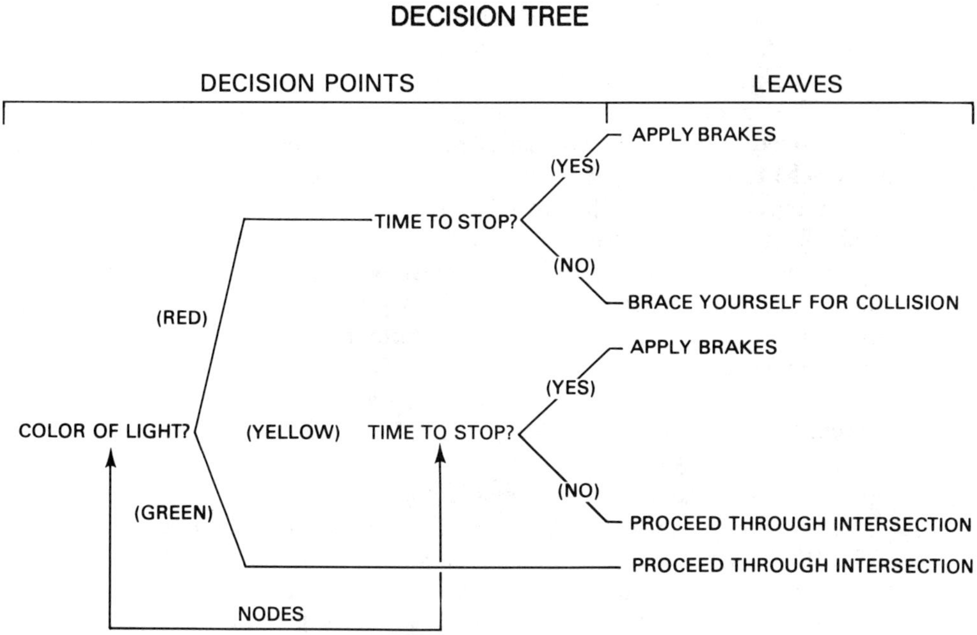

Figure 4.13

(programming and data-base design), but are also used in the fields of biology, information theory, and switching theory. In general, there are four main fields that use tree structures:

1. Taxonomy: diagnosis and pattern recognition.
2. Circuit (logic) design and reliability testing.
3. Algorithm analysis.
4. Programming and data-base design.

Tree structures are also used in many areas as documentation aids primarily because they are so simple and self-explanatory. Their ability to organize complex logic finds use in computer software and electronics hardware design. They have been used in the identification of unknown objects in biology and in system faults, and thus are also known as *diagnostic keys*.

One important characteristic of decision trees is that they limit redundant testing. With decision trees one has a map of the logic to use in constructing test plans; therefore a test for every logic branch can be constructed. The tree structure forms an algorithm that indicates which variable to test next, based on the previously tested variables and their results. We continue to evaluate conditions, left to right, until we reach the final value or action. In general, we can expand the role of a decision tree and view it as a means of partitioning a problem. Each node (*decision point*) or leaf (*action point*) can be a tree in and of itself.

We can illustrate the structure of a structured specification by using a decision tree. We partition a problem through successive decomposition when constructing data flow diagrams. The context diagram is the root node on the tree. Each process on level 1 is a node off the root node. Once we have reached a functional primitive, a single decision tree is effective in completing the specification. Decision trees serve well when there are several outcomes from a condition. They are most applicable when the number of conditions is relatively small. If the number is large, we can structure the conditions to provide a binary result (yes or no, true or false). In this case, a decision table may serve as a better means of documenting the policy of a functional primitive.

Decision Tables

In cases where the policy statements or logic for the process is more complex than can be comfortably handled by **If . . . then . . . Else** statements, use a *decision table.* **A DECISION TABLE is a tabular expression of procedural decision situations that is characterized by one or more sets of actions that are executed based upon the state of the conditions.**[C-5] The basic ingredient for a decision table is a *decision rule.* **A DECISION RULE describes a set of condition alternatives and a series of actions to be performed.**[C-5] There are two basic parts of a decision table. The first part is the statements of condition. The condition part is divided into two sections: the *condition stubs,* which contain all the conditions that apply to a particular problem, and the *condition entries,* which state a particular set of results of the statements of condition (yes or no, true or false). See Fig. 4.14. The second part of the decision table is the statements of action. The action part is likewise divided into two sections: the *action stubs,* which contain all the statements of action that are possible within the system,

FORMAT OF A DECISION TABLE

<Condition Stub>	*Decision Rule 1*	*Decision Rule 2*	*Decision Rule 3*	*Decision Rule 4*	*Decision Rule 5*	*Decision Rule 6*	*Decision Rule 7*	*Decision Rule 8 . . .*
Condition 1 :								
Condition 2 :					*<--- Condition Entries --->*			
Condition 3 :								
<Action Stub>								
Action 1 :								
Action 2 :								
Action 3 :					*<--- Action Entries --->*			
Action 4 :								

Figure 4.14

and the *action entries*, which indicate the selected actions that apply to a particular combination of results from the condition part.

For ease in constructing a decision table, phrase the conditions in the condition stub to accept a binary answer (i.e., yes or no, true or false). Sometimes we may have to divide the condition into several conditions before achieving a simple set of yes or no answers. Once the conditions have been phrased to accept a binary answer, we can complete the decision rules.

To determine how many decision rules are possible when there are three statements of condition, we can compute the maximum number as follows:

> The first statement of condition can have a yes or no answer (a total of two possible outcomes). The second statement of condition, can have a yes or no answer while condition 1 is yes, and it can have a yes or no answer while condition 1 is no (a total of four possible outcomes). Thus, there are four possible combinations with two statements of condition. These four combinations can exist when a third condition is yes, and the four can exist again when the third condition is no. Thus, when there are three statements of condition, the total number of possible outcomes is eight.
>
> Continuing in this manner, we find that for each new condition that accepts a yes or no answer, the total number of possible outcomes doubles. The table of Fig. 4.15 provides a means of computing the

MAXIMUM NUMBER OF DECISION RULES

CONDITIONS	OUTCOMES	MAXIMUM NUMBER OF POSSIBILITIES
Condition 1 :	Y N	2 or 2^1
Condition 1 :	Y Y N N	
Condition 2 :	Y N Y N	4 or 2^2
Condition 1 :	Y Y Y Y N N N N	
Condition 2 :	Y Y N N Y Y N N	8 or 2^3
Condition 3 :	Y N Y N Y N Y N	
Condition 1 :	Y Y Y Y Y Y Y Y N N N N N N N N	
Condition 2 :	Y Y Y Y N N N N Y Y Y Y N N N N	16 or 2^4
Condition 3 :	Y Y N N Y Y N N Y Y N N Y Y N N	
Condition 4 :	Y N Y N Y N Y N Y N Y N Y N Y N	
⋮	⋮	
Condition 1 :	Y Y Y Y Y Y Y Y... N N N N N N N N	
⋮		2^n
Condition *n* :	Y N Y N Y N Y N... Y N Y N Y N Y N	

Figure 4.15

maximum number of decision rules for any number of statements of condition.

Now that we have determined all possible combinations of *n* number of statements of condition, we can consider the action part of the decision table. In that part, we list all of the possible alternative actions. An appropriate mark (such as an "*" or "X") indicates the action(s) taken for each decision rule.

Consider the following example of computing the shipping and handling costs for a mail-order business. This example was used to illustrate the various formats of a mini-spec earlier in the chapter. The statements of condition would be

Is CUSTOMER-NAME a "COD-Only" customer? *(COD means Cash On Delivery)*

Is the CUSTOMER-NAME's credit "Good"?

The possible alternative actions are

Set SHIPPING-$ = SHIPPING-COST.

or Set SHIPPING-$ = SHIPPING-COST + COD-CHARGES.

or Set SHIPPING-$ = SHIPPING-COST + SPECIAL-HANDLING-FEE.

and the customer's order is either

Sent to the Transportation department.

or Sent to the Cash Sales department.

These are shown in the decision table of Fig. 4.16.

DECISION TABLE

	←----- DECISION RULES -----→			
"COD-Only" Customer?	Y	Y	N	N
Is the Customer's credit "Good"?	Y	N	Y	N
Set SHIPPING-$ = SHIPPING-COST			X	
Set SHIPPING-$ = SHIPPING-COST + COD-CHARGES	X	X		
Set SHIPPING-$ = SHIPPING-COST + SPECIAL-HANDLING-FEE				X
Send order to Transportation Dept.	X	X	X	
Send order to Cash Sales Dept.				X

Figure 4.16

If the customer is a "COD-Only" customer, then it does not matter if the customer's credit is "Good". When a statement of condition does not change the alternative actions, we can "compress" the decision table into one having fewer decision rules, as shown in Fig. 4.17.

COMPRESSED DECISION TABLE

	←---- DECISION RULES ----→		
"COD-Only" Customer?	Y	N	N
Is the Customer's credit "Good"?	-	Y	N
Set SHIPPING-$ = SHIPPING-COST		X	
Set SHIPPING-$ = SHIPPING-COST + COD-CHARGES	X		
Set SHIPPING-$ = SHIPPING-COST + SPECIAL-HANDLING-FEE			X
Send order to Transportation Dept.	X	X	
Send order to Cash Sales Dept.			X

Figure 4.17

If we construct a decision table properly, it can be used not only to define the logic of a process performed by a program module, but actually used to generate the code for the program module. For more detailed information on decision tables, see the report prepared by the Conference on Data Systems Languages (CODASYL).[C-5]

Action Diagrams

An ACTION DIAGRAM is a diagramming notation and specification tool that uses nested layers of brackets that represent the structure of the logic used to transform input data into output data. One advantage of an action diagram is that it is useful throughout the development life cycle. During the analysis, the action diagrams are prepared in a very general manner to specify only the users' policies of transforming the input data into output. During the design stage, we can expand these diagrams to include a more detailed level of specification of both the structure of the process and the processing logic. These diagrams are refined even further during the low-level design to plan and document the structure and programming logic of each program or to specify the procedures used in each process in the system. We can use the action diagram as the foundation to specifying the actual code during the implementation stage. Action diagrams cater to the top-down development approach as there is a natural movement from the high-level structure of the system to the low-level details of the procedures used to accomplish the intended results.

A second advantage in using action diagrams is that this tool has a very tight correlation between the structure of the logic and the specific logic statements used to accomplish a task. Thus, if we use only the three basic constructs of Bohm and Jacopini (easily verified by the layered brackets in an action diagram), the resulting logic conforms to *good* structured programming standards.

Action diagrams are also easy to understand and use. They were originally designed to assist end users in using report-generation software packages (see fourth-generation languages in Chapter 12). Because of the high concern for user understanding and verification of the structured specification, action diagrams are an excellent means of specifying the transformation of input data into output data in functional primitives.

In an action diagram, the entire functional primitive has a large square bracket down its left side. The square bracket is the basic building block of an action diagram. Within the bracket is a sequence of operations that always starts at the top and progresses down to the bottom. Brackets can be nested to indicate a hierarchical logic structure, as shown in Fig. 4.18.

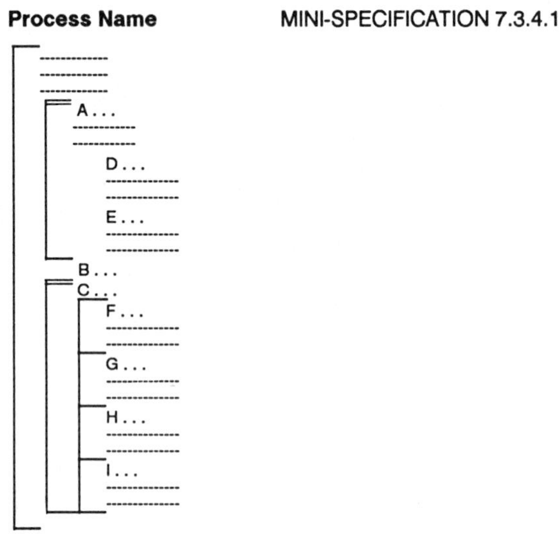

Figure 4.18

There are brackets to illustrate each of the three basic constructs of program logic: SEQUENCE (series of actions), REPETITION (repeating actions), and SELECTION (mutually exclusive actions).

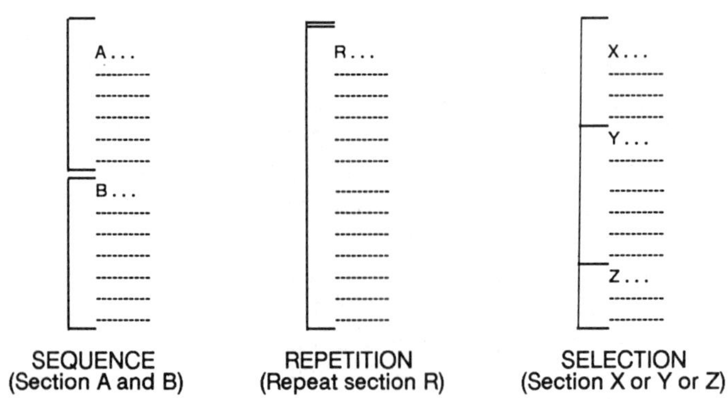

Figure 4.19

If a condition determines whether or not a bracket of logic applies, state the condition at the head of the bracket, as shown in Fig. 4.20.

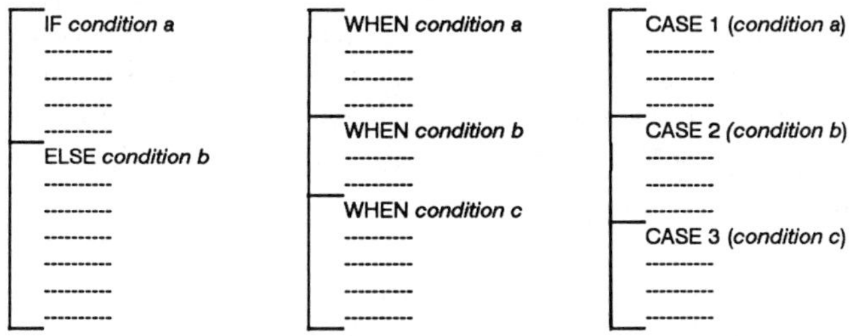

Figure 4.20

If a condition determines the repetition of a bracket, state the condition either at the top or bottom of the bracket, as shown in Fig. 4.21.

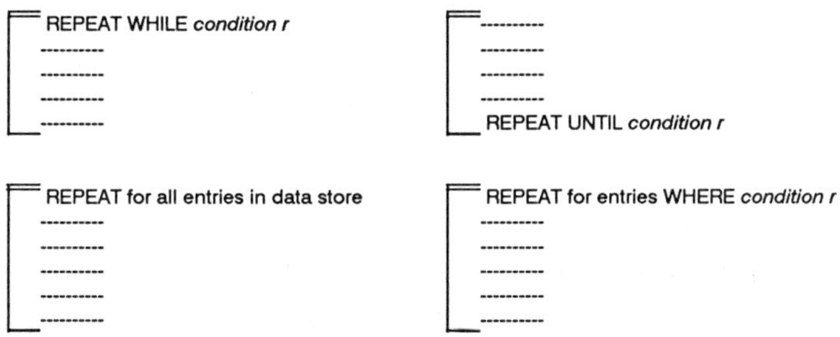

Figure 4.21

If a condition causes the process to end somewhere within a bracket, a left arrow that penetrates the bracket indicates that the processing has ended, as shown in Figs. 4.22a to 4.22c.

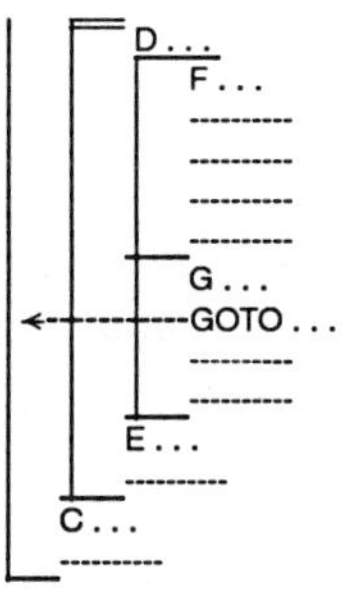

After GOTO, processing resumes at C.

Figure 4.22a

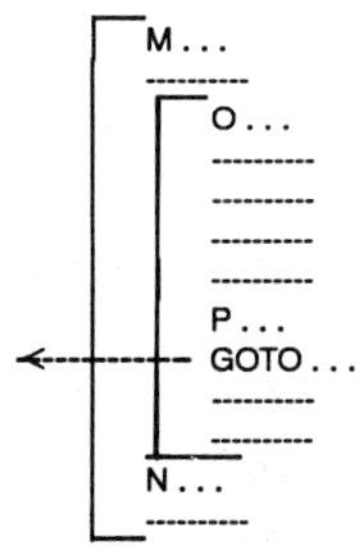

After GOTO, all processing ends.

Figure 4.22b

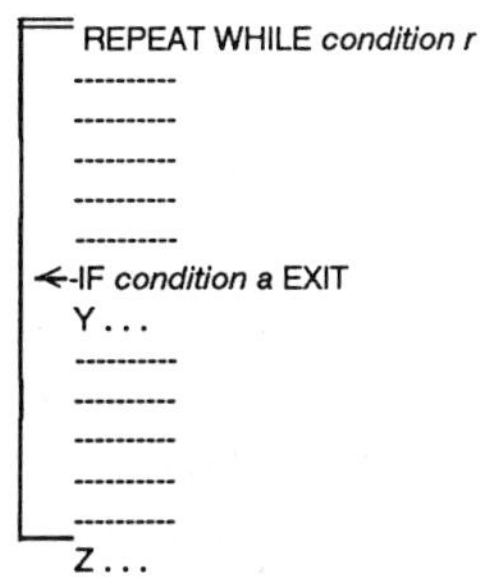

Upon condition a, processing of current iteration ends.

Figure 4.22c

In keeping with good structured programming techniques, the escape structure only allows a forward skip to the end of the bracket penetrated by the arrow. If the logic of the mini-specification requires a direct breakout, such as a transfer to an error-handling process, a dotted-line arrow indicates a direct escape that does not resume processing anywhere in the functional primitive (shown in Fig. 4.22b). The escape indicated in Fig. 4.22a breaks out of the SELECTION logic and the REPETITION logic. Control resumes at point C. Figure 4.22c indicates an escape that ends the current iteration of a repetition bracket and resumes at the next iteration of the repetition bracket. If we wanted to end the repetition logic completely and resume at point Z, the arrow in Fig. 4.22c would have to penetrate the repetition bracket.

If we only wish to transfer control to a common processing routine, use a Call statement and identify the statement by surrounding it with a round-cornered box with a double line on the left side, as shown in Fig. 4.23.

Figure 4.23

For real-time processes that need to show concurrency of operations, the symbols of Fig. 4.24 attach two (or more) concurrent processes.

Figure 4.24

To illustrate where and when inputs and outputs are used or generated, extend the bracket to form a rectangle. List all the inputs along the top right-hand corner outside the rectangle. List the corresponding outputs along the bottom right-hand corner outside the rectangle (see Fig. 4.25).

Action diagrams can be linked to a data model and can be generated automatically from decision trees, decision tables, and other diagrams. They are particularly well suited for use in mini-specifications because of their simplicity and effectiveness in communicating with users, as well as software engineers and database designers.

ACTION DIAGRAM FORMAT

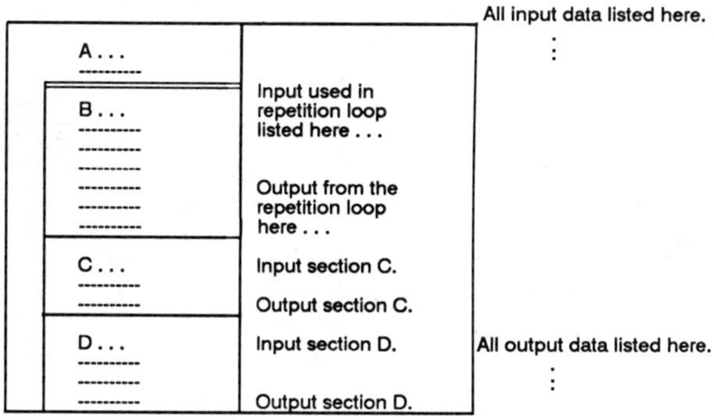

Figure 4.25

Charts and Graphs

The policies of transforming input data into output data, as well as other types of requirements, are included in a mini-spec. Occasionally, the analyst specifies other requirements by using normal narrative text. However, when using narrative text, be careful about ambiguity. The following are some guidelines to keep in mind:

1. Accompany narrative text with graphs, tables, diagrams, and other graphical aids that can help in the explanation of the policy whenever possible.
2. Use mathematical algorithms whenever possible.
3. Use indentations, outline and case structure formats as aids in organizing narrative text.
4. List all inputs and outputs of the process at the beginning of the narrative text as an aid to the reader in setting the stage for the basic transformations described.
5. Use examples to illustrate how the results are derived.
6. Provide a layout or example of the reports or screens produced by a process.
7. Use simple sentences.

Above all, *consider your audience*, and always submit your documentation to a peer review.

Mathematical formulas and graphs are especially helpful in specifying performance and capacity requirements. Use graphs to show the normal operating ranges for processes, as shown in Fig. 4.26.

PERFORMANCE RANGES IN PROCESSING INCOMING RECORDS

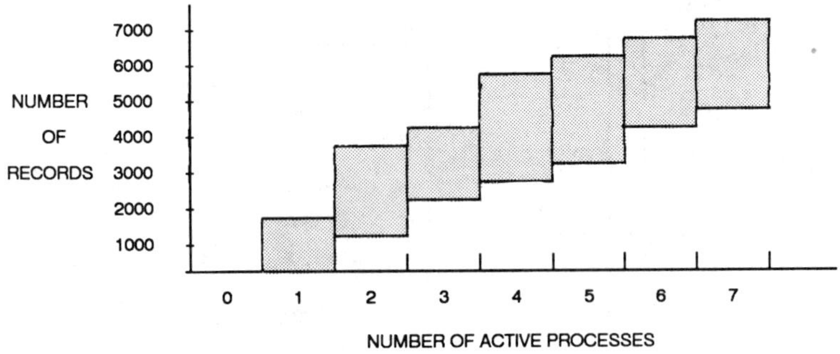

Figure 4.26

Charts and graphs are helpful in specifying time requirements such as delays between starting and stopping of processes in real-time applications. *Time-dependency charts*, like the one shown in Fig. 4.27, can pictorially represent the timing considerations in synchronizing the start and ignition activities of a heating system. **A TIME-DEPENDENCY CHART is a graphical illustration of the sequence, concurrency, and duration requirements, and/or restrictions among processes in a system.**

Time-dependency charts may often apply to the system as a whole and not to a specific process. This occurs most often in real-time system specifications. The chart can be used to specify the time constraints of a set of processes, as shown in Fig. 4.28,

TIME-DEPENDENCY CHART
(PROCESS 7.2.1)

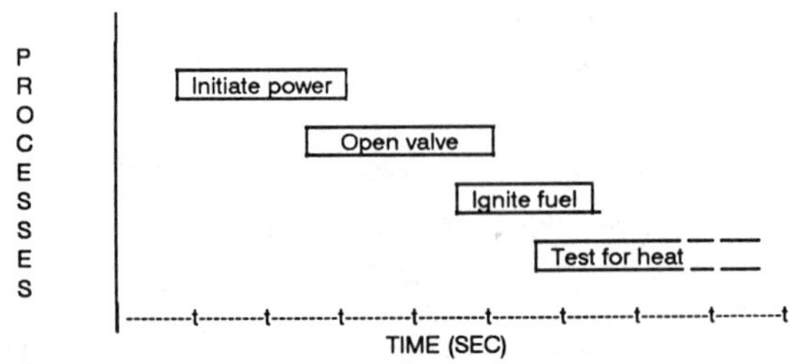

If heat sensor does not detect heat, . . . *such requirements specified here.*

Safety Actions:

1. If no heat is detected, valve must be closed within 15 sec.
2. . . . *other similar requirements specified here.*

Figure 4.27

or the overall time requirements for specific activities to take place in the system. When the time-dependency chart pertains to a single process, it is included in the mini-spec for the process. When it is used to specify performance requirements for activities in the system, it can be found in several places, depending on how the analyst organizes the structured specification. If there is a narrative description of the system, the time-dependency charts are useful to declare the performance constraints or requirements. Others may choose to include them in an appendage section of the mini-specs and cross-reference them in the process mini-specs that are involved in meeting time requirements. Still others may treat them as specifications for common processing routines and place them in the data dictionary with appropriate requirement names. The requirement names are then cross-referenced in the appropriate mini-specs for the processes involved in meeting the time requirements.

TIME-DEPENDENCY CHART
(SYSTEM REQUIREMENTS)

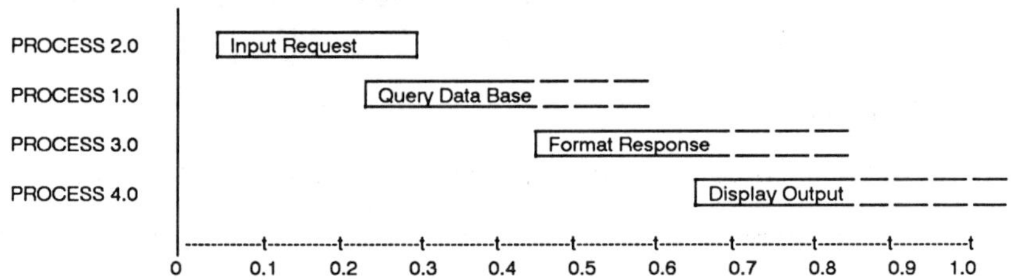

Output must be sent to the sink within 1.0 sec elapsed time for 99% of the input requests.

Figure 4.28

SUMMARY
. . . writing is intended to be understood, not solved

The mini-spec is where we specify what is to be done to solve the problem or produce the desired result. Often much of the contents of the mini-spec is narrative text. The text is intended for use by both the technical community of developers and the nontechnical user community of office workers who are ultimately responsible for making the system meet its objectives. The structured specification has two primary goals: *communication* and *comprehension*. In preparing any written material, remember that office workers read and comprehend best material written at the seventh-grade level. As a result many magazines such as *Time* and *Newsweek* are written at the seventh-grade level. *The New York Times* is written at the tenth-grade level. *Scientific American* is written at the eleventh-grade level. However, when technical people prepare written specifications they tend to write at the twentieth-grade level. Frequent iterative reviews with the users together with a heightened awareness on the part of the analyst of the primary goals of the requirements document are two valuable ingredients in preparing a structured specification.

The American Management Association offers the following worst-bad-better versions of the same narrative text as an example of clarity in documentation:

Worst:

At this point in time the incremental use generated by the various functional areas interfacing with the Data Processing Department has resulted in a cumulative, multifaceted demand that far exceeds the total working capacity of our present hardware.

Bad:

Increasing use by various functional areas within the organization has resulted in a demand that exceeds our present data processing capacity.

Better:

The equipment we have does not meet our present needs.

Aside from structured English or pseudocode that is used to explain the data transformations, good, clear, concise narrative text is used to specify other requirements. The following requirements categories are suggested for inclusion in the mini-specification for each process. Requirements for these categories are usually written in simple narrative text.

Processor Requirement:

Security Requirements:

Time Requirements:

Processing Priorities:

Performance Requirements:

Safety Requirements (Software and Manual) :

Other Requirements:

It is a good practice to include these items in all mini-specs even if some are not applicable (N/A). These sections or paragraph headings can serve as a reminder to both the analyst and user to specify requirements that would otherwise be forgotten until much later in the development life cycle.

QUESTIONS

1. Where is a common processing routine documented in a structured specification?
2. What are the ways to specify conditional statements in a mini-spec?
3. Give an example of an arithmetic operator, a logical operator, and a relational operator.
4. What are the basic components of structured English?
5. What is the purpose of the "Stop" statement in a mini-spec?
6. List the ways to represent a repetitive sequence of logic and explain the differences.
7. List some of the most common organizational styles of mini-specs.
8. Show the four forms of the assignment statement.
9. What are the purpose and goal of a mini-spec?
10. What are the guidelines or recommended ways of constructing a mini-spec?
11. How does one number a mini-spec?
12. In what case (upper or lower) are the following to be specified: local variables, data stores, commands, functional primitive names, common processing routine names, prepositions, and other verbs such as "to be" verbs?
13. Compare how input data and output data are handled in a mini-spec for a functional primitive and a common processing routine.
14. What type of requirements are documented in a mini-spec?
15. When would one use a decision table over a decision tree?
16. What distinct advantages do decision tables have over structured English statements, such as the nested If . . . then . . . Else statements or the Case structure.
17. a. What is the maximum number of decision rules in a decision table that has seven statements of condition?
    b. What is the maximum number of leaves in a decision tree that has four conditions, each having three possible outcomes?
18. What are the advantages of using action diagrams?
19. When would one use a time-dependency chart?
20. When requirements are specified using narrative text, what things should be kept in mind?

PROBLEMS

1. Rewrite the example in Fig. 4.7b (Balance Checking Account) using an action diagram.

2. Create a decision tree and a decision table for the following policy statement:

   Normal admission price is $3.00. If the customer is a senior citizen, a 10 percent discount applies. Students with school ID can be admitted for $2.00. All couples can be admitted for $5.00 (senior citizens' discount also applies to couple rate). All seats in the balcony are $1.50 each, and the senior citizens' discount does not apply.

5

PHYSICAL MODELS

The Rent-em Video store buys new video tapes from wholesalers and rents them to the general public. There is no membership fee required. However, before a customer can rent any tape, he or she must fill out a customer information card and pay a $50.00 deposit. Upon payment, the customer receives a list of the tapes that the store has for rent. The customer may rent up to three tapes from the list at the rate of $3.00 per tape for the first 24-hour rental period. If the customer wishes to keep the tape(s) longer, a rate of $1.00 per tape for each additional 24-hour period applies. A clerk at the counter will determine if the tape is available, and if it is, will issue the tape to the customer.

The customer is not permitted to rent more than three tapes until the previously rented tapes are returned and payment is made. The customer returns the tapes to the clerk who collects the amount owed and returns the tape(s) to the tape library.

All data in the system is handled manually, either verbally or on paper, and data files are maintained in a file drawer.

CONTEXT DIAGRAM
. . . defines the boundaries of the system

The context diagram indicates that the system obtains input from two sources (Customer and Video Tape Wholesaler) and provides information and/or outputs to three sinks (Bank, Customer, and Video Tape Wholesaler), as shown in Fig. 5.1.

The data flows indicate all the possible data that can flow into or out of the system from the various sources/sinks without regard to time or sequence. Every data flow that appears on the context diagram must appear on diagram 0, which illustrates the next level of detail about the system.

The data dictionary contains:

1. All input and output data flows found on the context diagram defined in terms of elementary and group data elements
2. All elementary data elements defined in terms of a descriptive comment
3. All data stores defined in terms of the elementary and group data elements with the primary key underlined or the organizational element(s) declared in an associated comment.

CONTEXT DIAGRAM

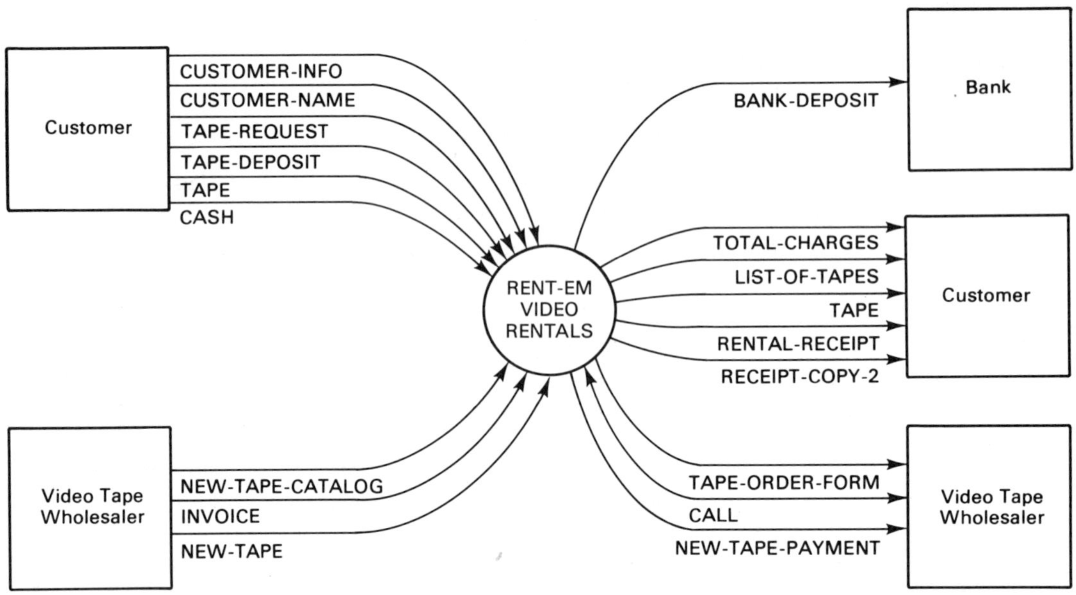

Figure 5.1

DATA FLOW DIAGRAMS
. . . a graphical illustration of the flow of data

The main bubble on the context diagram (Fig. 5.1) is decomposed into five major processes. The number assigned to each process is only for reference and *does not* indicate any sequence of execution. Those processes that are not broken down on a lower-level data flow diagram are functional primitives. A process number with an asterisk (1.0*, 4.0*, and 5.0*) indicates a functional primitive. Further detailed information regarding the functional primitives are obtained from the mini-specs.

The only data stores shown on level 1 (see Fig. 5.2) are those shared between two or more processes or those accessed by a functional primitive. Every data store and data flow is in the data dictionary. The context diagram of Fig. 5.1 and diagram 0 in Fig. 5.2 are balanced because every input and output data flow shown on the context **diagram** is shown as an input or output to some process on diagram 0. CUSTOMER-NAME is correctly shown *once* on the context diagram even though it is shown *twice* on diagram 0, once as input to Rent Tapes 3.0 and again as input to Receive Rental Tapes 4.0.

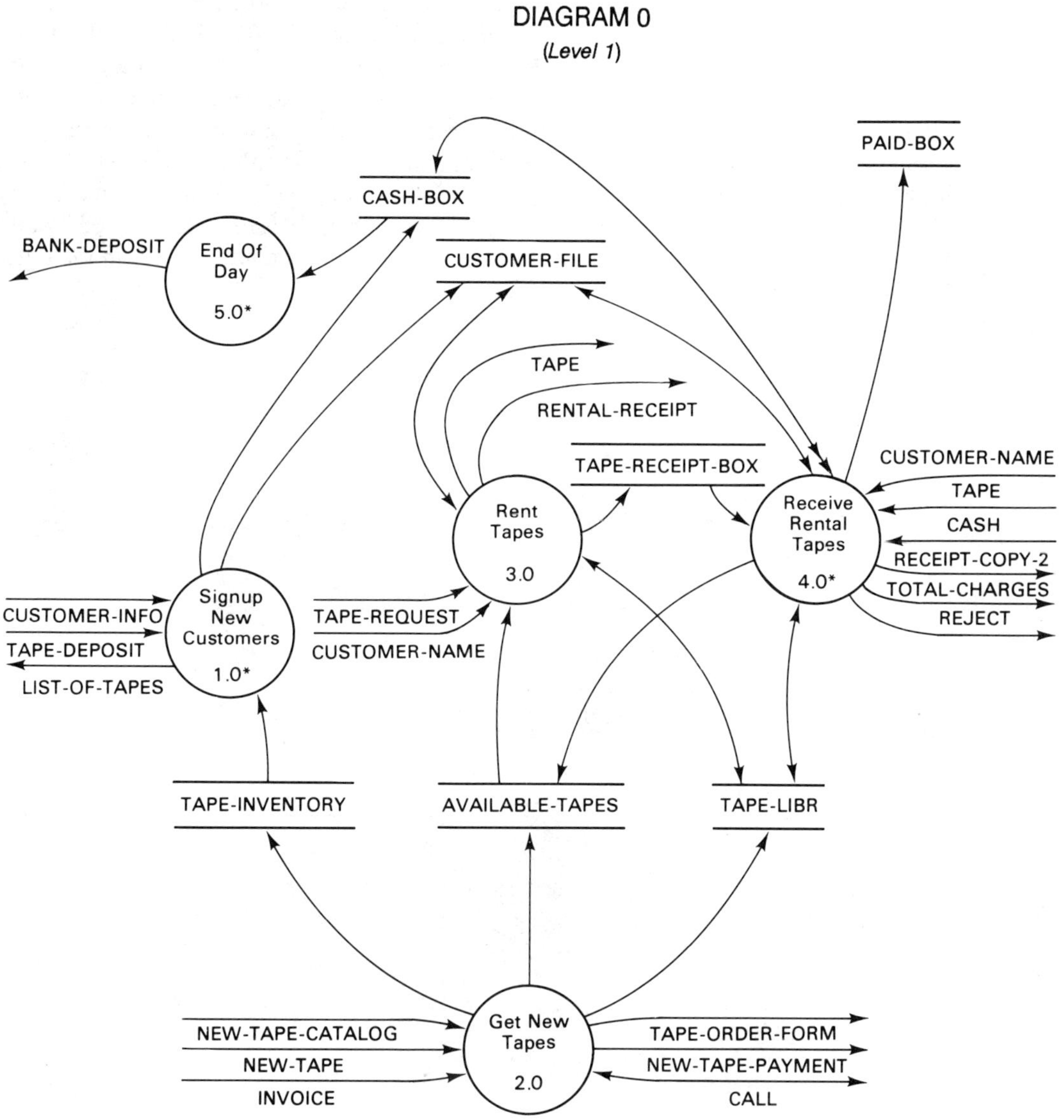

Figure 5.2

In Fig. 5.2, there are two processes that are decomposed further: Get New Tapes 2.0 and Rent Tapes 3.0. The detailed data flow diagrams are labeled diagram 2.0 (Fig. 5.3) and diagram 3.0 (Fig. 5.4), respectively.

Again, you will note that all of the data flows that enter or leave the process Get New Tapes 2.0 on diagram 0 (Fig. 5.2) are shown entering or leaving one of the processes on diagram 2.0 (Fig. 5.3). The data store INVOICE-BOX is used to pass information between two processes entirely contained within the "parent" process 2.0. As a result, the data store is not shown on diagram 0, but is instead shown on diagram 2.0. The data stores that are shown on diagram 0 (Fig. 5.2), TAPE-INVENTORY, TAPE-LIBR, and AVAILABLE-TAPES, are shown only as simple data flows on diagram 2.0 (Fig. 5.3). This convention of not repeating data store notation on two different levels of the data flow diagram reduces redundancy and facilitates change. If we change the system design whereby the data store becomes a direct data flow, only one data flow diagram and the corresponding data dictionary definition would change. Likewise, if a direct data flow has to be stored temporarily before it is retrieved by the receiving process, only one data flow diagram and the corresponding entry in the data dictionary is updated.

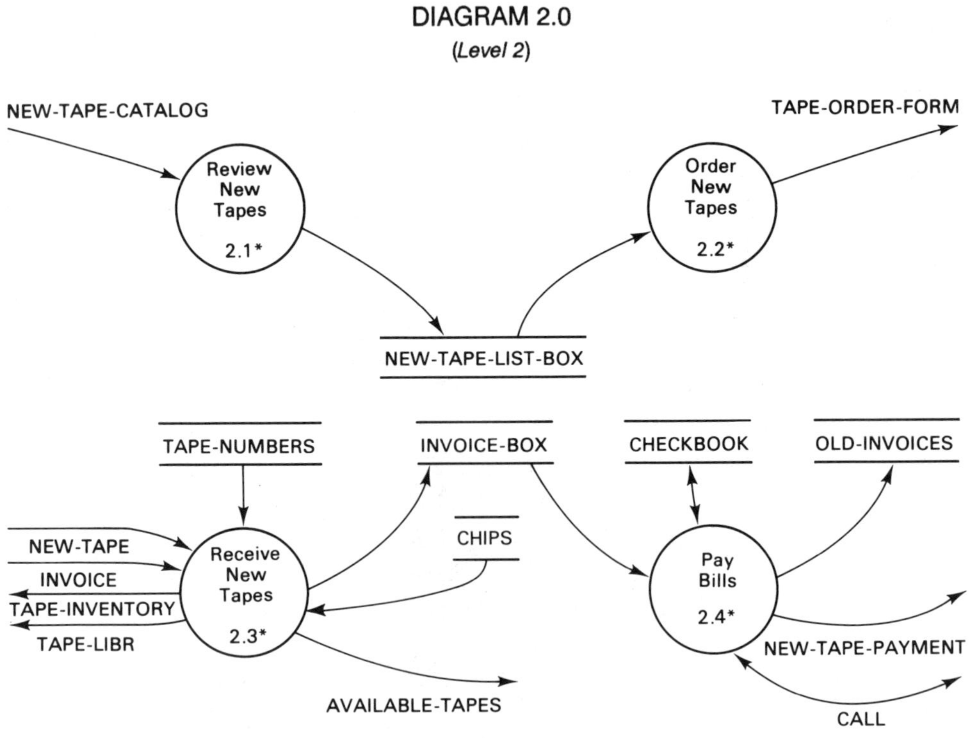

Figure 5.3

On diagram 2.0 in Fig. 5.3, we see there is a distinct discontinuity between processes 2.1 and 2.2, and processes 2.3 and 2.4. No data passes between these two segments of the system. So why associate them together in a "parent" process 2.0? Because that is the way the current system operates. The systems analyst must take great care to model the system *exactly* the way it operates, and not to model the system the way the analyst *thinks* it should operate. The only association these processes can have is that the same person does all of these activities, or that they all take place in one area of the store. Regardless of the physical, political, or technical reasons, if the processes are in fact associated by the user, then that is the way the analyst must model the system in preparing the physical structured specification.

DIAGRAM 3.0
(*Level 2*)

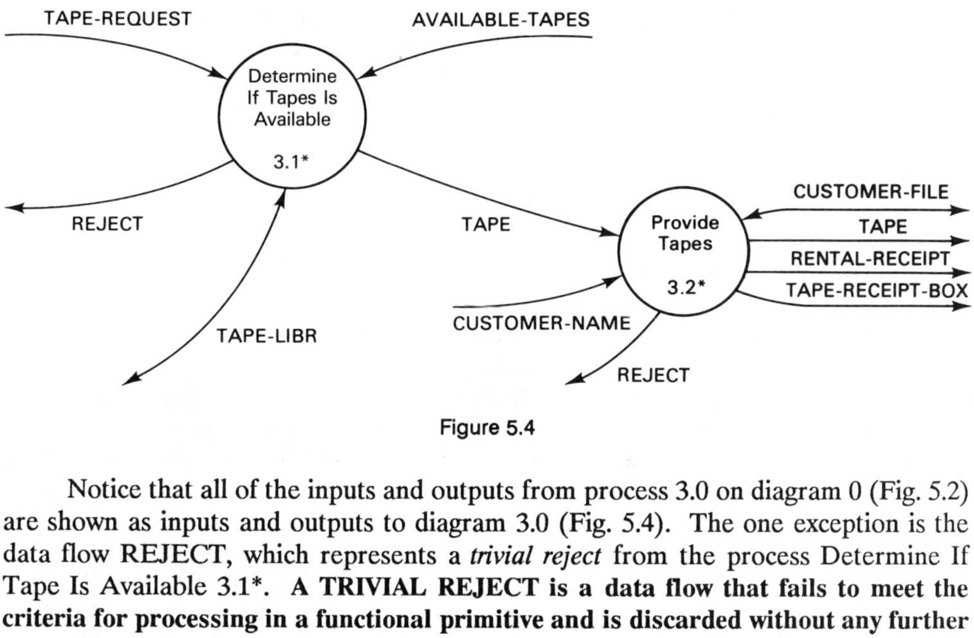

Figure 5.4

Notice that all of the inputs and outputs from process 3.0 on diagram 0 (Fig. 5.2) are shown as inputs and outputs to diagram 3.0 (Fig. 5.4). The one exception is the data flow REJECT, which represents a *trivial reject* from the process Determine If Tape Is Available 3.1*. **A TRIVIAL REJECT is a data flow that fails to meet the criteria for processing in a functional primitive and is discarded without any further processing.** We do not show trivial rejects at any higher level than the functional primitive that rejects the data. The data flow REJECT is not defined in the data dictionary and can be used many times throughout the specification to indicate that processes have rejected data. REJECT is only used for trivial rejects, those that do not require any further processing. Nontrivial rejects are shown as normal data flows with unique names and appear in the data dictionary. We can consider the dialogue data flow CALL as a nontrivial reject. If there is a problem with the data on the INVOICE, process 2.4* cannot process the INVOICE in the usual manner and initiates a dialogue (CALL) with the Video Tape Wholesaler. Nontrivial reject processing is a standard part of the structured specification and is documented as a normal data flow.

Because processes Signup New Customers 1.0*, Receive Rental Tapes 4.0*, and End Of Day 5.0* are functional primitives, there is no need to have a diagram 1.0, diagram 4.0, or diagram 5.0. Instead, details for these processes are found in mini-spec 1.0, mini-spec 4.0, and mini-spec 5.0. Reading through the data flow diagrams, we can easily tell whether we should look for further details on another data flow diagram, or in a mini-spec by the asterisk (*) following the process number. The asterisk indicates that the process is a functional primitive and further details can be found in the mini-spec.

In the data flow diagram of Fig. 5.4, we can also see how physical items s as the actual video tapes (TAPE) are modeled in the system. TAPE is retrieved by the process Determine If Tape Is Available 3.1* from the data store TAPE-LIBR, and is passed to process Provide Tapes 3.2*. That process inputs TAPE and outputs TAPE to the Customer. Usually, a process does not input and output the exact same data flow. However, when dealing with physical items, processes will handle them in exactly the manner shown in Fig. 5.4.

DATA DICTIONARY

. . . data dictionary adds rigor to the system model

The data dictionary is an alphabetical list of all of the inputs, outputs, internal data flows, data stores, and any common processing routines or functions found in the system. When an existing system that deals with physical items is modeled, occasionally there are data elements available within the system that have no processes to use them. Some analysts choose to omit such data elements completely from the structured specification. Others believe that because the data elements are physically present in the system, they should be included in the structured specification. The reason to include them is that they could be used in a future enhancement. To know all of the data that are readily available in the system is a definite advantage when studying proposed changes. In the Rent-em Video system, there are data elements that are available in the system but not used by any process in the system. Such data elements are included in this example and are shown in small italic type, as shown in Fig. 5.5.

DATA DICTIONARY

ACTOR	= ** Name of the actor in the movie. **
ACTRESS	= ** Name of the actress in the movie. **
AUTH-SIGNATURE	= * Signature of Rent-em Video Rental store manager. *
AVAILABLE-TAPES	= {TAPE-BOX + (TAPE-CHIP)}
	= * Organized by TYPE + TITLE. *

Figure 5.5

DATA DICTIONARY (Cont.)

BANK-DEPOSIT	= * All CASH from the CASH-BOX less STARTING-AMT. *
BASE-RATE	= * Rental rate for the first 24 hours ($3.00). *
	= * Rental rate for a TAPE for the first 24 hour period. *
CALL	= * Conversation with Video Tape Wholesaler. *
CASH	= * U.S. currency. *
CASH-BOX	= {CASH}
CHECK	= CHECK-DATE + RECIPIENT + CHECK-AMT + AUTH-SIGNATURE
	= * Check written to pay Video Tape Wholesalers for NEW-TAPEs. *
CHECK-AMT	= * Amount in U.S. dollars payable to RECIPIENT. *
CHECKBOOK	= {CHECK# + CHECK-NAME + CHECK-AMT + CHECK-DATE + REMAINING-BAL}
CHECK-DATE	= * Date the CHECK was issued. *
CHIPS	= * A collection of new blue or yellow TAPE-CHIPs. *
CUSTOMER-ADDRESS	= * Customer's home address. *
CUSTOMER-FILE	= {CUSTOMER-RECORD}
	= * Organized by CUSTOMER-NAME + CUSTOMER-ADDRESS. *
CUSTOMER-INFO	= CUSTOMER-NAME + CUSTOMER-ADDRESS + CUSTOMER-PHONE
CUSTOMER-NAME	= * Customer's first and last name. *
CUSTOMER-PHONE	= * Home phone number of CUSTOMER-NAME. *
CUSTOMER-RECORD	= CUSTOMER-INFO + {TAPE-INFO}
DATE-RENTED	= * Date tape was rented. *
DATE-RETURNED	= * Date tape was returned. *
DAY-RATE	= * Daily rental rate after the first 24 hours ($1.00). *
	= * Rental rate for a TAPE for each 24 hour period after the first 24 hrs. *
INVOICE	= WHOLESALER-NAME + WHOLESALER-PHONE + INVOICE-DATE + {TITLE + TYPE + TAPE-NUMBER + QUANTITY + PRICE + (NOTE)} + TOTAL-INVOICE-AMT
	= * One original and two copies. *
INVOICE-BOX	= {INVOICE}
	= * Organized by INVOICE-DATE. *
INVOICE-DATE	= * Date the invoice was created by the Video Tape Wholesaler. *
LIST-OF-TAPES	= {TYPE + TITLE + TAPE-RATING}
	= * A list of all the tapes the store has to rent and organized by TYPE + TITLE. *
MARKED-NEW-TAPE-LIST	= {(["1" | N-COPIES]) + TAPE-NUMBER + TITLE + TYPE + TAPE-RATING + PRICE} + TAPE-ORDER-FORM
	= * NEW-TAPE-CATALOG with quantity besides the TAPE-NUMBER of the tape(s) to be purchased. *
N-COPIES	= * Number of copies of a particular movie ordered. *
NEW-TAPE	= TAPE + TAPE-BOX
	= * A new video tape movie received from the Video Tape Wholesaler. *
NEW-TAPE-CATALOG	= {TAPE-NUMBER + TITLE + TYPE + TAPE-RATING + PRICE + TAPE-ORDER-FORM}
	= * List of new tapes available from the Video Tape Wholesaler. *
NEW-TAPE-LIST-BOX	= {MARKED-NEW-TAPE-LIST}
	= * Organized by TAPE-NUMBER. *
NEW-TAPE-PAYMENT	= "RENT-EM VIDEO RENTALS" + "123 South Main St. Videoville, USA" + CHECK
NOTE	= * Explanation as to the discrepancy found between the INVOICE and the actual TAPEs received from the Video Tape Wholesaler. *
OLD-INVOICES	= {INVOICE}
	= * Organized by INVOICE-DATE.*

Figure 5.5 (cont.)

DATA DICTIONARY (Cont.)

PAID-BOX = {RECEIPT-COPY-1}
= * Organized by DATE-RETURNED. Paid up rental RECEIPT-COPY-1 kept for auditing purposes.*

PLOT = ** Description of the story in the movie. **

PRICE = * Price of one tape from the Video Tape Wholesaler. *

QUANTITY = * The number of TAPEs of a specific TAPE-NUMBER. *

RECEIPT-COPY-1 = RENTAL-RECEIPT + ("PAID")
= * Put into PAID-BOX and kept for auditing purposes. *

RECEIPT-COPY-2 = RENTAL-RECEIPT + ("PAID")
= * Given to the customer after rental is completed and paid. *

RECIPIENT = * Person or business paid by CHECK. *

RENTAL-RECEIPT = CUSTOMER-INFO + {TAPE-INFO + TAPE-CHARGE} + TOTAL-CHARGES
= * One original and two copies: RECEIPT-COPY-1 and 2. *

REQUEST-TAPE = TITLE + TYPE
= * Customer asks for a specific tape. *

STARTING-AMT = * The amount of CASH left in the CASH-BOX for start of new day ($20.00).*

STARS = *{[ACTOR | ACTRESS]}*

STUDIO = ** Name of the studio that produced the movie. **

TAPE = TYPE + TITLE + TAPE-RATING
= * A video tape of a movie. *

TAPE-BOX = TYPE + TITLE + TAPE-RATING + (TAPE-NUMBER) *+ STUDIO + TAPE-VIEWING-TIME + {STARS} + (PLOT)*

TAPE-CHARGE = * Amount owed for renting a tape rented. *

TAPE-CHIP = TAPE-NUMBER + TYPE
= * TYPE is indicated by the color of the Chip: Blue = "BETA," Yellow = "VHS." *

TAPE-DEPOSIT = * Deposit of $50.00 used as a security deposit. *

TAPE-INFO = TYPE + TITLE + DATE-RENTED + (DATE-RETURNED)

TAPE-INVENTORY = {LIST-OF-TAPES}
= * Stacks of printed lists of all movies available for rent at Rent-em Video Rentals. *

TAPE-LIBR = {TAPE-NUMBER + TYPE + TITLE + [TAPE | TAPE-CHIP]}

TAPE-NUMBER = * A unique number for each TYPE and TITLE of tape used as reference by the Rent-em Video Rental System. *

TAPE-NUMBERS = {TAPE-NUMBER}
= * A sequence of available TAPE-NUMBERs in ascending order.*

TAPE-ORDER-FORM = {TITLE + TYPE + TAPE-NUMBER + PRICE + QUANTITY}

TAPE-RATING = ["G" | "PG" | "PG-13" | "R" | "X"]

TAPE-RECEIPT-BOX = {RENTAL-RECEIPT}
= *Original RENTAL-RECEIPT organized by CUSTOMER-NAME.*

TAPE-REQUEST = [TITLE + TYPE | TAPE-CHIP]

TAPE-VIEWING-TIME = ** Viewing time in minutes of the movie on the TAPE. **

TITLE = * Title of the movie. *

TOTAL-CHARGES = TAPE-CHARGE
= * Total amount owed for all tapes rented. *

TOTAL-INVOICE-AMT = * Total amount of invoice for all movies purchased from the Video Tape Wholesaler. *

TYPE = ["VHS" | "BETA"]
= * VHS and BETA are two distinctly different formats and size of tape available for home Video Cassette Recorders.*

WHOLESALER-NAME = * Name of the Video Tape Wholesaler who sells tapes.*

WHOLESALER-PHONE = * Phone number of the Video Tape Wholesaler. *

Figure 5.5 (Cont.)

There are several points to note regarding a data dictionary:

1. CURRENT-DATE, CURRENT-TIME, and REJECT are special elementary data elements that are always available within a manual or automated system. These elements are used repeatedly and we **do not** need to define them explicitly in the data dictionary.

2. Although aliases should always be avoided, if they exist in a current system they must be modeled just as they are used, so that the user can verify the model as an accurate representation of the current system.

3. TAPE-BOX is defined to be TYPE, TITLE, TAPE-RATING, and TAPE-NUMBER. These are the data elements used by the processes in the system. The other data elements denoted in small type do in fact enter the system on the TAPE-BOX but are not used anywhere in the system. The analyst can choose to ignore all data elements that the system does not use, such as ACTOR, ACTRESS, TAPE-VIEWING-TIME, STARS, STUDIO, and PLOT, and not include them in the data dictionary.

4. Examples of literals are found in the data elements NEW-TAPE-PAYMENT and TAPE-RATING. A literal is some specific set of data values used in the system, such as "G," "PG," "PG-13," "R," and "X" for the rating of a particular movie.

5. As we do in good programming practices, we should not "hard code" constant values that are subject to change in a structured specification. An example would be the rental rate (BASE-RATE) of $3.00 for the first 24-hour rental of a TAPE. When a system is modeled, all such constants (BASE-RATE, DAY-RATE, STARTING-AMT, etc.) are details that should be avoided or postponed to the implementation stage. Such data items should be defined as elementary data elements and given specific values later in the development life cycle. The logic in the mini-spec should only reference the data element names. The data elements are given example values as a comment in this structured specification for illustration purposes only.

6. Three ways of defining data stores are illustrated in this structured specification:

> PAID-BOX, whose primary key is documented by a comment statement that specifies the organization of the data store.
>
> TAPE-LIBR, whose primary key is indicated by the underlined data element TAPE-NUMBER. If the primary key is composed of more than one data element, it is assumed that the data element listed first is of primary importance and the data element listed second is of less

significance. If TYPE + TITLE were the primary key for TAPE-LIBR, the data store would be organized first by TYPE (whether it is "VHS" or "BETA") and then alphabetically by TITLE.

TAPE-INVENTORY and CHIPS, which have no primary keys because they are repeating groups of identical data elements. In fact, the data definitions represent a stack of printed materials stored until needed for distribution to customers or prospective customers or a collection of blue and yellow identification chips. A comment specifies when the data is used, the first, or topmost, LIST-OF-TAPES entry in the data store TAPE-INVENTORY is always selected.

Notice that because there is no organization indicated for NEW-TAPE-CATALOG, it is not a data store in this system. If we follow the logic carefully in the mini-spec for process Review New Tapes 2.1*, we will see that the NEW-TAPE-CATALOG becomes the data store NEW-TAPE-LIST-BOX, whose organizational data element is specified as TAPE-NUMBER.

MINI-SPECIFICATIONS WITH ACTION DIAGRAMS
. . . mini-specs specify the transformations of data

The mini-specs are organized numerically for each functional primitive in the system, as shown in Figs. 5.6a to 5.6i. In order to permit quick access to the logic, mini-specs are organized one to a page and arranged numerically. The logic is specified using structured English (pseudocode) as presented in Chapter 4. Action diagram notation is used to illustrate the structure **as an example only**, because this system contains all manual processes, the mini-specs do not necessarily need to show the structure of the procedures or be rigorously defined in pseudocode. Action diagrams are used in mini-specs to illustrate the structure of the logic for the software used in processes, however, even in manual processes action diagrams aid the reader by

1. Segmenting sections of procedure specifications and designating those input data and output data associated to each section
2. Illustrating the segments of procedure specifications that repeat
3. Illustrating choices among various segments of procedure specifications (If . . . then . . . Else . . . , or Case structures)
4. Segmenting sections of sequential procedure specifications
5. Illustrating segments of procedures that operate simultaneously
6. Designating hand-offs (Calls) to alternative sections of procedure specifications.

The following mini-specs are presented with action diagrams.

Signup New Customers MINI-SPECIFICATION 1.0

Input CUSTOMER-INFO from Customer. Copy CUSTOMER-INFO onto the CUSTOMER-RECORD. Output CUSTOMER-RECORD to CUSTOMER-FILE. Input TAPE-DEPOSIT from Customer. Output TAPE-DEPOSIT to CASH-BOX. Input LIST-OF-TAPES from TAPE-INVENTORY. Output LIST-OF-TAPES to Customer. Stop.	CUSTOMER-INFO TAPE-DEPOSIT LIST-OF-TAPES (TAPE-INVENTORY) CUSTOMER-RECORD (CUSTOMER-FILE) TAPE-DEPOSIT (CASH-BOX) LIST-OF-TAPES

Processor Requirements: Manual process performed by the cashier.

Security Requirements: The cashier is responsible for the money in CASH-BOX.

Time Requirements: N/A.

Processing Priority: Customer's shall be processed in order of their arrival time. This process has a high priority.

Performance Requirements: Customers shall be processed as quickly as possible to minimize the customer's inconvenience.

Safety Requirements: N/A.

Other Requirements: None.

Figure 5.6a

Review New Tapes

MINI-SPECIFICATION 2.1

NEW-TAPE-CATALOG

Input NEW-TAPE-CATALOG from Video Tape Wholesaler.
Select which movies to buy by placing a "1" or an integer (N-COPIES) indicating the number of copies to purchase along side the TAPE-NUMBER on the MARKED-NEW-TAPE-LIST.
Output MARKED-NEW-TAPE-LIST to NEW-TAPE-LIST-BOX.
Stop.

MARKED-NEW-TAPE-LIST
(NEW-TAPE-LIST-BOX)

Processor Requirements: Manual process performed by the manager.

Security Requirements: N/A.

Time Requirements: N/A.

Processing Priority: This process has a low priority.

Performance Requirements: The NEW-TAPE-CATALOG shall be processed within one week of its arrival.

Safety Requirements: N/A.

Other Requirements: None.

Figure 5.6b

Order New Tapes MINI-SPECIFICATION 2.2

MARKED-NEW-TAPE-LIST
(NEW-TAPE-LIST-BOX)

```
Input MARKED-NEW-TAPE-LIST from NEW-TAPE-LIST-BOX.
For each entry in MARKED-NEW-TAPE-LIST where there
        is indicated a "1" or a number N-COPIES:
    Copy TAPE-NUMBER, TITLE, TYPE, and PRICE from
        MARKED-NEW-TAPE-LIST to the TAPE-ORDER-FORM.
    If there is a N-COPIES, then
        Set QUANTITY to N-COPIES on TAPE-ORDER-FORM.
    Else Set QUANTITY to 1 on TAPE-ORDER-FORM.
Endfor.
Output TAPE-ORDER-FORM to Video Tape Wholesaler.
Discard MARKED-NEW-TAPE-LIST.
Stop.
```

TAPE-ORDER-FORM

Processor Requirements: Manual process performed by the counter clerk.

Security Requirements: N/A.

Time Requirements: This process shall be done at least once per month.

Processing Priority: This process has a low priority.

Performance Requirements: N/A.

Safety Requirements: N/A.

Other Requirements: None.

Figure 5.6c

Receive New Tapes MINI-SPECIFICATION 2.3

```
                                                              NEW-TAPE
Input NEW-TAPE from Video Tape Wholesaler.        NEW-TAPE     TAPE-NUMBER
   Input the next TAPE-NUMBER from TAPE-NUMBERS.  TAPE-NUMBER    (TAPE-NUMBERS)
   If TYPE is "VHS",                              new TAPE-CHIP new TAPE-CHIP
      Input a new Yellow TAPE-CHIP from CHIPS.                   (CHIPS)
      Copy TAPE-NUMBER onto a Yellow TAPE-CHIP.                INVOICE
   Endif.
   If TYPE is "BETA",
      Input a new Blue TAPE-CHIP from CHIPS.
      Copy TAPE-NUMBER onto a Blue TAPE-CHIP.
   Endif.
   Remove TAPE from the TAPE-BOX.
   Output TAPE to TAPE-LIBR.
   Mark TAPE-NUMBER on the TAPE-BOX.
   Output TITLE, TYPE, & TAPE-RATING
      to TAPE-INVENTORY.                          TAPE
   Output TAPE-BOX to AVAILABLE-TAPES             TITLE+TYPE
      (on the shelve in the store).                 +TAPE-RATING
   Output TAPE-CHIP to AVAILABLE-TAPES            TAPE-BOX
      (on the shelve beneath the TAPE-BOX).         +TAPE-CHIP
   Input INVOICE from Video Tape Wholesaler.      INVOICE
   Compare TAPE-NUMBER and QUANTITY on INVOICE
      with the NEW-TAPEs received.
   If the TAPE-NUMBER or QUANTITY does not match
         TAPE(s), then Circle the TITLE, TYPE,                 TAPE
         TAPE-NUMBER, QUANTITY, and PRICE on                     (TAPE-LIBR)
         INVOICE with a red ink pen.                           TITLE+TYPE
      Write an explanatory NOTE describing the                   +TAPE-RATING
         discrepancy on the INVOICE.                             (TAPE-INVENTORY)
   Endif.                                                      TAPE-BOX
   Output INVOICE to INVOICE-BOX.                 INVOICE        +TAPE-CHIP
Stop.                                                            (AVAILABLE-TAPES)
                                                               INVOICE
                                                                 (INVOICE-BOX)
```

Processor Requirements: Manual process performed by the counter clerk.

Security Requirements: N/A.

Time Requirements: N/A.

Processing Priority: This process has a medium priority and may be interrupted by process 3.1*.

Performance Requirements: NEW-TAPEs shall be made available to rent as soon as possible with respect to the other specified priorities.

Safety Requirements: N/A.

Other Requirements: None.

Figure 5.6d

Pay Bills MINI-SPECIFICATION 2.4

```
                                                                        INVOICE
                                                                         (INVOICE-BOX)
  Search INVOICE-BOX for INVOICE with                  INVOICE          CHECK
     DUE-DATE = CURRENT-DATE + 3 days,                                   (CHECKBOOK)
  Input INVOICE from INVOICE-BOX.
  If any TITLE, TYPE, TAPE-NUMBER, QUANTITY,
        and PRICE has a red ink circle,
     Output CALL to Video Tape Wholesaler and
        resolve the discrepancy as stated in the NOTE.
     Subtract any adjustment amount from
        TOTAL-INVOICE-AMT agreed to via CALL.
  Endif.                                               CALL

  Input next CHECK from CHECKBOOK.                     CHECK
  Set CHECK-AMT to TOTAL-INVOICE-AMT.
  Set CHECK-NAME to WHOLESALER-NAME .
  Set CHECK-DATE to CURRENT-DATE.
  Set AUTH-SIGNATURE to manager's signature.
  Subtract CHECK-AMT from REMAINING-BAL.               CHECK-AMT          CALL
  Output CHECK-AMT, CHECK-NAME, CHECK-DATE,              +CHECK-NAME      CHECK-AMT
     REMAINING-BAL to CHECKBOOK.                         +CHECK-DATE        +CHECK-NAME
  Output NEW-TAPE-PAYMENT to Video Tape                  +REMAINING-BAL     +CHECK-DATE
     Wholesaler.                                        NEW-TAPE-PAYMENT    +REMAINING-BAL
  Output INVOICE to OLD-INVOICES.                                            (CHECKBOOK)
  Stop.                                                                   NEW-TAPE-PAYMENT
                                                                          INVOICE
                                                                            (OLD-INVOICES)
```

Processor Requirements: Manual process performed by the manager.

Security Requirements: N/A.

Time Requirements: N/A.

Processing Priority: This process shall be completed by the end of the business day on every Tuesday and Thursday.

Performance Requirements: All INVOICEs due within the next 3 days shall be processed.

Safety Requirements: N/A.

Other Requirements: None.

Figure 5.6e

Determine If Tape Is Available MINI-SPECIFICATION 3.1

```
                                                                        TAPE-REQUEST
Input TAPE-REQUEST from Customer.                                       TAPE-CHIP
Case 1 (Customer provides TAPE-CHIP):                 TAPE-CHIP           (TAPE-LIBR)
    Search TAPE-LIBR for TAPE-NUMBER.                  +TAPE              (AVAILABLE-TAPES)
    If TAPE-NUMBER is not found,                                        TAPE
        Reject TAPE-REQUEST as REJECT (tell                               (TAPE-LIBR)
            customer we don't have movie).                              TYPE+TITLE
<-------Endcase 1.                                                        (AVAILABLE-TAPES)
    Endif.
    Input TAPE from TAPE-LIBR.
    Output TAPE to Provide Tapes 3.2*.
    Output TAPE-CHIP to TAPE-LIBR as a                REJECT
        replacement for TAPE.                         TAPE
Endcase 1.                                            TAPE-CHIP
Case 2 (Customer provides TYPE + TITLE):              TYPE+TITLE
    Search AVAILABLE-TAPES for TYPE + TITLE           TAPE-CHIP
    If TYPE + TITLE is not found,                     TAPE
        Reject TAPE-REQUEST as REJECT (tell
            customer we don't have movie).
<-------Endcase 2.
    Endif.
    Input TAPE-CHIP from AVAILABLE-TAPES.
    Input TAPE from TAPE-LIBR.
    Output TAPE-CHIP to TAPE-LIBR as a
        replacement for TAPE.                         REJECT
    Output TAPE to Provide Tapes 3.2*.                TAPE-CHIP         REJECT
Endcases.                                             TAPE
Stop.                                                                   TAPE-CHIP
                                                                          (TAPE-LIBR)
                                                                        TAPE
```

Processor Requirements: Manual process performed by the counter clerk.

Security Requirements: N/A.

Time Requirements: N/A.

Processing Priority: This process has the highest priority for a counter clerk and may interrupt lower-priority processes done by a counter clerk.

Performance Requirements: Response to a customer's TAPE-REQUEST shall be done within 3 minutes to avoid customer dissatisfaction.

Safety Requirements: N/A.

Other Requirements: None.

Figure 5.6f

Provide Tapes MINI-SPECIFICATION 3.2

```
     Input CUSTOMER-NAME from Customer.
     Input CUSTOMER-RECORD from CUSTOMER-FILE for
        CUSTOMER-NAME.
  ┌─ If there is no CUSTOMER-RECORD for CUSTOMER-NAME,
  │     Reject the customer's request to rent tape(s) as REJECT.
<-┼------Stop.
  ├─ Else,
  │ ┌─ If CUSTOMER-NAME has 3 or more tapes now being rented,
  │ │     Reject customer's request to rent tape(s) as REJECT.
<-┼-┼------ Stop.
  │ ├─ Else, Input TAPE from Determine If Tape Is Available 3.1*.
  │ │ ╔═ For each TAPE:
  │ │ │     Copy TITLE and TYPE from TAPE to TAPE-INFO
  │ │ │        on CUSTOMER-RECORD.
  │ │ │     Set DATE-RENTED to CURRENT-DATE.
  │ │ │     Output CUSTOMER-RECORD to CUSTOMER-FILE.
  │ │ │     Copy TAPE-INFO on CUSTOMER-RECORD
  │ │ │        to RENTAL-RECEIPT.
  │ │ │     Output TAPE to Customer.
  │ │ │     Output RENTAL-RECEIPT (original) to Customer.
  │ │ └─ Endfor.
  │ │    Output RECEIPT-COPY-1 and RECEIPT-COPY-2
  │ │       to TAPE-RECEIPT-BOX.
  │ └─ Endif.
  └─ Endif.
     Stop.
```

Inputs	Outputs
CUSTOMER-NAME	REJECT
CUSTOMER-RECORD (CUSTOMER-FILE)	CUSTOMER-RECORD (CUSTOMER-FILE)
TAPE	TAPE
	RENTAL-RECEIPT
	RECEIPT-COPY-1 + RECEIPT-COPY-2 (TAPE-RECEIPT-BOX)

Processor Requirements: Manual process performed by the cashier.

Security Requirements: N/A.

Time Requirements: N/A.

Processing Priority: This process has a medium priority.

Performance Requirements: A customer shall not wait longer than 5 minutes to be processed to avoid customer dissatisfaction.

Safety Requirements: N/A.

Other Requirements: None.

Figure 5.6g

Receive Rental Tapes MINI-SPECIFICATION 4.0

Mini-specification	Data elements	Inputs / Outputs
		CUSTOMER-NAME
Input CUSTOMER-NAME from Customer.	CUSTOMER-NAME	CUSTOMER-RECORD
Input CUSTOMER-RECORD from CUSTOMER-FILE	CUSTOMER-RECORD	(CUSTOMER-FILE)
for CUSTOMER-NAME.	TAPE	TAPE
Input TAPE from Customer.		CASH
For each TAPE:		RECEIPT-COPY-1
Set DATE-RECEIVED on CUSTOMER-RECORD		RECEIPT-COPY-2
to CURRENT-DATE.		(TAPE-RECEIPT-BOX)
Output CUSTOMER-RECORD to CUSTOMER-FILE.		TAPE-CHIP
Calculate NUMBER-OF-DAYS = CURRENT-DATE		(TAPE-LIBR)
- DATE-RENTED.		TAPE-BOX
Calculate AMOUNT-DUE = BASE-RATE +		(AVAILABLE-TAPES)
DAY-RATE * (NUMBER-OF-DAYS - 1)		
Endfor.		
Sum all AMOUNT-DUE into TOTAL-CHARGES.	CUSTOMER-RECORD	
Output TOTAL-CHARGES to Customer.	TOTAL-CHARGES	
Input CASH from Customer.	CASH	
If CASH does not equal AMOUNT-DUE,		
Reject CASH as REJECT.		
Endif.	REJECT	
Output CASH to CASH-BOX.	CASH	
Input RECEIPT-COPY-1 and RECEIPT-COPY-2 from	RECEIPT-COPY-1	
TAPE-RECEIPT-BOX for CUSTOMER-NAME.	+RECEIPT-COPY-2	
Set RECEIPT-COPY-1 and RECEIPT-COPY-2 to "PAID"		
Output RECEIPT-COPY-1 to PAID-BOX.	RECEIPT-COPY-1	CUSTOMER-RECORD
Output RECEIPT-COPY-2 to Customer.	RECEIPT-COPY-2	(CUSTOMER-FILE)
	TAPE	TOTAL-CHARGES
For each TAPE:	TAPE-CHIP	CASH
Replace TAPE-CHIP with TAPE in TAPE-LIBR.	RECEIPT-COPY-1	(CASH-BOX)
Match TAPE-NUMBER on TAPE-CHIP	TAPE-BOX	REJECT
to TAPE-NUMBER on TAPE-BOX		RECEIPT-COPY-1
in AVAILABLE-TAPES.		(PAID-BOX)
Output TAPE-CHIP to AVAILABLE-TAPES.	TAPE	RECEIPT-COPY-2
Endfor.	TAPE-CHIP	TAPE
Stop.		(TAPE-LIBR)
		TAPE-CHIP
		(AVAILABLE-TAPES)

Processor Requirements: Manual process performed by the cashier.

Security Requirements: N/A.

Time Requirements: N/A.

Processing Priority: This process has a relatively low priority for a cashier and may be interrupted by higher-priority processes. However, once the process begins to process a customer, it shall not be interrupted.

Performance Requirements: Response to a customer waiting to be processed shall not be longer than 6 minutes to avoid customer dissatisfaction.

Safety Requirements: N/A.

Other Requirements: None.

Notice that the data element NUMBER-OF-DAYS is created and used exclusively in this mini-spec. For such data elements, no entry is made in the data dictionary. These data elements are known as local variables *and are used within this mini-spec only.*

Figure 5.6h

End of Day MINI-SPECIFICATION 5.0

Input all CASH less STARTING-AMT from CASH-BOX. Set BANK-DEPOSIT to all CASH taken from CASH-BOX. Output BANK-DEPOSIT to Bank. Stop.	CASH (CASH-BOX) BANK-DEPOSIT

Processor Requirements: Manual process performed by the manager.

Security Requirements: The manager is responsible for the money in the CASH-BOX.

Time Requirements: This process is performed at the end of every day.

Processing Priority: This process is the last process to be performed in the business day.

Performance Requirements: N/A.

Safety Requirements: N/A.

Other Requirements: None.

Figure 5.6i

SUMMARY
. . . structured specifications model the physical aspects of systems

The structured specification, consisting of the SYSTEM DESCRIPTION, CONTEXT DIAGRAM, DATA FLOW DIAGRAMS, DATA DICTIONARY, and MINI-SPECIFICATIONS, is a physical model of the current system. In Chapter 9, we use this model for logical analysis, object analysis, and event analysis to produce the logical model of the current system. Once we complete the physical and logical models of the system, we have a good understanding of what the system does and how it works. Using these models as tools, we can review the structured specification to propose changes to meet new requirements or needs of the users. After the users review and approve the proposed changes, the revised structured specification becomes the requirements used to develop the new changes or enhancements agreed upon.

QUESTIONS

1. a. Explain the use of RENTAL-RECEIPT (original) in the system.

   b. Is RENTAL-RECEIPT (original) ever stamped "PAID?"
2. What data elements of RENTAL-RECEIPT are filled in by process 3.2?
3. If the manager considered the answer to question 2 to be insufficient information, what changes would you make to the mini-spec to rectify the problem?
4. If the manager found from experience that the CASH-BOX had to have $40.00 in it at the beginning of each day, indicate all of the changes one would have to make to the entire system specification.
5. In this system, who is responsible for ordering new movies?
6. In this system, who is responsible for balancing the CASH-BOX?
7. If you were the owner of the business and found that money was missing from the CASH-BOX, who would be the likely suspect(s) that has access to the CASH-BOX as part of his or her normal duties?
8. What ultimately happens to the RENTAL-RECEIPT and all of its copies (RECEIPT-COPY-1 and RECEIPT-COPY-2)?
9. Why are entries such as ACTOR, ACTRESS, STARS, STUDIO, and PLOT shown in small caps in the data dictionary?

10. How are the data elements MARKED-TAPE-LIST and NEW-TAPE-CATALOG related?
11. What are the two types of rejects that can exist within a system and how are they documented in the structured specification?
12. According to the structured specification, what is the result if a registered customer requests to rent four tapes?
13. According to the structured specification what is the result if:
    a. A registered customer who has no outstanding tapes being rented, requests to rent two tapes and then returns within an hour and requests to rent two additional tapes without returning the first two tapes?
    b. The same registered customer returns a third time after the action in question 13a and requests to rent one tape and does not return any of the tapes rented so far?
14. Does the policy implemented in the mini-specs of the system match the system's description at the beginning of the chapter regarding the maximum number of tapes a customer can rent?

PROBLEMS

1. Modify the Rent-em Video System to sell used video tapes for 50 percent of the price paid to the Video Wholesaler after the tapes have been rented more than 30 times.

2. Modify the Rent-em Video System to give a 10 percent discount to all senior citizens when they present a Senior Citizen's Card or a driver's license.

   SENIOR-CITIZEN-CARD = SENIOR-NAME + SENIOR-ADDR + "Senior Citizen"

PART II

Logical Specifications

This part presents the tools and techniques for preparing a logical structured specification. To construct this model, we must separate the physical aspects of the system from the logical aspects. Activities are separated into those that exist because of the technology used in the system and those that must exist in order to accomplish the objectives of the system. Data are also organized into forms whereby we can easily identify the data being stored and the relationships among the data, independent from the technology used to store the data. And, finally, the text presents a partitioning of a system from the users' perspective by organizing processes and data by the events to which they respond.

6

LOGICAL ANALYSIS

We have now reached a point where we have a complete, accurate, unambiguous, nonredundant structured specification of a working system as it currently exists. The model that we've constructed is able to be understood and verified by the users, namely those who work with or within the system. Although we may have spent many hours creating the physical model, we have not yet done any system analysis. We have just defined the scope of our analysis. Our goal is to specify the requirements for a new system. The new system will either rectify some outstanding problem(s) in the current system, increase functionality or capacity, or include some new constraint(s) as specified by management. As Steve McMenemin and John Palmer stated, "The purpose of a systems analysis is to produce a statement of the *true requirements* for the system to be built."[M-5] In order to produce *true requirements*, the systems analyst must provide a policy statement of the system's behavior. To begin to derive this statement of policy for the new system, we must understand the policy statement of the current system. We need to take the model of the current system and analyze it in such a way that we can distill the basic policy of the system and see how it compares to what the user wants in the new system.

Let's consider our structured specification of the physical system and examine what we have. Notice that in the physical model, we specified not only *what* was being done, but also *how* it was being implemented. The physical model is contaminated with many physical and technology-related activities and memories simply because *that is the way things are done* in the current system. We want to separate the *what* from the *how*. Namely, we want to separate the things that do the data transformations from the things that service the processes that do these transformations. Simply stated, we want to know what processes and data stores **serve the users** and which ones **serve the system**. In this way, we can distinguish between changes in the technology of *how* things are done from changes to the fundamental business practices implemented in the system.

Basically, there are four sets of activities in a system:

1. Those that carry *communications* among the processes
2. Those that perform *administrative functions* usually needed because of some processing requirement, inefficiency, or unreliability in the system
3. Those that service the *essential memories* in some way
4. Those that perform the *essential data transformations*, converting input data into output data.

Our goal is to identify those activities, data flows, and data stores that are necessary no matter what type of technology is used in the system. Such activities and data stores give the system its *value added*, or *reason for being*, from the users' perspective. These activities, data flows, and data stores that service the primary purpose of the system are known as *essential* components. The remaining activities and data stores that carry on the communications and administrative functions operate on a level below those responsible for performing the data transformations and providing the required output of the system. The prefix *infra* means below or within. In looking at the overall structure of the system, we wish to separate those functions and data that service the primary purpose of the system from those that are *subordinate* to the primary responsibilities but still operate *within* the system. Activities, data flows, and data stores that service the system instead of the users are said to operate at a lower level within the system. Subordinate functions and data stores are usually necessary because we have selected a particular type of technology for use in the system. Thus, we will refer to the communications and administrative functions and certain data stores as belonging to the *infrastructure* of the system. The activities, data flows, and data stores that carry on the primary responsibility of the system are referred to as the *essence* of the system. Let's discuss each of these categories separately.

SYSTEM'S INFRASTRUCTURE
. . . components that serve the system

Identify the Communications Infrastructure

Systems that use more than one process to accomplish an essential activity must have a means of transporting data among the processes. The means or network is known as an *infrastructure*. **INFRASTRUCTURE refers to a network of components that transports people, goods, and information from one location to another in a society.**[M-5] We want to identify all the activities and data stores whose only purpose is to keep data flowing in the system. Such a network of activities and memories forms the communications infrastructure. The activities can be manual or automated. For example, a receptionist that routes phone calls, a secretary that routes mail, or a process that inputs terminal transactions and routes them to the appropriate processing activity are all carrying out activities that are part of the *internal communications infrastructure*. Also included in this category are input routines and report-generating programs, which are fundamentally mechanical activities whose main purpose is to pass data in and out of the system. The activities classified as internal communications infrastructure not only transport the customer's requests or data, but can convert, translate, format, or organize the data. Within the communications category, we may also wish to include activities such as merge, sort, or collate data as it is being passed on to another process in the system. These types of functions are related solely to the type of technology used to implement the system.

All of the activities and processes found in the infrastructure of a system frequently relate to the level of technology used in the system. If we wish to send data to a person located in a distant building, there are several ways to do so:

1. Make a photocopy and mail it to the recipient.
2. Telefax a copy to the recipient.
3. Hire a carrier service to personally deliver the data.
4. Enter the data into a computer and transmit the data to the recipient's computer.
5. Commit the data to human memory and use mental telepathy to communicate the data to the mind of the recipient.

Each of the above alternatives uses a different type of technology to transmit the data. The technology varies from the simple (mailing a photocopy) to the very advanced (mental telepathy). Each involves different processes and/or data stores to carry out the activity of communicating the data to the recipient.

Data stores in and of themselves can also serve the function of transporting data and holding it until the receiving activity is ready to process the data. Examples of manual data stores that perform communication functions are "in-boxes," file drawers, stacks of paper, "out" files, notebooks, bulletin boards, etc. In automated systems, internal communication data stores can also appear as buffers, tables, queues, as well as temporary and permanent files. Any place where data is held to permit a passing of time until some process can use it as input is an internal communications memory. We want to identify all the processes and memories (data stores) that make up the communication functions and eliminate them in our search for the *essence* of the system.

So, let's look at what we have and give some specific definitions that we can work with.

INTERNAL COMMUNICATIONS ACTIVITY (ICA) is a process that communicates the results of one activity to another or links another process to a data store. The main purpose of the activity is to *transport* data, and, in some cases, it may even *translate* data from one form of representation into another or *organize* data by merging, sorting, or collating. These processes frequently pack, ship, carry, receive, and unpack data traveling about the system.

INTERNAL COMMUNICATIONS MEMORY (ICM) is a data store that acts as a temporary storage area to synchronize communications in the system by holding data until the receiving process is available to receive the data.

Notice that the ICAs can *translate and/or convert data* from one form to another while it is being routed to its intended destination. The sorting, merging, and collating activities are considered *organizing data* from one structure to another. Likewise, the processes that separate data flows into bins so that the appropriate set of processes can process them are also considered part of the internal communications infrastructure.

A function that would input a MAIL-ORDER-REQUEST into a computerized system can merely be converting the customer's order into another equivalent data flow usable by a computer. In the data flow diagram of Fig. 6.1, both the process Greet Customer 1.4* and the data store RETRIEVAL-ORDER are part of the internal communications infrastructure of the system.

The internal communications infrastructure consists of those subordinate activities, data flows, and data stores that manage the transportation of data within the system. Without these functions and memories, the system would have to have

COMMUNICATIONS INFRASTRUCTURE EXAMPLE

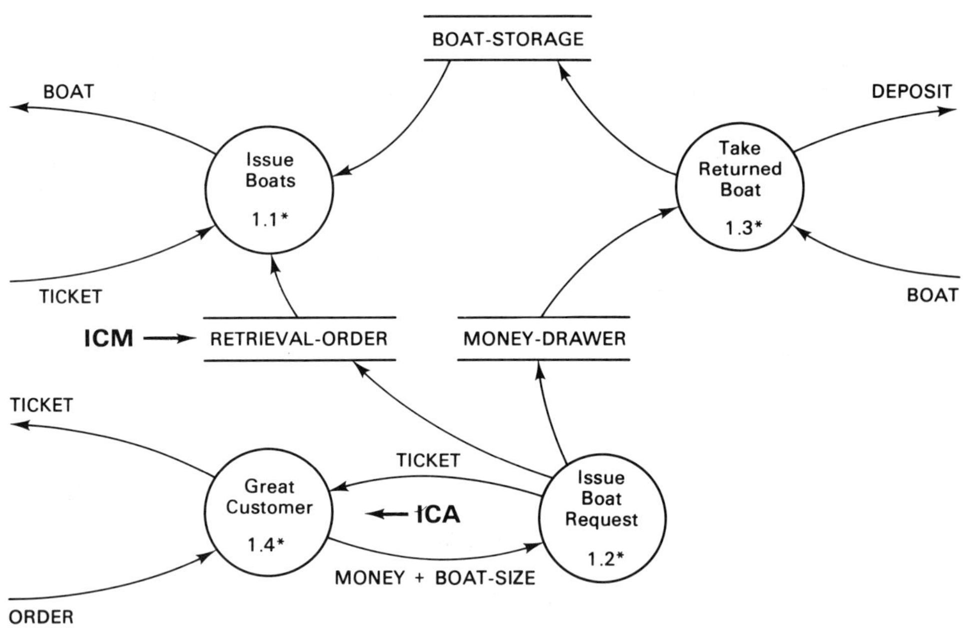

Figure 6.1

perfect telepathic communications, in which there would be no need to sort, merge, separate, translate data, or synchronize the flow of data between processes.

The activities and data flows that directly interface with the source of the input to the system are usually considered *essential*, particularly because we may not have any control over the technology used by the source. However, in determining the technology-free *logical* view of the system, we categorize these activities and data flows as *infrastructure* because they merely move data to a process that transforms input (ORDER) into output (TICKET), as seen in Fig. 6.1. After we obtain the pure logical essence of the system, many of these activities, data flows, and data stores of the infrastructure will have to be put back into place to implement the system in the *real world*. We may or may not have the opportunity to change the technology used in the system, but we will at least be able to identify those activities, data flows, and data stores that exist solely because of the implementation technology used in the system.

Identify the Administrative Infrastructure

There are several types of processes that **serve the system** rather than **serve the users** of the system. These activities are administrative in nature and include such functions as editing data, auditing results, approving requests, logging information, measuring system performance, backing up files, recovering data, restarting processes, monitoring and managing system resources, and maintaining an orderly flow of data. The central theme of these activities is to detect, eliminate, or prevent errors from producing erroneous results or otherwise affecting the outputs or results of the system. The administrative functions also form an infrastructure that operates subordinately to the system's primary responsibilities. It is the mechanism that protects the system from itself, as well as detects erroneous input received from outside the system.

Protecting the system from erroneous input data may seem to be an essential activity for a system to perform, and it is. However, just as in the case of the communications activities that service the boundaries of the system, all administrative activities that relate to technology need to be identified and removed from the logical model. Later, when it comes time to specify the requirements of implementing the system using available technology, many of these administrative activities, data flows, and data stores will be put back into the structured specification.

Also, included in the administrative components of a system are those processes and data stores that maintain an orderly flow of data. These types of processes, data flows, and data stores relate strongly to the kind of technology used in the system. Punch-card verification is a necessary administrative activity when one transcribes data onto 80- (or 96-) character punch cards. This activity identifies and filters out errors so that they would not be processed by the system. Activities such as this are part of the internal administrative infrastructure and are not the reason why the system exists. These activities are often related to the technology used to implement the system. The following are other examples of administrative activities and data stores that make up the infrastructure of the system.

- Activities that accumulate and maintain a transaction log
- Activities that maintain memory for restart and recovery processes in the event of system failure
- Activities that check for erroneous inputs given to or generated within the system
- Activities that monitor real-time processes within a system or manage limited resources of the system. (Activities that control real-time processes are considered part of the infrastructure if the process they control is part of the infrastructure)
- Manual or automated processes that verify input data
- Activities that provide approvals so that data can continue to flow through the system
- A data store containing identification or sequence numbers that a process assigns to a group data element in order to maintain an orderly flow of data
- Data stores that hold blank multipart forms used when a customer inputs data into the system
- Transaction logs that record all customer interactive requests
- Archive files to save data for historical reference, etc.

Processes that regulate or manage the system's resources (just as a receptionist routes incoming calls to the next available service representative) are either administrative or communications activities. Occasionally, accounting systems have a *write-only* file that is necessary to save some financial data, if for no other reason than the IRS may ask about it someday. Administrative activities and administrative memories would not be present if the system's input processes, data conversion processes, data flows, data stores, and output processes were infallible and could respond instantly to all requests with error-free results.

In a real-time system, we can find many types of audit processes that seek out problems caused by imperfect technology. The audit process may even output a data flow informing the system user of an anomaly in the system. Such processes, even though they output data, are still considered part of the administrative infrastructure and are classified as internal administrative activities (IAA). The specific definitions for the administrative infrastructure follows.

INTERNAL ADMINISTRATIVE ACTIVITY (IAA) is a process that maintains orderly flow of data, manages the affairs of editing, auditing, approving, or measuring the performance of the system activities, detects and possibly corrects operational errors, monitors and regulates the use of resources by the other processes, or keeps a historical record. These activities are frequently set up to protect processes from receiving erroneous data produced by another process and manages the affairs of how the system goes about producing the required outputs.

INTERNAL ADMINISTRATIVE MEMORY (IAM) is a data store that holds information about the system for either some current use of maintaining orderly flow of data, managing the system's resources, or for some future use that is not part of the normal activities or purpose of the system. Such data stores are identification numbers, forms, transaction logs, audit trails, historical archives, performance data, restart/recovery, and other files that service the system. If a process has to hold information temporarily until the system can continue processing the information, the process can place the data in an *internal administrative memory* (IAM) and retrieve it when processing can resume.

ADMINISTRATIVE INFRASTRUCTURE

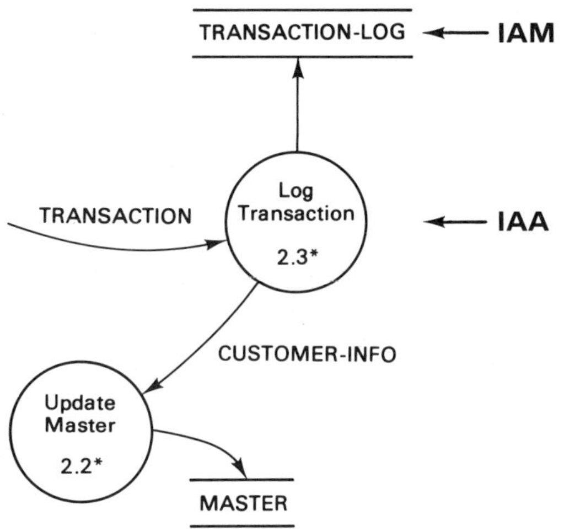

Figure 6.2a

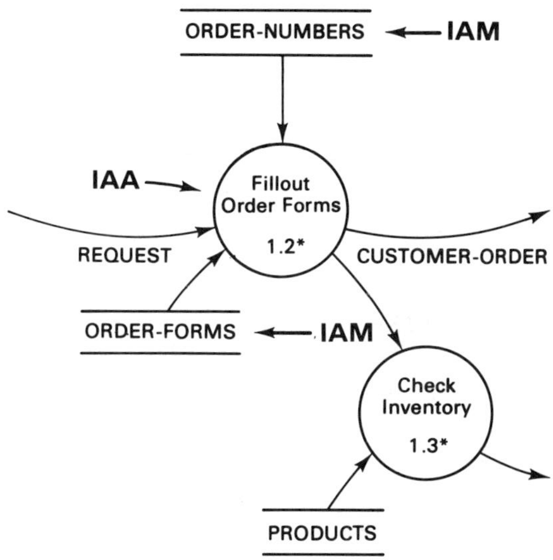

Figure 6.2b

The data flow diagrams in Figs. 6.2a to 6.2d illustrate some administrative functions that are part of the internal administrative infrastructure.

Loosely speaking, administrative activities are the subordinate processes that manage the orderly flow of data and the efficient use of system resources. They also maintain the integrity of the data within the system, as well as the integrity of the system itself. Although it is our intention to identify all administrative activities for the purpose of eliminating them from the logical model, care must be taken in classifying activities that interface directly with the sources and sinks of the system. Every data flow that crosses the system's boundaries must be traceable to a process that is responsible for servicing the input or providing the output. (There is one exception: the case where an IAA outputs a message to the user to inform them of a detected anomaly. This usually applies only to real-time systems.) In the example in Fig. 6.2b, process 1.2* seems to be accepting a customer's REQUEST for a PRODUCT. The process assigns an order number to a form obtained from ORDER-FORMS, translates the verbal REQUEST onto the form, and passes it on to process 1.3*. All the tasks performed in process 1.2* qualify as administrative activities. Even though this process interacts directly with a source, it is only preparing the order for the process that is responsible for processing the order. If perfect technology were available, the input data flow would go directly to process 1.3* and there would be no need for transcribing an order onto paper.

ADMINISTRATIVE INFRASTRUCTURE

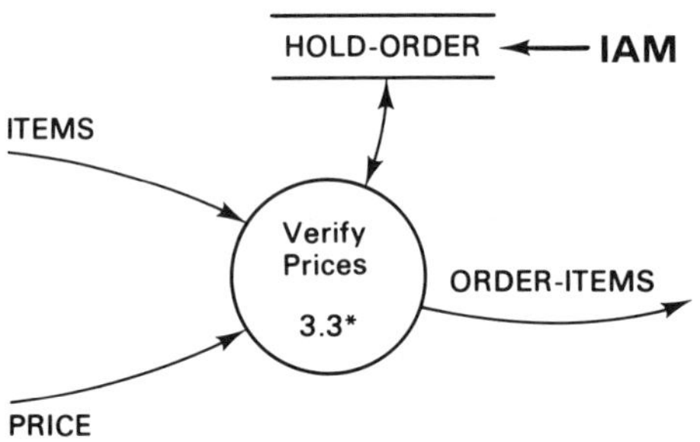

Figure 6.2c

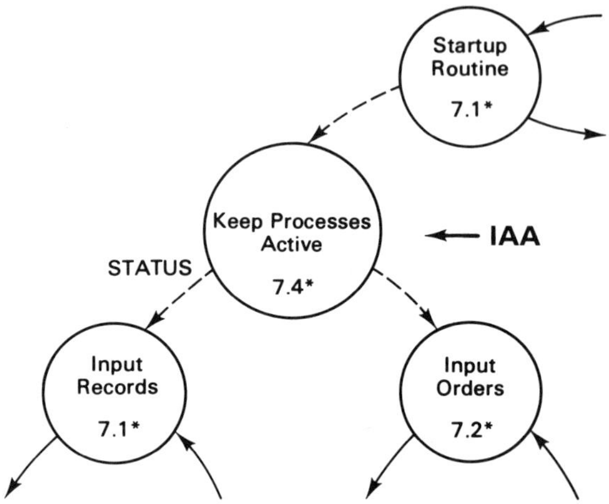

Figure 6.2d

If instead of receiving a REQUEST from a source outside the system, suppose process 1.2* received a REQUEST from a previous process that took phone orders from a source. Process 1.2* would assign the proper order number to the form selected from ORDER-FORMS and would send it on to one of several departments for processing. Even in this case, process 1.2* would be considered an internal administrative activity because it performs an administrative task totally internal to the system. If, on the other hand, process 1.2* received mail, opened the mail, and sent all checks to the sink Accounting Office, while all other mail was sorted and routed to the appropriate process for further processing, the process would then basically be performing an administrative function. However, because it would be the only process responsible for providing one of the system's outputs, namely checks, it would be performing an essential fundamental activity. As a result, the more important responsibility of producing one of the system's defined outputs classifies the activity as an essential activity. Care must be taken when the data flow crosses the system's boundaries to make sure that there is at least one essential process ultimately responsible for processing every input data flow and producing every output data flow. The responsible process is part of the essence of the system and not the infrastructure. The essential process need *not* be the first process that receives the input data flow nor the final process that sends the output data flow to the sink.

SYSTEM'S ESSENCE
. . . components that serve the user

Identify the Essential Elements of a System

Now that we have identified all of the activities and memories that form the communications and administrative infrastructures of a system, we can turn our attention to the most important elements, the *essential components*. The **ESSENTIAL COMPONENTS of a system are those essential activities and essential memories that must be present no matter what type of technology is used to implement the system.** There are two types of essential activities: the fundamental activities that perform the actual data transformations in the system and those that maintain the essential memory of the system.

ESSENTIAL FUNDAMENTAL ACTIVITY (EFA) is a process that carries out the responses that justify the existence of the system. Every output data flow must be able to be traced to an EFA that is responsible for producing the output.

ESSENTIAL CUSTODIAL ACTIVITY (ECA) is a process that establishes and maintains the essential memory, such that it is sufficiently correct to support the essential fundamental activities. Every input data flow must ultimately be processed by either an EFA or an ECA.

ESSENTIAL MEMORY (EM) consists of the information that must be remembered to support the essential fundamental activities. Essential memories contain anything that the system must remember or have in order to do its job. In addition to keeping data, the essential memories can contain any products or physical items that a user might request. Examples of physical items kept in an essential memory are brochures, catalogs, merchandise, etc. Any process that keeps the data up to date or replenishes the supply of physical items is an essential custodial activity.

The objective in establishing such classifications of activities and memories is to separate the infrastructure from the processes that perform the transformation of input data into output data in a system. Having the ability to make such a distinction allows us to better understand the scope and effect of the changes needed to satisfy new constraints or requirements. Most of the changes required in information systems are to the infrastructure of the system. This is because the infrastructure is closely tied to technology and it is subject to the many technological changes constantly taking place in our industry. The essential fundamental activities rarely change, and, if they do, the impact is usually broader in scope and has greater effect on the processes and data stores of a system.

How many times have companies rewritten their payroll systems? I submit that in all of those changes, no one has changed the fundamental process of computing the amount paid by applying the duration of time-worked to some specified rate, and deducting the appropriate taxes, deductions, and exemptions. However, there have been many changes in the technology that presents the input data to the system. Originally, the data probably came in the form of cardboard "time cards," then on keypunched 80-column cards. Later, data entered the system by keyboard-entry terminals, or on-line automatic "time clocks," or perhaps even on-line magnetic card readers, where employees use cards that resemble credit cards and pass them through a magnetic-strip reader. To be partitioned properly, the support activities of the infrastructure and fundamental activities should be linked via simple and well-defined interfaces. Under that type of system architecture, we can accommodate massive technological changes with limited trauma to the system.

Figure 6.3 is an example of a data flow diagram that indicates essential fundamental activities (EFAs), an essential custodial activity (ECA), and an essential memory (EM) in a small system that maintains a subscription mailing list for a monthly publication. The reasons the system exists are

1. To register new subscriptions
2. To send reminders when subscriptions are about to expire
3. To generate mailing labels for the publication
4. To make sure a customer has an up-to-date subscription and to handle any complaints when the customer does not receive the publication

Notice that when the data first enters the system, a process separates it into one of two data flows: APPLICATIONs and all other mail (which is ultimately rejected by the process File Mail 1.5*). Process 1.5* passes all APPLICATIONs to the Register New Subscriber 1.1* process, which has the responsibility of keeping the essential memory (EM) SUBSCRIBERS up to date so that the remaining activities can take place. In addition to performing an essential custodial activity (ECA), process 1.1* performs an internal administrative activity (IAA) and files the customer's APPLICATION in the APPLICATIONS-FILE in case someone needs to see it. If a process performs at least one essential function (either an EFA or an ECA), it will be carried over in the logical model as an essential component. The main distinction we are trying to make is between those activities or data stores that are part of the infrastructure and those that are essential components. Whether an activity or data store is part of the communications or administrative infrastructure is of less importance. The main distinction is between the infrastructure and the essential components of a system. In the example of Fig. 6.3, the process File Mail 1.5* is an internal administrative activity (IAA) that separates MAIL into APPLICATIONs and all other mail, which it sends out of the process as REJECT. We can consider

LOGICAL ANALYSIS OF PROCESS 1.0

DIAGRAM 1.0 (LEVEL 0)

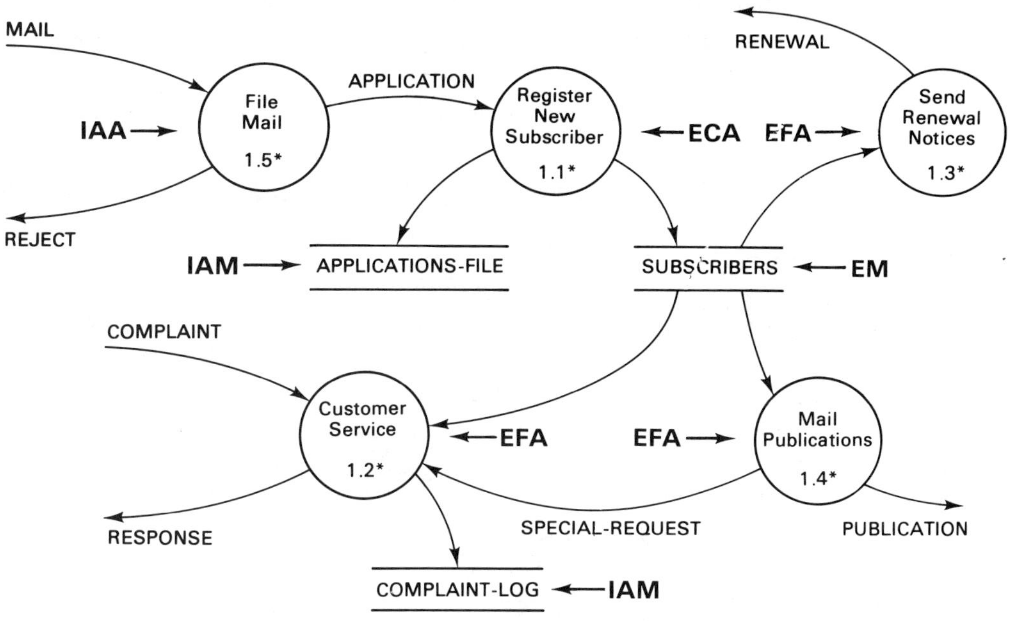

Figure 6.3

process 1.5* as an internal communications activity (ICA); either way, it is part of the infrastructure and will not be carried over into the logical model.

The only other input to the system is the data flow COMPLAINT. The process Customer Service 1.2* is responsible for satisfying the complaint and providing the necessary RESPONSE. When a process is totally responsible for providing an output of the system, it is an essential fundamental activity (EFA). Also, there is a log where all complaints are filed after appropriate action has been taken. The COMPLAINT-LOG data store is an internal administrative memory (IAM). The data placed in COMPLAINT-LOG is not used anywhere else in the system and is probably referenced by management at some later time by some other system (or at least we would like to think so).

Example of Logical Analysis

We begin the process of logical analysis by going through the data flow diagrams of the physical model and identifying each functional primitive as an ICA, IAA, ECA, or EFA, and each data store as an ICM, IAM, or EM. Consider the TYPICAL BILLING SYSTEM of Fig. 6.4. The system registers new customers, opens new accounts, applies charges to existing accounts, removes charges from existing accounts, and provides three other outputs:

1. Displays all current account charges upon request.
2. Issues a BILL to customers, which includes a statement of account and a self-addressed envelope for return payment.
3. Reports on the level of business to management.

CONTEXT DIAGRAM OF A TYPICAL BILLING SYSTEM

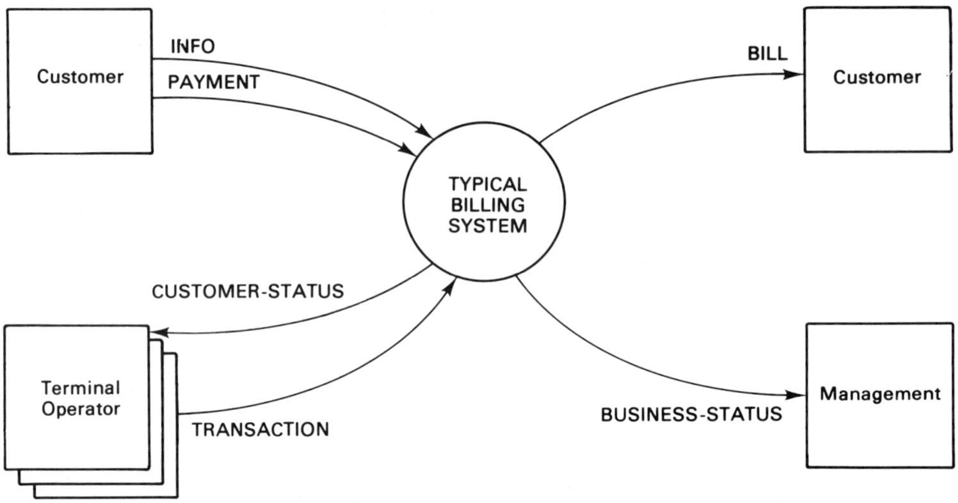

Figure 6.4

In Fig. 6.5, we can see all of the functional primitives that make up the TYPICAL BILLING SYSTEM as they appear in the physical model.

Formally, **LOGICAL ANALYSIS is the process of separating the physical characteristics of a system (physical model) into the supporting infrastructure and the essential activities and memories that are required for the system to achieve its intended purpose.** From the logical analysis, we obtain the essential memories upon which we perform object analysis (see Chapter 7) and we come to understand which processes and data serve the system (infrastructure) and which serve the users.

To begin, logical analysis consists of the following four steps:

1. Identify and remove the communications infrastructure.
2. Identify and remove the administrative infrastructure.

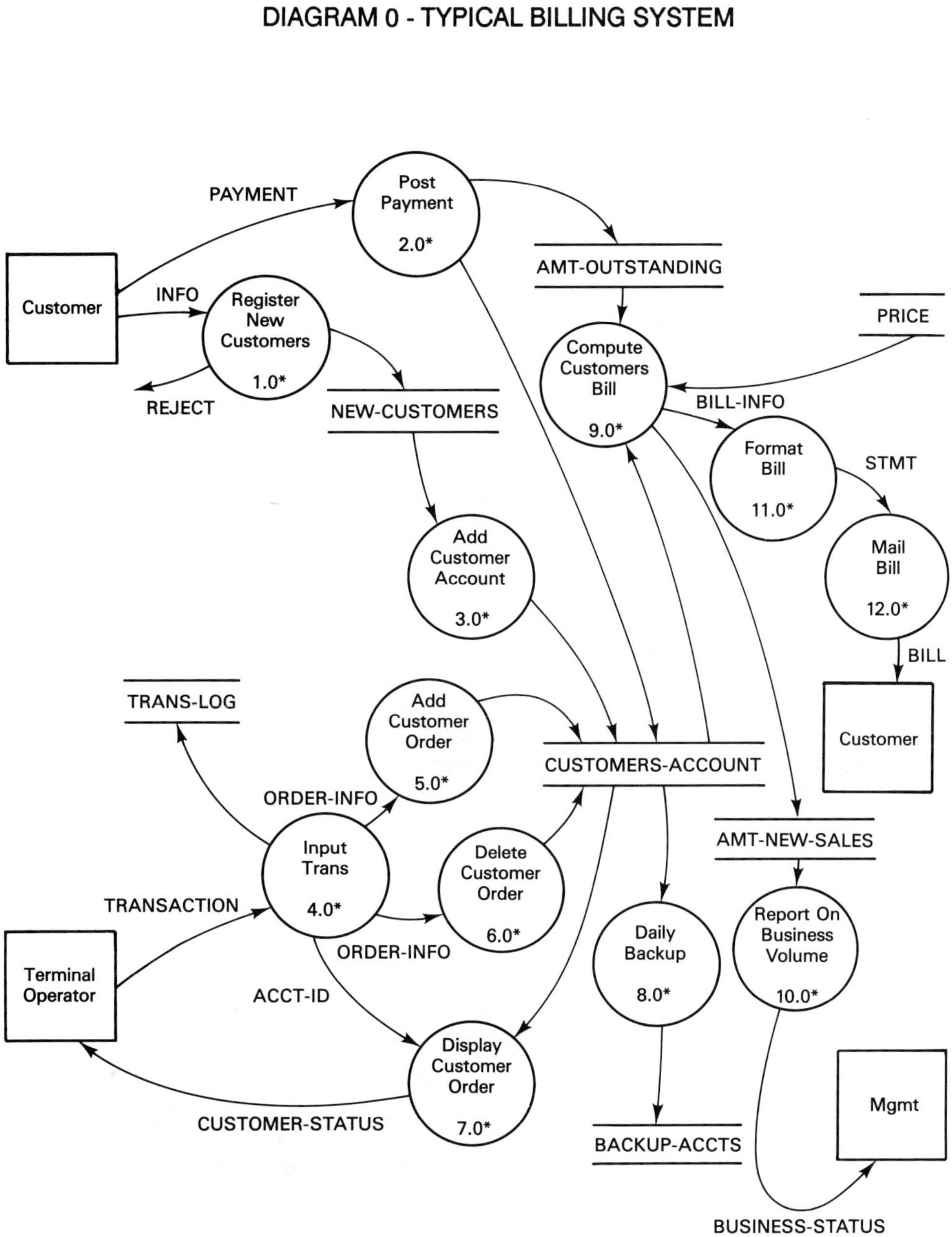

DIAGRAM 0 - TYPICAL BILLING SYSTEM
PAYMENT
Post Payment 2.0*
Customer
INFO
Register New Customers 1.0*
REJECT
AMT-OUTSTANDING
PRICE
NEW-CUSTOMERS
Compute Customers Bill 9.0*
BILL-INFO
Format Bill 11.0*
STMT
Add Customer Account 3.0*
Mail Bill 12.0*
BILL
TRANS-LOG
Add Customer Order 5.0*
Customer
CUSTOMERS-ACCOUNT
ORDER-INFO
AMT-NEW-SALES
Input Trans 4.0*
Delete Customer Order 6.0*
TRANSACTION
Report On Business Volume 10.0*
Daily Backup 8.0*
ORDER-INFO
Terminal Operator
ACCT-ID
Display Customer Order 7.0*
Mgmt
CUSTOMER-STATUS
BACKUP-ACCTS
BUSINESS-STATUS

Figure 6.5

Now, consider the remaining essential activities and memories:

3. Partition the essential memories into the basic objects for which the system is storing data, and the data attributes that describe those objects (see Chapter 7, Object Analysis).
4. Reorganize the essential activities and object data stores (determined in step 3) according to the events to which they respond (see Chapter 8, Event Analysis).

When we complete these four steps, we can:

1. Clearly differentiate between the activities and memories that effect the fundamental purpose of the system from those that are directly related to the type of technology used in implementing the system.
2. Clearly understand what *data* are stored and used in the system to create each response the system makes.
3. Clearly relate the basic activities and memories of the system to the *events* that are responsible for evoking responses from the system.

INPUT/OUTPUT PROCESSING CHART
. . . tool to aid logical analysis

To begin classifying each process and data store in the system, we should create a chart of all the inputs, outputs, and data stores. The input/output processing chart shown in Fig. 6.6 will aid in organizing the logical analysis of processes and data stores in the system. This chart is also helpful during event analysis when we are trying to identify the events of the system.

We create the chart by listing all **inputs**, **outputs**, and **data stores** along the left margin and listing the process numbers of all **functional primitives** across the top of the chart. Two summary areas consolidate the analysis and determine the final disposition of each process and data store:

1. Each entry on the PROCESS SUMMARY line consists of the highest-priority classification for each functional primitive process. The process column lists all of the classifications of all the tasks performed by the process as it relates to each of the various input and output data flows. The process is then given a single classification in the PROCESS SUMMARY line according to the highest-priority function the process provides. The priorities are as follows:

highest ESSENTIAL FUNDAMENTAL ACTIVITY (EFA)
↕ ESSENTIAL CUSTODIAL ACTIVITY (ECA)
INTERNAL ADMINISTRATIVE ACTIVITY (IAA)
lowest INTERNAL COMMUNICATIONS ACTIVITY (ICA)

The same process that is an IAA for one input data flow and an ECA for another input data flow is officially classified by the second highest-priority function, namely ECA, as shown above.

2. As each input or output data flow is considered, the data stores that are involved in the processing are also classified in a priority scheme as follows:

    highest ESSENTIAL MEMORY (EM)
    ↕ INTERNAL ADMINISTRATIVE MEMORY (IAM)
    lowest INTERNAL COMMUNICATIONS MEMORY (ICM)

After the analysis is complete, we consider each data store and determine its final disposition by assigning it the highest-priority classification to the far right beyond the equal sign (=) in the data store section of the input/output processing chart.

INPUT/OUTPUT PROCESSING CHART

	<-- ALL FUNCTIONAL PRIMITIVE PROCESSES --> 1.0.1 . . . . . . 9.7.9 . . .	
Inputs: (List each input data flow here alphabetically.)	(Classify the functional primitives according to the type of function they perform with the input data flow.)	
Outputs: (List each output data flow here alphabetically.)	(Classify the functional primitives according to the type of function they perform in providing the output data flow.)	
PROCESS SUMMARY:	(Summarize each functional primitive by listing the highest-priority classification here.)	
Data Stores: (List each data store here alphabetically.)	(Classify each data store according to how it is used by the functional primitive in processing the above data flows.)	= (List the highest = priority classification : for each data store.)

Figure 6.6

From the example of Fig. 6.5, consider each of the input data flows originating from some source outside the system.

Input Section

INFO. A customer provides INFO when registering as a new customer. Following that data flow through the system, we see that it passes through two processes and two data stores: Register New Customers 1.0*, Add Customer Account 3.0*, NEW-CUSTOMERS, and CUSTOMERS-ACCOUNT. Analysis of mini-specs 1.0* and 3.0* (not included here), reveals that process 1.0* merely verifies that all of the necessary data is present in the data flow INFO. The process then places the INFO into the data store NEW-CUSTOMERS until process 3.0* can add the customer to the data store CUSTOMERS-ACCOUNT.

From the above analysis, we can conclude that process 1.0* is basically doing a verification activity (IAA), passing the data to the data store NEW-CUSTOMERS (ICM), where it resides until process 3.0* is ready to place it into the data store CUSTOMERS-ACCOUNT. CUSTOMERS-ACCOUNT is the data store that the system must remember to do its job, so we classify it as an essential memory (EM). Because process 3.0* is partially responsible for keeping an essential memory up to date, we determine that it is an essential custodial activity (ECA).

So we have:

Process 1.0* (IAA), using NEW-CUSTOMERS (ICM)
Process 3.0* (ECA), using NEW-CUSTOMERS (ICM) and
CUSTOMERS-ACCOUNT (EM)

The input/output processing chart of Fig. 6.7 is updated to contain the information derived from the preceding steps in the logical analysis.

INPUT/OUTPUT PROCESSING CHART

	← ALL FUNCTIONAL PRIMITIVE PROCESSES →												
	1.0	2.0	3.0	4.0	5.0	6.0	7.0	8.0	9.0	10.0	11.0	12.0	
Inputs:													
INFO	**IAA**		**ECA**										
PROCESS SUMMARY:	**IAA**		**ECA**										
Data Stores:													
CUSTOMERS-ACCOUNT			**EM**										= **EM**
NEW-CUSTOMERS	**ICM**		**ICM**										= **ICM**

Figure 6.7

Continuing our analysis, let's consider the next input data flow PAYMENT and add our results to the input/output processing chart shown in Fig. 6.8.

PAYMENT. The system receives a PAYMENT in process Post Payment 2.0*, which, according to the mini-spec (not included here), verifies the status of the customer's account in CUSTOMERS-ACCOUNT and updates CUSTOMERS-ACCOUNT to reflect the payment. If the payment did not cover the entire outstanding amount, the process posts an entry in the AMT-OUTSTANDING data store for the customer. Because CUSTOMERS-ACCOUNT has been determined to be an EM, any process that updates that data store is an ECA. From analyzing the mini-spec for process 2.0*, we cannot determine how the data store AMT-OUTSTANDING is used. Therefore, we will postpone our classification of that data store. However, we do want to record the fact that process 2.0* does interact with the data store AMT-OUTSTANDING while processing the input data flow PAYMENT. We can do this by placing "***" in the input/output processing chart for AMT-

OUTSTANDING under process 2.0*. Thus, from this analysis, we can update our chart as shown in Fig. 6.8.

So we have:

Process 2.0* (ECA), using AMT-OUTSTANDING (***) and
CUSTOMERS-ACCOUNT (EM)

INPUT/OUTPUT PROCESSING CHART

	<---- ALL FUNCTIONAL PRIMITIVE PROCESSES ---->												
	1.0	2.0	3.0	4.0	5.0	6.0	7.0	8.0	9.0	10.0	11.0	12.0	
Inputs:													
INFO	IAA		ECA										
PAYMENT		**ECA**											
PROCESS SUMMARY:	IAA	**ECA**	ECA										
Data Stores:													
AMT-OUTSTANDING		***											= ***
CUSTOMERS-ACCOUNT		**EM**	EM										= EM
NEW-CUSTOMERS	ICM		ICM										= ICM

Figure 6.8

There is one more input data flow to consider: TRANSACTION. From the data dictionary, we can see that TRANSACTION can be one of three types of input:

1. A new customer order
2. A return or credit applied to a customer order
3. An inquiry as to the status of a customer's account

Our analysis continues by considering each of the three types of input TRANSACTIONs separately.

TRANSACTION (1). When the TRANSACTION is a new customer order, process 4.0* inputs the data flow and carries out two functions. The first function makes a log entry in the data store TRANS-LOG for restart/recovery purposes because an on-line computer system is used to carry out this portion of the system. The data store TRANS-LOG is not used by any process in the system and is therefore an IAM. Process 4.0* performs an IAA in recording all TRANSACTIONs in the TRANS-LOG data store. The second function of process 4.0* is a communications function (ICA) that passes the ORDER-INFO to process 5.0*. Because an IAA task is of a higher priority than an ICA task, process 4.0* will be listed as an IAA. Process 5.0* updates the essential memory (EM) CUSTOMERS-ACCOUNT data store to reflect the new purchase. Because CUSTOMERS-ACCOUNT is an essential memory, process 5.0* is an ECA.

So far we have:

Process 4.0* (IAA), using TRANS-LOG (IAM)
Process 5.0* (ECA), using CUSTOMERS-ACCOUNT (EM)

TRANSACTION (2). When the TRANSACTION is a return of merchandise, process 4.0* inputs the data flow and carries out the following two functions. First, it makes a log entry to the data store TRANS-LOG for restart/recovery purpose which is an IAA task. Process 4.0* then passes the ORDER-INFO to process 6.0*, which is a communications function (ICA). An IAA task outranks an ICA task with respect to priority, and thus process 4.0* is classified as an IAA. Process 6.0* receives its data from ORDER-INFO and updates the essential memory (EM) CUSTOMERS-ACCOUNT to reflect the returned purchase. Because CUSTOMERS-ACCOUNT is an essential memory, process 6.0* is an ECA.

So we have:

Process 4.0* (IAA), using TRANS-LOG (IAM)
Process 6.0* (ICA and ECA), using CUSTOMERS-ACCOUNT (EM)

TRANSACTION (3). When the TRANSACTION is a request to display the status of a customer's account, receiving process 4.0* does not make a log entry for this type of input data flow. Thus, it only performs a communications function (ICA) of passing the ACCT-ID to process 7.0*. Process 7.0* retrieves data from the essential memory CUSTOMERS-ACCOUNT and provides the system output CUSTOMER-STATUS. Process 7.0* is the only process responsible for providing the system output and therefore is an essential fundamental activity (EFA). Every system output must be the responsibility of some essential fundamental activity because without it, the system could not meet one of its objectives: to provide CUSTOMER-STATUS.

And finally we have:

Process 4.0* (ICA) to Process 7.0*
Process 7.0* (EFA), using CUSTOMERS-ACCOUNT (EM)

With the above analysis, we can further update our chart, as shown in Fig. 6.9. In the above analysis, process 4.0* is an IAA while processing TRANSACTION (1) and TRANSACTION (2), but is an ICA while processing TRANSACTION (3). To determine the PROCESS SUMMARY entry for process 4.0*, we consult the priority

INPUT/OUTPUT PROCESSING CHART

	ALL FUNCTIONAL PRIMITIVE PROCESSES												
	1.0	2.0	3.0	4.0	5.0	6.0	7.0	8.0	9.0	10.0	11.0	12.0	
Inputs:													
INFO	IAA		ECA										
PAYMENT		ECA											
TRANSACTION				**IAA**	**ECA**	**ECA**	**EFA**						
PROCESS SUMMARY:	IAA	ECA	ECA	**IAA**	**ECA**	**ECA**	**EFA**						
Data Stores:													
AMT-OUTSTANDING		***											= ***
CUSTOMERS-ACCOUNT		EM	EM		**EM**	**EM**	**EM**						= **EM**
NEW-CUSTOMERS	ICM		ICM										= ICM
TRANS-LOG				**IAM**									= **IAM**

Figure 6.9

of activities listed at the beginning of this section and always classify the process according to the highest-priority task performed. In the case of process 4.0* processing a TRANSACTION, it is an IAA. The highest-priority task in each column is carried down to the PROCESS SUMMARY line. The PROCESS SUMMARY classification is the official classification for the functional primitive.

Now that we have finished all of the input data flows, we should look at the chart for each input data flow (row) and verify that one of the two following classifications exist: there is at least one essential fundamental activity (EFA) process ultimately responsible for producing an output data flow, or the data is used to update an essential memory (EM) by an essential custodial activity (ECA) process. If not, we have to go back and repeat the analysis for that input data flow because every input/output data flow must be processed somewhere along the line by an essential activity, either custodial or fundamental. Likewise, after we complete the analysis for all of the output data flows in the output section, we will also verify that every output data flow is the result of some EFA somewhere in the system.

When one classifies a process as an ECA, there must be at least one essential memory listed in the process column in the data store section of the input/output processing chart. (There is one exception to this; see Question 11 at the end of the chapter.)

After examining all of the input data flows, we continue our analysis by considering each output data flow. We trace each output data flow back through the system to analyze the activities and memories that contribute in providing the output data flow.

Output Section

BILL. Backtracking BILL through the system, we see that it is a result of processes 9.0*, 11.0*, and 12.0*. Analyzing the mini-specs (not included here) for these processes, we can conclude that process 9.0* uses information from the AMT-OUTSTANDING, CUSTOMERS-ACCOUNT, and PRICES data stores to compute the amounts owed by the customer. Because process 9.0* determines the amount of the customer's bill, it is the EFA responsible for producing the output data flow BILL. The data stores AMT-OUTSTANDING and PRICES serve the EFA (process 9.0*) and are essential memories (EM). A copy of the amount of the bill is stored in the data store AMT-NEW-SALES. Assume the mini-spec for process 9.0* does not provide enough information for us to determine what classification is appropriate for the data store AMT-NEW-SALES, so we will just indicate that it is used by process 9.0* by using "***". Information is also used from the essential memory (EM) CUSTOMER-ACCOUNT to produce the data flow BILL-INFO.

So we have:

Process 9.0* (EFA), using CUSTOMERS-ACCOUNT (EM)
AMT-OUTSTANDING (EM)
PRICES (EM)
AMT-NEW-SALES (***)

Process 11.0* takes the BILL-INFO, reformats the data into a human-readable statement (STMT), confirms the totals, and passes the data flow STMT to process 12.0*. Thus, process 11.0* is an IAA that serves the system by preventing erroneous data from being propagated. Process 12.0* merges the statement with an envelope and produces the BILL that is mailed to the customer. The task performed by process 12.0* simply combines data and passes it on, which makes the process an ICA.

And finally we have:

Process 11.0* (IAA)
Process 12.0* (ICA)

Figure 6.10 shows the above analysis added to the input/output processing chart.

INPUT/OUTPUT PROCESSING CHART

	←—— ALL FUNCTIONAL PRIMITIVE PROCESSES ——→												
	1.0	2.0	3.0	4.0	5.0	6.0	7.0	8.0	9.0	10.0	11.0	12.0	
Inputs:													
INFO	IAA		ECA										
PAYMENT		ECA											
TRANSACTION				IAA	ECA	ECA	EFA						
Outputs:													
BILL									**EFA**		**IAA**	**ICA**	
PROCESS SUMMARY:	IAA	ECA	ECA	IAA	ECA	ECA	EFA		**EFA**		**IAA**	**ICA**	
Data Stores:													
AMT-NEW-SALES									*******				= *******
AMT-OUTSTANDING		**EM**							**EM**				= **EM**
CUSTOMERS-ACCOUNT		EM	EM		EM	EM	EM		**EM**				= EM
NEW-CUSTOMERS	ICM		ICM										= ICM
PRICES									**EM**				= **EM**
TRANS-LOG				IAM									= IAM

Figure 6.10

BUSINESS-STATUS. Management receives a report from this system indicating the level of sales, etc. Process 10.0* provides the BUSINESS-STATUS report and is therefore an EFA. Assume that analysis of the mini-spec for process 10.0* tells us that the data store AMT-NEW-SALES is only a temporary repository of data. It is used to synchronize process 9.0*, which produces the raw data, with process 10.0*, which provides the BUSINESS-STATUS report. Thus, the data store AMT-NEW-SALES is an ICM that passes data between processes 9.0* and 10.0*.

So we have:

Process 10.0* (EFA), using AMT-NEW-SALES (ICM)

Figure 6.11 shows the above analysis added to the input/output processing chart. With the above information on process 10.0* using AMT-NEW-SALES as an internal communications memory, we can now update process 9.0* use of the AMT-NEW-SALES data store to an ICM.

INPUT/OUTPUT PROCESSING CHART

	←—— ALL FUNCTIONAL PRIMITIVE PROCESSES ——→												
	1.0	2.0	3.0	4.0	5.0	6.0	7.0	8.0	9.0	10.0	11.0	12.0	
Inputs:													
INFO	IAA		ECA										
PAYMENT		ECA											
TRANSACTION				IAA	ECA	ECA	EFA						
Outputs:													
BILL									EFA		IAA	ICA	
BUSINESS-STATUS										**EFA**			
PROCESS SUMMARY:	IAA	ECA	ECA	IAA	ECA	ECA	EFA		EFA	**EFA**	IAA	ICA	
Data Stores:													
AMT-NEW-SALES									**ICM**	**ICM**			**= ICM**
AMT-OUTSTANDING		EM							EM				= EM
CUSTOMERS-ACCOUNT		EM	EM		EM	EM	EM		EM				= EM
NEW-CUSTOMERS	ICM		ICM										= ICM
PRICES									EM				= EM
TRANS-LOG				IAM									= IAM

Figure 6.11

CUSTOMER-STATUS. CUSTOMER-STATUS is an output of process 7.0* when process 7.0* receives the input data flow ACCT-ID. Process 4.0* passes ACCT-ID to process 7.0* when a TRANSACTION (3), display customer status, is received. Because process 4.0* only passes ACCT-ID onto process 7.0*, it is an ICA. Process 7.0* provides the output data flow CUSTOMER-STATUS by retrieving the requested data from the essential memory (EM) CUSTOMERS-ACCOUNT. Thus, process 7.0* performs an EFA by providing one of the system's outputs.

So we have:

Process 4.0* (ICA)
Process 7.0* (EFA), using CUSTOMER-ACCOUNT (EM)

INPUT/OUTPUT PROCESSING CHART

	←—— ALL FUNCTIONAL PRIMITIVE PROCESSES ——→												
	1.0	2.0	3.0	4.0	5.0	6.0	7.0	8.0	9.0	10.0	11.0	12.0	
Inputs:													
INFO	IAA		ECA										
PAYMENT		ECA											
TRANSACTION				IAA	ECA	ECA	EFA						
Outputs:													
BILL									EFA		IAA	ICA	
BUSINESS-STATUS										EFA			
CUSTOMER-STATUS				**ICA**			**EFA**						
PROCESS SUMMARY:	IAA	ECA	ECA	IAA	ECA	ECA	EFA		EFA	EFA	IAA	ICA	
Data Stores:													
AMT-NEW-SALES									ICM	ICM			= ICM
AMT-OUTSTANDING		EM							EM				= EM
CUSTOMERS-ACCOUNT		EM	EM		EM	EM	**EM**		EM				= EM
NEW-CUSTOMERS	ICM		ICM										= ICM
PRICES									EM				= EM
TRANS-LOG				IAM									= IAM

Figure 6.12

Figure 6.12 shows the above analysis added to the input/output processing chart. In the figure notice that data stores classified as ICMs must have at least two entries in the data store row: one for the process that writes the data and the second for the process that reads the data. Only a read-only EM and a read-only or write-only IAM can have a single entry. Read-only EMs can provide essential data to the system. Read-only IAMs provide blank forms, identification or sequence numbers, and other data to control the flow of data in the system. Write-only IAMs hold data for historical purposes or as a precaution against system failure. An example of a read-only EM is the price list (PRICES) used to compute the amount the customer owes. An example of a write-only IAM is a log or archive file (such as TRANS-LOG) that keeps a historical log of all transactions processed by the system.

The approach of using inputs and outputs as a means of going through the data flow diagrams and classifying every process and data store does not necessarily cover all processes and data stores. Looking back at the chart in Fig. 6.12, we see that the process Daily Backup 8.0* and BACKUP-ACCTS were not covered. In analyzing the mini-spec for process 8.0*, we can conclude that process 8.0* is simply an IAA that is necessary because of the technology used to save customer's account information in the essential memory (EM) CUSTOMERS-ACCOUNT. BACKUP-ACCTS is the write-only data store used as an IAM for backup/recovery purposes. No other process in the system uses the data from this data store. It serves only as a backup file.

So, we have:

Process 8.0* (IAA), using CUSTOMERS-ACCOUNT (EM) and BACKUP-ACCTS (IAM)

Our final input/output processing chart is shown in Fig. 6.13.

INPUT/OUTPUT PROCESSING CHART

	←---- ALL FUNCTIONAL PRIMITIVE PROCESSES ----→												
	1.0	2.0	3.0	4.0	5.0	6.0	7.0	8.0	9.0	10.0	11.0	12.0	
Inputs:													
INFO	IAA		ECA										
PAYMENT		ECA											
TRANSACTION				IAA	ECA	ECA	EFA						
Outputs:													
BILL									EFA		IAA	ICA	
BUSINESS-STATUS										EFA			
CUSTOMER-STATUS				ICA			EFA						
PROCESS SUMMARY:	IAA	ECA	ECA	IAA	ECA	ECA	EFA	**IAA**	EFA	EFA	IAA	ICA	
Data Stores:													
AMT-NEW-SALES									ICM	ICM			= ICM
AMT-OUTSTANDING		EM							EM				= EM
BACKUP-ACCTS								**IAM**					**= IAM**
CUSTOMERS-ACCOUNT		EM	EM		EM	EM	EM	**EM**	EM				= EM
NEW-CUSTOMERS	ICM		ICM										= ICM
PRICES									EM				= EM
TRANS-LOG				IAM									= IAM

Figure 6.13

Our purpose for classifying all of the functional primitives in the system was to identify clearly the essential components. Figure 6.14 applies the information from the input/output processing chart to the data flow diagram of Fig. 6.5. The light type

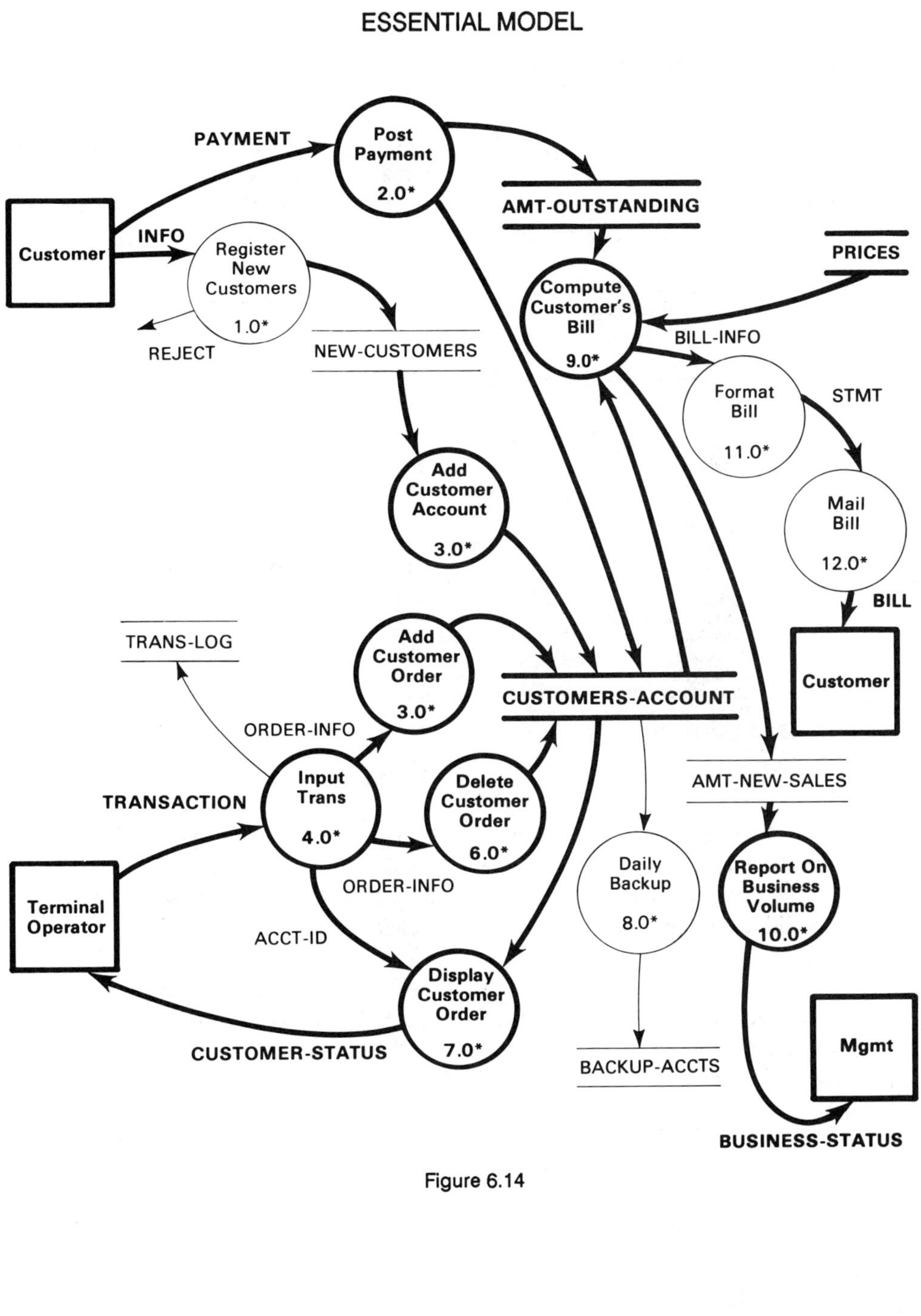

Figure 6.14

areas illustrate the infrastructure of the system with the essential components shown in bold type. After establishing the essential model, we can proceed to do object analysis (Chapter 7) and event analysis (Chapter 8). The result is a complete, logical, structured specification of the system.

VERIFYING INPUT/OUTPUT PROCESSING CHARTS
. . . aid in recognizing errors in analysis

From an input/output processing chart, we can identify several suspicious characteristics about the system. The characteristics can either be problems in the system itself, errors in the logical analysis of the processes and/or data stores, or errors in constructing the chart.

INCORRECT INPUT/OUTPUT PROCESSING CHART

	←------ ALL FUNCTIONAL PRIMITIVE PROCESSES ------→												
	1.0	2.0	3.0	4.0	5.0	6.0	7.0	8.0	9.0	10.0	11.0	12.0	
Inputs:													
INFO	IAA		ECA										
NEW-PRICE								ICA					
PAYMENT		ECA											
REQUEST				IAA	ICA	ECA	EFA						
Outputs:													
BILL									EFA		IAA	ICA	
BUSINESS-STATEMENT										EFA			
CUSTOMER-LETTER							EFA						
PROCESS SUMMARY:	IAA	ECA	ECA	IAA	ECA	ECA	EFA	ICA	EFA	EFA	IAA	ICA	
Data Stores:													
ACCOUNTS-PAYABLE		IAM					EM						= IAM
ACTIVE-SALES									ICM				= ICM
DAILY-LOG*				IAM									= IAM
MEMBER-FILE			EM		ICM	EM	EM		EM				= EM
NEW-CUSTOMERS	ICM		ICM										= ICM
PRICES								EM	EM				= EM
SALES-HISTORY								IAM					= IAM

Figure 6.15

Examine the incorrect input/output processing chart of Fig. 6.15 and see if we can find any problems or inconsistencies in the analysis. The following lists the erroneous items from the chart, and, in each case, a convenient rule is stated that we should keep in mind when constructing charts.

1. Process 2.0* receives PAYMENT and is designated as an ECA. However, upon examining column 2.0* we find that the process does not interface with any EM data store. If this process is truly saving data to be used by another process (as in the case of process 7.0*), ACCOUNTS-PAYABLE is erroneously denoted as an IAM.
Rule 1:

*** ALL ECAs MUST BE ASSOCIATED TO AT LEAST ONE EM.***

(There is one exception to this rule, see Rule 4.)

2. The data store ACTIVE-SALES is shown as an ICM for process 9.0*. However, there is no other process that uses ACTIVE-SALES and thus the data store cannot be an ICM.

Rule 2:

*** ALL ICM DATA STORES MUST SERVE AT LEAST TWO PROCESSES.***

3. Process 5.0* receives REQUEST as an input and performs the internal communications activity of moving the data into a designated ICM data store MEMBER-FILE. However, MEMBER-FILE is designated as an essential memory of the system and therefore cannot be classified as an internal communications memory for any process.

Rule 3:

*** IF A DATA STORE IS AN ESSENTIAL MEMORY FOR ONE PROCESS, IT IS AN ESSENTIAL MEMORY FOR ALL PROCESSES IN THE SYSTEM.***

4. Process 8.0* receives the input data flow NEW-PRICE and moves the data to the data store PRICES. However, because PRICES is an essential memory, processes that update essential memories are essential custodial activities and not internal communications activities.

Rule 4:

*** ONLY ESSENTIAL ACTIVITY (EFA or ECA) PROCESSES UPDATE OR WRITE TO AN EM.***

(See the exception that follows)

Caution: In this example, process 8.0 is the only one that handles the arrival of input *and* the updating of the essential memory; consequently, it is an ECA. However, there are cases where an accountant or bookkeeper updates a journal ledger (or some data from an essential memory) and gives the ledger to a clerk process. The clerk simply "transports" the ledger to a filing cabinet. The accountant or bookkeeper process is the ECA, whereas the clerk process is simply an ICA that places the updated ledger into the filing cabinet.

5. Verify that there is at least one process designated as an EFA in the system. If the system's sole purpose is to update data stores, the designated essential custodial activities would be considered EFAs because updating data stores is the system's primary purpose.

Rule 5:

*** THERE MUST BE AT LEAST ONE EFA IN THE SYSTEM.***

6. Examine every row in the input section. An EFA or an ECA must appear under some functional primitive process for every input data flow in the chart.

Rule 6:

*** EVERY INPUT MUST BE PROCESSED BY AN EFA OR AN ECA.***

7. Examine every row in the output section. Every output data flow must be processed by at least one EFA. The functional primitive classified as an EFA is the one that is responsible for producing that output data flow.

Rule 7:

*** EVERY OUTPUT MUST BE PROCESSED BY AT LEAST ONE EFA.***

8. *It is possible* to have the following characteristics in an input/output processing chart:

a. A data store classified as an IAM can be associated to only one process.
b. IAA and ICA processes do not necessarily have to be associated to any data store.
c. A process can be associated to only one data store and receives no other inputs or produces no other outputs other than the interactions with that one data store.
d. A process does not have to have any association with a data store. It can only receive input data and produce output data, or simply just produce output data.

9. *It is not possible* to have the following characteristics in an input/output processing chart:

a. It is an error if a process exists and does not receive any inputs or produce any outputs, associates to any data stores, or interfaces to any other process.
b. It is an error if a process only receives inputs and does not produce any output data or interface to any other process or data store.

In reviewing structured specifications, or input/output processing charts, remember two things:

1. All input data flows received from a source outside the system must be ultimately processed by either an essential custodial activity (ECA) or an essential fundamental activity (EFA).

2. All output data flows that are sent to a sink outside the system must be ultimately produced by some essential fundamental activity (EFA) within the system. (Essential custodial activities (ECA) can send outputs as a dialogue data flow to obtain information to update an essential memory (EM). The output dialogue data flow does not in itself warrant the process being classified as an essential fundamental activity (EFA).)

Consider a system that received an input data flow and an internal administrative activity (IAA) that simply placed the data into an internal administrative memory (IAM) for historical purposes without any other processing. Has the analysis correctly classified the processes as an administrative infrastructure? No, all input data flows must be processed by an essential activity (EFA or ECA). It may be the case that the system's primary purpose is to update a historical data store. Thus, the activity that updates performs an essential fundamental activity (EFA). The data store itself would then be considered an essential memory (EM) because the primary purpose of the system is to save data in the historical data store. Only when a write-only file is used by a process for purposes of maintaining an historical archive, or for the system's restart/recovery capability, is the process and the data store considered as internal administrative components.

SUMMARY
. . . now we can focus on the essential components of the system

Another benefit from classifying a system's activities in the manner described in this chapter is to be able to understand what the essential processes are that transform the data into information. The human mind best deals with complexity by partitioning things into subsets and continues to partition the subsets until each component is easily understood. Then, the subsets are organized into groups, where each group pertains to a particular service or objective of the system. When dealing with a system whose function and transformation of data is complex, it is helpful to first partition the system's activities into those responsible for communication, administration, and maintenance of the system's memory. By identifying and separating the system's communications and administrative activities, the processes that perform the essential data transformations are clearly exposed for further investigation.

Once the system's processes and data are classified in the manner described in this chapter, we can begin to organize the essential components into sets of processes and data that produce a desired result or perform a particular service. One method used to reorganize activities and data stores is to group them by the events to which they respond. I will discuss organizing the system's processes and data around events in Chapter 8. First, we must discuss data analysis (the study of data objects and their relationships, Chapter 7) before discussing events.

QUESTIONS

1. What is the purpose of a system analysis?
2. What types of activities or processing takes place in a system?
3. What is a communications data store?

4. Define an internal communications activity and explain its purpose?
5. Define an internal communications memory and explain its purpose?
6. Define an internal administrative activity and explain its purpose?
7. Define an internal administrative memory and explain its purpose?
8. What are the essential components of a system and what are their purposes?
9. a. What are the two types of essential activities?

   b. What is an essential memory and what is its purpose?
10. Identify the classification of each of the following types of activities:

a.	Routing data to selected destinations	f.	Merging data flows
b.	Verifying that data flows meet a criteria	g.	Sorting data entries
c.	Assigning ID numbers to data flows	h.	Collating data flows
d.	Compressing data to occupy less storage	i.	Editing data elements
e.	Converting data flow into ASCII protocol	j.	Maintaining a log file

11. Under what conditions can we classify a functional primitive as an essential custodial activity and not have it directly interface to an essential memory?
12. Give two examples when tracing back an output data flow where we cannot find an essential fundamental activity responsible for producing the output data flow.

PROBLEMS

1. After someone analyzed a system the following conclusions were made:

   Process 2.1 accepts input data A and passes it onto process 3.1. Process 1.1 accepts input data B, verifies it for completeness, and gives it to process 1.2 that updates the MASTER file and places the data into an IN-BASKET which will be processed by process 1.3. some time later.

   Process 1.3 periodically takes the data form the IN-BASKET, inquires via an on-line terminal from the MASTER file to verify the update was made correctly, initials the data sheet and places it into the PROCESS-LOG file in case someone asks about it later.

   Near the end of the day, process 2.2 selects all of the data records from the MASTER file that must be sent to clients, places them in alphabetical order, and passes them to process 3.2. Process 3.2 takes each data record and generates a data sheet C. Process 3.2 updates the total amount for each data record it processes in the MASTER file and passes the data sheet C to process 3.3. Process 3.3 searches the CLIENT-FILE to locate the client's address and mails the data sheet C to the client. The process also makes a note in the MAIL-OUT notebook that the data sheet C was mailed out on the given date.

From the preceeding statements construct an input/output processing chart in proper format illustrating all inputs, outputs, and data stores with the appropriate classifications (ICA, IAA, ECA, EFA, ICM, IAM and EM).

2. Given the following input/output processing chart identify any discrepancies, inconsistencies or suspicious classifications of processes or data stores:

PROPOSED INPUT/OUTPUT PROCESSING CHART

	← ALL FUNCTIONAL PRIMITIVE PROCESSES →												
	1.0	2.0	3.0	4.0	5.0	6.0	7.0	8.0	9.0	10.0	11.0	12.0	
Inputs:													
IOU-S	IAA		ECA				ICA						
PRICE-LIST								ICA					
PAYMENT		ECA											
REQUEST				IAA	ICA	ECA	ECA						
Outputs:													
BILL									EFA		IAA	ICA	
BUSINESS-STATEMENT										EFA			
CUSTOMER-LETTER							EFA						
COMPLAINT				IAA	ICA		ECA	ICA					
PROCESS SUMMARY:	IAA	ECA	ECA	IAA	ECA	ECA	ECA	ICA	EFA	EFA	IAA	ICA	
Data Stores:													
OUTSTANDING-LOADS		IAM					EM						= IAM
ACTIVE-BORROWS		ICM											= ICM
TO-DO-LIST				IAM					IAM				= IAM
MEMBERS		EM		ICM	EM	EM		EM					= EM
NEW-MEMBERS	ICM		ICM										= ICM
COMPLAINT-LIST							IAM						= IAM

Figure 6.16

3. Perform a logical analysis and construct an input/output processing chart for problem #2 discribed at the end of Chapter 2 (Purchasing System).

7

OBJECT ANALYSIS

In analyzing the stored data in a system, our strategy will be to concentrate on the essential memories and factor the data into objects and attributes. To begin, let's first establish some definitions. There have been many definitions offered for the concept of an *object*.

> "An ENTITY (OBJECT) is a 'thing' which can be distinctly identified."[C-2]
>
> - Peter P. Chen

> "An OBJECT is a class of individuals that is characterized by a distinct functional identity, which is preserved as a whole over time in the system being modeled."[F-1]
>
> - Matt Flavin

For our purposes, ENTITY and OBJECT are synonymous. In this text, we will use the following definitions:

An OBJECT is some distinct identifiable thing for which the system must store data in order to perform the fundamental activities in the system. The concept of an object is easier to use than it is to describe.

An ATTRIBUTE is an elementary data element that describes an object or is a fact about an object.

A DATA VALUE is a particular value of an attribute.

An example of an object is a PERSON. A system may need to store descriptive data about a person such as the attributes NAME, Address (ADDR), Social Security Number (SS-#), OCCUPATION, BIRTH-DATE, Height (HT), Weight (WT), etc. An example of a data value for the attribute NAME is "J.C. JONES."

Another example is the object AUTO. A system may need to store descriptive data about an auto such as the attributes MAKE, YEAR, MODEL, COLOR, SERIAL-#, etc. The data values for the attributes of a particular occurrence of the object AUTO are "OLDS," "1985," "98-SEDAN," "GREY," etc.

Now consider forming two data stores: one named PERSON which contains all of the attributes that describe a person, and a second named AUTO, which contains all of the attributes that describe an auto.

PERSON = {NAME + ADDR + SS-# + OCCUPATION + BIRTH-DATE + HT + WT
+ SEX + EYE-COLOR}

AUTO = {MAKE + YEAR + MODEL + EXT-COLOR + INT-COLOR + INT-FABRIC
+ AUTO-SERIAL-# + PRICE}

Each attribute in the above data stores is a single, elementary data element. The data elements are descriptions or facts about their respective objects. Within each of the data stores, we need a means of identifying a particular entry. **An attribute (or set of attributes) in a data store that uniquely identifies a single entry from all others in the data store is known as a CANDIDATE KEY.** There can be several candidate keys for a particular data store. The only requirement is that it uniquely identifies the entries in the data store. The candidate key that we select as "*the*" identifier for entries in a particular data store is called the *primary key*. **The PRIMARY KEY is the particular candidate key chosen to identify a single entry from all others in the data store.** The candidate keys that are not selected are the *alternate keys*. **An ALTERNATE KEY is a candidate key that is not chosen to be the primary key.** So, the set of all candidate keys consists of the primary key and all alternate keys.

Data Store Integrity Rule #1 (Object Integrity)

The primary key is used to locate and retrieve the attributes associated with a particular entry in a data store. To do this effectively, no part of the primary key must ever take on a *null value*. In other words, a *null value* is not permitted for any attribute that comprises the primary key. **A NULL VALUE means that no data value is assigned to the attribute.** A null value does *not* mean that the data value is a "*space*," a "*blank*," or "*zero*." It means that **no value** has been assigned to the attribute.

If no part of the primary key can be null, it cannot have a data element that is optional. The following example is a data store with an **invalid primary key**:

STOCK = {COLOR + SIZE + SLEEVE-LENGTH + (COLLAR-TYPE) + PRICE
+ FABRIC + ["Dry clean only" | "Washable"]

For the entries where COLLAR-TYPE is not provided, the value is NULL and the data store violates the Object Integrity Rule (Integrity Rule #1).

The Data Store Integrity Rule #1 permits every **data store entry** in the system to have a unique address. The unique address is

< Data-Store-Name, Primary-Key value >

Furthermore, we can address every **data value** stored in the system by specifying

< Data-Store-Name, Primary-Key value, Attribute-Name >

Consider, for example, the following data store:

AUTO = {SERIAL-# + MFG + MAKE + MODEL + YEAR + COLOR + PRICE + . . . }

If you think of a data store as a table of entries, with each entry a fact or description (*attribute*) of the item (*object*) it describes, the table looks like that of Fig. 7.1.

TABLE OF ENTRIES

	←——— *Primary Key*	———→	←———	———	*Attributes*	———	———→	
Data Store: **AUTO**	SERIAL-#	MFG	MAKE	MODEL	YEAR	COLOR	PRICE	. . .
Entries:	ZK123-4567	GM	OLDS	CUTLASS	1985	GRAY	$14,864	. . .
	⋮	⋮	⋮	⋮	⋮	⋮	⋮	
	XY789-5432	FORD	FORD	LTD	1980	WHITE	**$12,756**	. . .
	⋮	⋮	⋮	⋮	⋮	⋮	⋮	

Figure 7.1

The address of the entry is denoted by

< AUTO, 'XY789-5432, FORD' >

Likewise, there is a unique address for every *data value* stored in the data store. The address of the highlighted data value '$12,756' is

< AUTO, 'XY789-5432, FORD' , PRICE >

where AUTO is the name of the data store, 'XY789-5432, FORD' is the *value* of the primary key used to locate the specific row in the table, and PRICE is the attribute name (column) of the data value that we wish to locate.

The object of the data store AUTO is **auto**[1]. All of the attributes, including the primary key attribute(s), describe or are facts about an auto. The primary key

1 For simplicity we have named the object data store the same as the object itself, although these names do not have to be the same.

contains the data value(s) that identify exactly one auto that has a collection of descriptive attributes saved in the data store.

When preparing specifications that serve as the requirements for system development, it is important that analysts have an addressing capability for stored data that is independent of the technology used to implement the data structure in the system. The analyst's focus is on problem solving (i.e., providing solutions), rather than in the technical details of providing a plan for the necessary linkage to the data. Therefore, there is a need to be able to simply and directly refer to any value of a data element in a system while being independent of the implementation technology.

NORMALIZATION
. . . associate attributes to the objects they describe

When we prepared the physical model of the system in the previous chapters, we modeled the data stores exactly as we found them. Now, we want to take a closer look at the data stores. We want to associate all of the attributes to the object that they describe in the system. To identify the objects in a system, we can apply the process of *normalization*. **NORMALIZATION is the decomposition of complex data structures into the normal forms of data, based on the rules of data dependency.** By using normalization, we transform the physical essential memories into *third normal form* and ultimately *Boyce/Codd normal form* (BCNF). Normalization is a data-base design tool. However, we will find that once the data stores are in BCNF, each resulting data store will automatically associate with a unique *object* of the system.

In a causal sense, the set of objects for a system can be determined by the sets of descriptive attributes. From a practical viewpoint, there is little value in defining an object in a system if there is no information to store about the object. We would not store AUTO-LICENSE-NUMBER in a system if it did not relate to a vehicle, person, lease contract, or some other object in the system. Therefore, every attribute in the system pertains to some particular object. This is to say that we can have a data store in third normal form with only one data element, the primary key. The following data store contains only one data element, the primary key, and it defines an object of the system:

LOST-STOLEN-CREDIT-CARDS = {CARD-NUMBER}

When we use a credit card to pay for a purchase, the merchant looks in a small book to see if the card number is listed among all of the cards that have been lost or stolen. In this case, the object is CREDIT-CARD, and there is an implied attribute associated with the card if it is listed in the data store of lost and stolen credit cards.

We want to reduce each of the essential memories into data stores such that each data store contains one or more attributes that describe a single object in the system. To find the objects of the system, we will use the data-base design tool of normalization to place the essential memories into Boyce/Codd normal form

(BCNF). Then each BCNF data store is associated to one and only one object. Let's begin.

The following example presents the normal forms of data by considering the data store CLIENT. Let's assume NAME is not unique. "NAME + ADDR," where ADDR is the client's home address, is unique. Likewise CHILD-NAME is not unique, however, two children with the same name are *never* assigned to the same ROOM# in a school. Furthermore, assume *a TEACHER can teach in more than one school* in the district. An example would be if Mr. Gray teaches sixth grade in one school during the morning shift (7:00 A.M. to 11:00 A.M.), and teaches junior high science in another school in the district during the afternoon shift (noon to 4:00 P.M.).

UNNORMALIZED DATA STORE

```
CLIENT   = {NAME + ADDR + OCCUPATION + BIRTH-DATE + OFFICE-ADDR
           + OFFICE-PHONE + OFFICE-SIZE + {CHILD-NAME
           + CHILD-BIRTH-DATE + SCHOOL-NAME + ROOM# + TEACHER}
           + ["HOUSE"|"APT"|"CONDO"] + DATE-CONSTRUCTED
           + HOUSE-STYLE}
```

Figure 7.2 graphically illustrates the embedded repeating group of data elements that are associated to the child of a client.

UNNORMALIZED FORM

UNNORMALIZED FORM

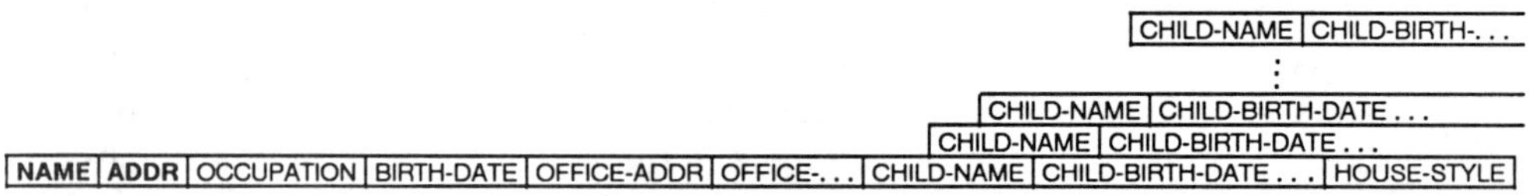

Figure 7.2

First Normal Form

For the **FIRST NORMAL FORM (1NF), entries of the data store contain only elementary data elements and each element has precisely one value.** To convert a data store in the first normal form, we do three things: first, replace all group data elements with the equivalent elementary data elements; second, remove all repeating groups; third, expand the primary key to include a unique identifier from each of the repeating group that was removed. The compound primary key then continues to uniquely identify entries in the 1NF data store. In the data store CLIENT, each person can have a different number of children. To place this data store in first normal form, concatenate all of the data attributes about the parent and the house with the data attributes about each child. To be able to identify a single entry in the

1NF data store, we need to expand the primary key by including a unique identifier of the repeating groups that has been removed from the data definition. The resulting data store is

FIRST NORMAL FORM

CLIENTS = {NAME + ADDR + OCCUPATION + BIRTH-DATE + OFFICE-ADDR
+ OFFICE-PHONE + OFFICE-SIZE + CHILD-NAME + CHILD-BIRTH-DATE
+ SCHOOL-NAME + ROOM# + TEACHER + ["HOUSE" | "APT"
| "CONDO"] + DATE-CONSTRUCTED + HOUSE-STYLE}

Figure 7.3 graphically illustrates the data structure before and after the embedded repeating group of data elements is removed.

FIRST NORMAL FORM

UNNORMALIZED FORM

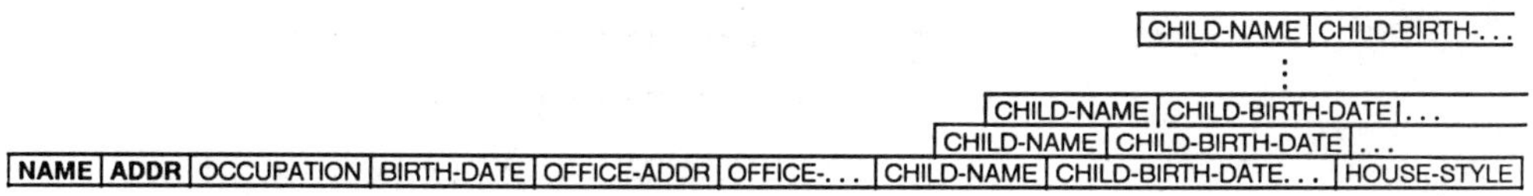

FIRST NORMAL FORM

NAME	**ADDR**	OCCUPATION	BIRTH-DATE	OFFICE-ADDR	OFFICE-. . .	CHILD-NAME	CHILD-BIRTH-DATE. . .	HOUSE-STYLE

Example of Data in First Normal Form

<-PRIMARY-| |KEY------>

ABBEY JIM	123 MAIN ST.	ENGR	6-04-47	543 MILL ST.	. . .	JOHNNY	7-18-70	. . .	2 STORY
ABBEY JIM	123 MAIN ST.	ENGR	6-04-47	543 MILL ST.	. . .	SUSAN	3-12-78	. . .	2 STORY
BASON AL	4 WATER ST.	MGR	5-09-45	17 EAST AVE.	. . .	MATTHEW	8-15-71	. . .	RANCH
CHILDS JOE	45 HIGHLAND	CPA	7-19-59	25 NORTH ST.	. . .	BRIAN	8-15-71	. . .	RANCH
MALEC W.	954 OAK ST.	MGR	5-09-45	80 EAST AVE.	. . .	MANDY	8-15-71	. . .	RANCH
SNOW BOB	4 WATER ST.	ENGR	2-29-49	75 FRANKKLIN	. . .	JENNIFER	1-17-71	. . .	RANCH
⋮	⋮	⋮	⋮	⋮		⋮	⋮		⋮

Figure 7.3

After removing the repeating group from an unnormalized data store, the original primary key is insufficient to uniquely identify the new data entries. For the primary key to identify each entry in the CLIENT's data store in first normal form, we must include an additional data element (or elements) from the set of attributes in the repeating group. Because parents give unique names to their children, adding CHILD-NAME to NAME + ADDR forms a primary key that will do the job in this example.

Second Normal Form

For the **SECOND NORMAL FORM (2NF), all nonkey* attributes must provide a fact about the whole key and not just part of the key.** To place a data store into second normal form, we focus our attention on all of the attributes that make up the primary key. First of all, if the primary key is not a composite key, that is, a key consisting of more than one elementary data element, the data store is automatically in second normal form even if it is in first normal form. If the primary key is a composite key, then we need to examine the data store to verify that every nonkey attribute describes, or is a fact about, the whole key and not just part of the key.

In our example, there are attributes associated to the **client** (NAME), and there are other attributes that describe just the **residence** (ADDR). There are still other attributes that pertain to just a **child** (CHILD-NAME) of the client. The **client**, **residence**, and the **child** form three distinct sets of attributes. As a result, the original data store has to be factored into the following three data stores:

SECOND NORMAL FORM

```
CLIENT      = {NAME + ADDR + OCCUPATION + BIRTH-DATE + OFFICE-ADDR
               + OFFICE-PHONE + OFFICE-SIZE + CHILD-NAME}

RESIDENCE   = {ADDR + ["HOUSE"|"APT"|"CONDO"] + DATE-CONSTRUCTED
               + HOUSE-STYLE}

CHILD       = {CHILD-NAME + ADDR + CHILD-BIRTH-DATE + SCHOOL-NAME
               + ROOM# + TEACHER}
```

Because NAME is assumed not to be unique, the attribute ADDR is needed in the primary key even though ADDR is the primary key of the data store

SECOND NORMAL FORM

FIRST NORMAL FORM

NAME	**ADDR**	OCCUPATION	BIRTH-DATE	OFFICE-ADDR	OFFICE-. . .	CHILD-NAME	CHILD-BIRTH-DATE . . .	HOUSE-STYLE

SECOND NORMAL FORM

NAME	**ADDR**	OCCUPATION	BIRTH-DATE	OFFICE-ADDR	OFFICE-PHONE	OFFICE-SIZE	CHILD-NAME

ADDR	["HOUSE"|"APT"|"CONDO"]	DATE-CONSTRUCTED	HOUSE-STYLE

CHILD-NAME	**ADDR**	CHILD-BIRTH-DATE	SCHOOL-NAME	ROOM#	TEACHER

Figure 7.4

* A nonkey data element is that which is not the *primary* key.

RESIDENCE. If NAME is not unique, it also follows that CHILD-NAME is not unique. Likewise, ADDR has to be part of the primary key for the data store CHILD. In this example, it is assumed that ADDR is unique. CHILD-NAME is kept as a descriptive nonkey attribute of CLIENT. Figure 7.4 is an illustration of the data structure as it appears in first normal form and second normal form.

Third Normal Form

The third normal form is violated when a nonkey data element describes another nonkey data element and not the primary key. If two nonkey attributes are dependent on one another, we call the relationship a transient dependency. To be in third normal form all such transient dependencies much be removed. More formally, for the **THIRD NORMAL FORM (3NF), all nonkey attributes are functionally dependent on the primary key and are independent of each other.**

To ensure that each of the data stores are in third normal form, examine all the nonkey attributes. Each must describe the primary key, or be a fact about **the key, the whole key, and nothing but the key.** If the attribute describes another nonkey element, then we have another object embedded in the data store. Remove the embedded object and create a separate data store for the newly found object and all of the attributes that are associated with it.

In our example, we see that OFFICE-PHONE and OFFICE-SIZE are attributes of the nonkey attribute OFFICE-ADDR and not the primary key NAME + ADDR. Therefore, we need to create a fourth data store (OFFICE) that describes the object **office.**

The following data stores are in third normal form:

THIRD NORMAL FORM

CLIENT = {NAME + ADDR + OCCUPATION + BIRTH-DATE + CHILD-NAME + OFFICE-ADDR}

RESIDENCE = {ADDR + ["HOUSE"|"APT"|"CONDO"] + DATE-CONSTRUCTED + HOUSE-STYLE}

CHILD = {CHILD-NAME + ADDR + CHILD-BIRTH-DATE + SCHOOL-NAME + ROOM# + TEACHER}

OFFICE = {OFFICE-ADDR + OFFICE-PHONE + OFFICE-SIZE}

Figure 7.5 is a graphical illustration of the data structure in second and third normal forms. Notice that all of the data attributes that describe the client's office are put into a separate data structure to remove the transient dependency found in second normal form.

In a practical sense, once we have achieved third normal form, we have met our objective of organizing all of the data attributes according to the object that they describe. However, we will continue and discuss Boyce/Codd normal form.

THIRD NORMAL FORM

SECOND NORMAL FORM

NAME	**ADDR**	OCCUPATION	BIRTH-DATE	OFFICE-ADDR	OFFICE-PHONE	OFFICE-SIZE	CHILD-NAME

ADDR	["HOUSE" | "APT" | "CONDO"]	DATE-CONSTRUCTED	HOUSE-STYLE

CHILD-NAME	**ADDR**	CHILD-BIRTH-DATE	SCHOOL-NAME	ROOM#	TEACHER

THIRD NORMAL FORM

NAME	**ADDR**	OCCUPATION	BIRTH-DATE	CHILD-NAME	OFFICE-ADDR

ADDR	["HOUSE" | "APT" | "CONDO"]	DATE-CONSTRUCTED	HOUSE-STYLE

CHILD-NAME	**ADDR**	CHILD-BIRTH-DATE	SCHOOL-NAME	ROOM#	TEACHER

OFFICE-ADDR	OFFICE-PHONE	OFFICE-SIZE

Figure 7.5

Boyce/Codd Normal Form

The concept of normalization, that is, defining data in normal forms, was first introduced by E.J. (Ted) Codd[C-6] when he defined the first three normal forms of data. Normalization provides a means for the analyst to organize the data attributes stored in the system. This type of data organization meets the objective of having the data elements expressed in a way that is independent of any data-base technology. It is a view of the data requirements from a purely logical viewpoint.

However, the wording of Ted Codd's original definition of third normal form did not accomplish the intended objective in *all* cases. The inadequacy was pointed out by Raymond Boyce in 1974. The case not covered in Codd's original definition only occurred when the following three conditions were met:

1. The data store had multiple candidate keys.
2. The candidate keys contained more than a single attribute.
3. The candidate keys shared a least one common attribute.

Codd published a subsequent paper later in 1974[C-7] that stated a more restrictive definition for third normal form. It became known as the Boyce/Codd normal form to distinguish it from Codd's original definition of third normal form. To state the definition, it is convenient to define another term known as a *determinant*. **A DETERMINANT is any attribute on which some other attribute is (fully functionally) dependent.** Another way to define a determinant is to say there exists a one-to-one relationship between two attributes in both directions. This relationship exists if every value of one attribute is associated, or related, to exactly one value of the other attribute. Now the definition of the *Boyce/Codd Normal Form* can be stated: **A data store is in BOYCE/CODD NORMAL FORM (BCNF) if and only if every determinant is a candidate key.** Notice that the definition is stated in terms of candidate keys and not just the primary key.

Let's consider the following data store CHILD from our example that is in accordance to the original definition of third normal form:

THIRD NORMAL FORM

CHILD = {<u>CHILD-NAME + ADDR</u> + CHILD-BIRTH-DATE + SCHOOL-NAME + ROOM# + TEACHER}

Recall the assumption that a TEACHER could be assigned to more than one SCHOOL-NAME and ROOM#. Under these conditions, both SCHOOL-NAME + ROOM# and SCHOOL-NAME + TEACHER are determinants, because if you know the SCHOOL-NAME + TEACHER, you can *determine* the SCHOOL-NAME + ROOM#. Likewise, if you know SCHOOL-NAME + ROOM#, you can *determine* SCHOOL-NAME + TEACHER.

This example fulfills all three conditions mentioned earlier, namely:

1. **The data store has multiple candidate keys.**

   The candidate keys for the CHILDREN data store are CHILD-NAME + ADDR, CHILD-NAME + ROOM#, and CHILD-NAME + TEACHER.

2. **The candidate keys contain more than a single attribute.**

   All three candidate keys have more than one attribute.

3. **The candidate keys share a least one common attribute.**

   The candidate keys CHILD-NAME + ADDR, CHILD-NAME + ROOM#, and CHILD-NAME + TEACHER all share the attribute CHILD-NAME.

In the above example, TEACHER is *not* fully functionally dependent on ROOM#, nor is ROOM# fully functionally dependent on TEACHER. TEACHER is really fully functionally dependent on CHILD-NAME + ROOM#. (Or ROOM# is fully functionally dependent on CHILD-NAME + TEACHER.) Therefore, to be in BCNF, the CHILD data store has to be broken into two data stores: one that contains all of the attributes that describes the child, and a second that contains the attributes that designate which teachers are assigned to the particular homerooms in a school.

THIRD NORMAL FORM

CHILD = {CHILD-NAME + ADDR + CHILD-BIRTH-DATE + SCHOOL-NAME
+ ROOM# + TEACHER}

BOYCE/CODD NORMAL FORM

CHILD = {CHILD-NAME + ADDR + CHILD-BIRTH-DATE + SCHOOL-NAME
+ ROOM#}

ROOM# = {SCHOOL-NAME + ROOM# + TEACHER}

Once we have all of the essential memories defined in BCNF, we can name the object associated with each of the resulting object data stores.

CLIENT = {NAME + ADDR + OCCUPATION + BIRTH-DATE + CHILD-NAME
+ OFFICE-ADDR}

The object of the data store CLIENT is a **client.**

RESIDENCE = {ADDR + [HOUSE | APT | CONDO] + DATE-CONSTRUCTED
+ HOUSE-STYLE}

The object of the data store RESIDENCE is a client's **residence.**

CHILD = {CHILD-NAME + ADDR + CHILD-BIRTH-DATE + SCHOOL-NAME
+ ROOM#}

The object of the data store CHILD is a client's **child.**

HOME-ROOM = {SCHOOL-NAME + ROOM# + TEACHER}

The object of the data store HOME-ROOM is a child's **homeroom.**

OFFICE = {OFFICE-ADDR + OFFICE-PHONE + OFFICE-SIZE}

The object of the data store OFFICE is a client's **office.**

Data stores in Boyce/Codd normal form provide the systems analyst a means of associating all data elements of a system to the object that they describe. Using this information the analyst can define the object data stores, and all possible relationships that exist among the data elements can be easily and unambiguously represented. This information is important when stating the requirements for the stored data in a system. The problem that we are left with, if the object data stores are not in Boyce/Codd normal form, can be best illustrated with an example.

Let's consider the hypothetical case where the systems analyst defines the object data stores in third normal form. Imagine that a data-base designer built a data base of data elements exactly to the structure specified in the CHILD data store.

THIRD NORMAL FORM

CHILD = {CHILD-NAME + ADDR + CHILD-BIRTH-DATE + SCHOOL-NAME + ROOM# + TEACHER}

OBJECT DATA STORE - CHILD

CHILD-NAME	ADDR	CHILD-BIRTH-DATE	SCHOOL-NAME	ROOM#	TEACHER
ALLEN, JIM	**123 MAIN ST.**	**11-29-74**	**JEFFERSON**	**104**	**MR. GRAY**
BAKER, SUSAN	456 WATER ST.	12-21-74	JEFFERSON	213	MRS. SMITH
BAKER, SALLY	456 WATER ST.	12-21-74	JEFFERSON	213	MRS. SMITH
DANIELS, ED	**750 EAST AVE.**	**02-18-75**	**JEFFERSON**	**104**	**MR. GRAY**
DOLL, KIMBERLY	87 VALLEY RD.	11-10-70	WASHINGTON	204	MS. JONES
FOSTER, BOB	555 LILAC LANE	09-30-75	JEFFERSON	213	MRS. SMITH
GOLDBERG, JO	**567 MAIN ST.**	**07-18-78**	**LINCOLN**	**104**	**MR. GRAY**
HILL, DAVID	123 VALLEY RD.	04-06-74	JEFFERSON	213	MRS. SMITH
MICHAELS, CHAD	432 MARKET DR.	10-17-79	JEFFERSON	112	MS. RODGERS
RYAN, MARK	15 NORTH AVE.	09-30-73	WASHINGTON	107	MS. HILLS

Figure 7.6

Now visualize what would have to happen if there was a change in TEACHER for a particular SCHOOL-NAME + ROOM#, namely, Mr. Gray was changed from Room 104 to Room 110 at the Jefferson school. There are two possibilities:

1. An extensive search would have to be made of every entry in the data base in order to update the TEACHER (Mr. Gray) for every CHILD-NAME + ADDR entry in the file.
2. There would be an inconsistency in the data base if only one entry in the file was updated to reflect that the teacher (Mr. Gray) was assigned to a new SCHOOL-NAME + ROOM# (Jefferson, 110).

Consider the structured specification using the object data stores in the original third normal form that do not contain an explicit break-down of the SCHOOL-NAME + ROOM# having the attribute TEACHER. If after the system was implemented, it suffered from inconsistency or poor system performance when updating teachers assigned to homerooms, the deficiency would be traced to incomplete or poor system requirements.

Fourth and Fifth Normal Forms

For **FOURTH AND FIFTH NORMAL FORM**, see an article by William Kent.[K-1] To summarize normalization, we can view each form of data as follows:

Unnormalized data: One or more related data elements with repeating groups of data.

First normal form: One or more related data elements where each data element has a single value (no repeating groups of data).

Second normal form: One or more related data elements in first normal form where all nonkey data elements are fully functionally dependent on the whole primary key.

Third normal form: One or more related data elements in second normal form where all nonkey data elements are fully functionally dependent on the primary key *and* are independent of each other.

Boyce/Codd normal form: A data store where every determinant is a candidate key.

With the example data stores now in Boyce/Codd normal form, we have five distinct objects defined in our system that have data associated with them, namely, CLIENT, RESIDENCE, CHILD, HOME-ROOM, and OFFICE. Each data store has a primary key that uniquely identifies an occurrence of the respective objects. In addition to reorganizing the data elements as we have, there are relationships that exist among the object data stores.

Normalization breaks down the data stored in a system into its basic, most logical structure. From this basic structure, a systems designer can construct the physical data base that best suits the needs of the users with the available technology.

During the requirements stage, the systems analyst defines the problem and system solution. It is at this time that the analyst should present the data in a form that is not prejudicial toward any particular data-base structure or type of data-base technology. Object data stores present the necessary data organized according to the basic inherent structure of the data itself. Normalization, although a data-base design tool, aids the analyst in determining the objects and attributes of a system. All implementation considerations are left for the designers and those implementing the system.

Normalization of Physical Things

The question often comes up about how to represent physical items that are stored in a data store, such as BLOOD in a BLOOD-BANK, SKATE in a STORAGE-RACK, VIDEO-TAPES on the SHELVES, etc. First of all, we must list all of the data that can be derived from the physical item itself. In the case of a container of BLOOD, the label on the container would most likely contain the BLOOD-TYPE, DATE-RECEIVED, name of the BLOOD-BANK, etc. Likewise, in the case of SKATE, one can determine from the physical item itself whether the skates are for men or women, the SIZE, COLOR, CONDITION, etc. Once all of the data elements used by the system are identified, the data entry is formulated to contain the data derived from the physical item as well as the physical item itself, WHOLE-BLOOD or SKATE. Much data can be derived from the physical appearance of an item, but we are only concerned with the data used by the system. Consider the data store that follows.

UNNORMALIZED DATA STORE

BLOOD = {BLOOD-TYPE + DATE-RECEIVED + BLOOD-BANK + . . . + WHOLE-BLOOD}

SKATE = {["MALE"|"FEMALE"] + SIZE + COLOR + CONDITION + . . . + PAIR-OF-SKATES}

When the primary key is selected, the physical item is typically included as a single occurrence together with the attributes that permit the user to locate the physical item. In considering normal form, the primary key is usually just the physical item itself with all of the other attributes listed as descriptions or as facts about the physical item derived from observing the item itself.

FIRST NORMAL FORM

BLOOD = {BLOOD-TYPE + DATE-RECEIVED + BLOOD-BANK + . . . + WHOLE-BLOOD}

SKATE = {["MALE"|"FEMALE"] + SIZE + COLOR + CONDITION + . . . + PAIR-OF-SKATES}

When the primary key is the physical item itself and the nonkey attributes are derived data from observance of the physical item, the nonkey attributes are clearly descriptions of the primary key, the whole key, and nothing but the key. Because every candidate key is also a determinant, the relationship satisfies the criteria for the Boyce/Codd normal form.

DATA RELATIONSHIPS
. . . relationships among objects

Now that we have reorganized all of the essential memories into object data stores, we can consider the relationships that exist between the entities of one object data store with the entities of another object data store. An *entity-relationship (E-R) diagram* can be created to show these relationships. **ENTITY-RELATIONSHIP (E-R) DIAGRAM is a graphical illustration of the objects (also known as entities), and attributes, and the relationships that exist between them.** The E-R diagrams were introduced by Peter Chen[C-2] and are independent of any data-base technology used to implement the necessary data structures that serve the needs of the system. E-R diagrams provide a means of logically specifying data and relationships that can easily map into any data structures of the four basic technologies of data-base systems: inverted list, hierarchical, network or relational. Because E-R diagramming is technologically independent and can service all four data-base models, it is an ideal tool for systems analysis.

In creating an E-R diagram, we show objects as rectangles, relationships as small diamonds, and attributes as circles. Also, the notation **R#**[object-data-store-A/object-data-store-B, **1-1]** represents a one-to-one (1-1) relationship associating one occurrence of an entry from object-data-store-A to exactly one occurrence of an entry from object-data-store-B.

For all of the following examples of E-R diagrams, the following PERSON and AUTO data stores are used:

PERSON = {NAME + ADDR + SS-# + OCCUPATION + BIRTH-DATE + HT + WT + SEX + EYE-COLOR}

AUTO = {MAKE + YEAR + MODEL + EXT-COLOR + INT-COLOR + INT-FABRIC + AUTO-SERIAL-# + {OPTIONS}}

An E-R diagram permits us to model the relationships between the entries in one object data store and the entries in another object data store. There are four basic relationships that can exist:

1. **ONE-TO-ONE** (1-1) relationship exists when one associates one entry in an object data store with only one entry in another object data store.
2. **ONE-TO-MANY** (1-M) relationship exists when one associates one entry in an object data store with one or more entries in another object data store.
3. **MANY-TO-ONE** (M-1) relationship exists when one associates one or more entries in an object data store with exactly one entry in another object data store.
4. **MANY-TO-MANY** (M-M) relationship exists when one associates one or more entries in an object data store with one or more entries in another object data store.

In all of the definitions of relationships, "**1**" denotes exactly one, and "**M**" denotes *one or more*. In the entity-relationship models originally proposed by Peter Chen, the question as to whether or not "**M**" is permitted to be zero is unclear. As far as specifying system requirements in a structured specification, we will use "**M**" to mean zero, one, or more than one. If the relationship requires there be at least one, and also more than one, then we will specify the requirement by means of "**M***" (* = **minimum of one**). In doing the data-base design for a particular implementation of a system, it may be necessary to know exactly if "**M**" is permitted to be zero or not. In such cases, it may be convenient to modify the graphical notation for the E-R structures to distinguish between those one-to-many (**1-M**) relationships that permit "**M**" to be zero and those that require "**M**" to be at least one, but I will leave that to the data-base designers.

ENTITY-RELATIONSHIP DIAGRAMS
. . . how does all the data relate?

Let's show how one could illustrate each of the basic types of relationships using graphical notation in an E-R diagram. Figure 7.7a is an E-R diagram that illustrates the relationship R1[PERSON/AUTO, 1-1], which permits only one owner to own only one auto.

ONE-TO-ONE RELATIONSHIP

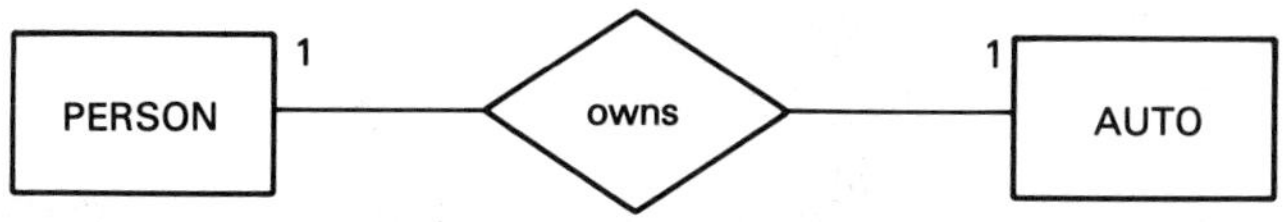

Figure 7.7a

We use the rectangle to designate the object data store or entity. The diamond designates the relationship between the object data stores. Some analysts choose to leave the diamond blank, others prefer to provide a descriptive relationship name or title, and still others prefer to insert the relationship designator **R#**, which corresponds to the following textual reference to the relationship:

R#[object-data-store-1/object-data-store-2, 1-1]

where # is an integer value used to uniquely number the relationships within the system. Choose whatever way you wish, but the purpose is to document the relationships among the entries in the various data stores in the system. Figure 7.7a is an illustration of the relationship R1[PERSON/AUTO, 1-1], whereas Fig. 7.7b illustrates the relationship R2[PERSON/AUTO, 1-M], which permits one owner to own one or more autos.

ONE-TO-MANY RELATIONSHIP

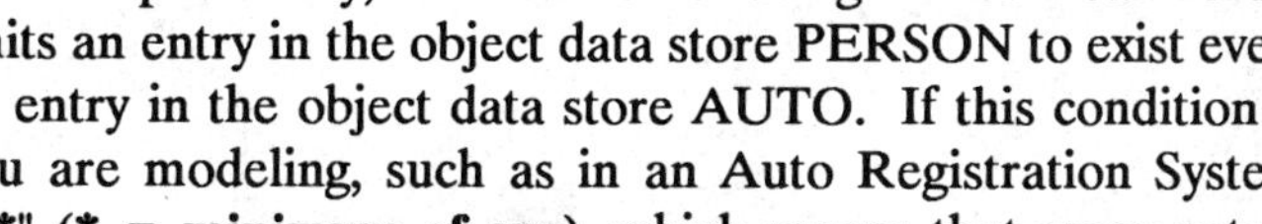

Figure 7.7b

As mentioned previously, our notation of using "**M**" to mean zero, one, or more than one permits an entry in the object data store PERSON to exist even if there is no corresponding entry in the object data store AUTO. If this condition is not true for the system you are modeling, such as in an Auto Registration System, the analyst would use "**M***" **(* = minimum of one)**, which means that every entry in the object data store PERSON must own at least one auto.

Figure 7.7c illustrates the relationship R3[PERSON/AUTO, M-1], where there is one or more owners for every auto and the owner(s) can own only one auto.

MANY-TO-ONE RELATIONSHIP

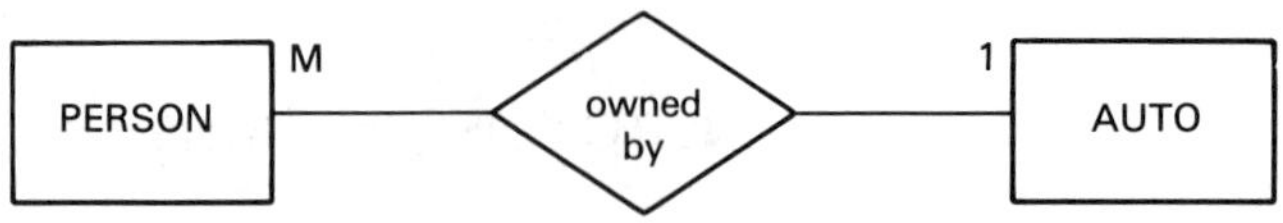

Figure 7.7c

Figure 7.7d shows the relationship R4[PERSON/AUTO, M-M], where every owned auto can have one or more owners and the owner(s) is permitted to own one or more autos.

MANY-TO-MANY RELATIONSHIP

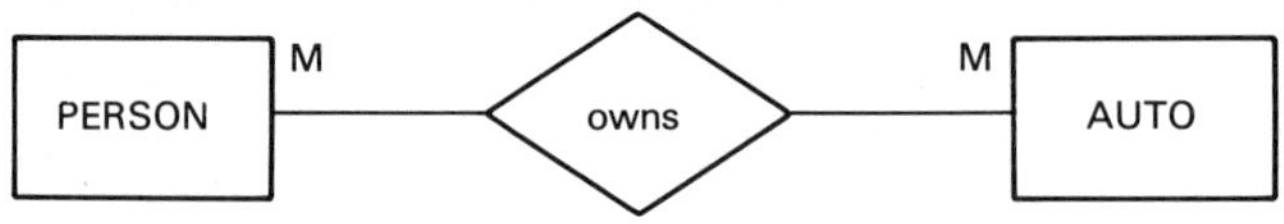

Figure 7.7d

Intersecting Data

There is a set of attributes that can exist only when a particular person is associated with a particular auto. Such attributes are LENDING-CO, AMT-OF-LOAN, LOAN-PAYMENTS, etc. These attributes only exist when there is a relationship of NAME + ADDR and AUTO-SERIAL-#.

The E-R diagram of Fig. 7.8 illustrates the relationship among the object data stores PERSON and AUTO with the object data store LOAN:

LOAN = {NAME + ADDRESS + AUTO-SERIAL-# + LENDING-CO + AMT-OF-LOAN
+ LOAN-PAYMENTS + DATE-OF-LOAN + (CO-SIGNER)}

In the example, R5[(PERSON/AUTO)/LOAN, (M-M)-1] of Fig. 7.8, the object data store LOAN contains entries that are dependent on entries in the two object data stores PERSON and AUTO. An entry in the data store LOAN exists only if there is an entry from the PERSON data store associated with an entry from the AUTO data store. The data elements in the object data store LOAN are known as *intersecting data*. **INTERSECTING DATA is an object data store or a data element(s) that describes or contains fact(s) about the relationship that exists between two object data stores.** Object data stores, as well as single data elements (attributes) that describe or contain facts about a relationship, can be intersecting data.

INTERSECTING DATA

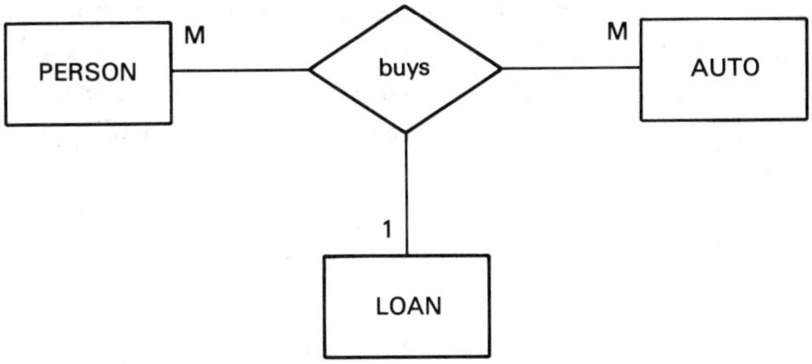

Figure 7.8

Other Special Relationships Between Objects

There can be cases where we have a MANY-TO-MANY-(TO-MANY . . ., etc.) relationship, such as in a PART, PROJECT, SUPPLIER relationship. Here we have an object data store that shows which suppliers are supplying which parts for which projects. With the square-bracket notation, the relationship can be specified as R6 [(PART/PROJECT/SUPPLIER)/PIECE-PART, (M-M-M)-1] and is illustrated in Fig. 7.9.

MANY-TO-MANY-TO-MANY RELATIONSHIP

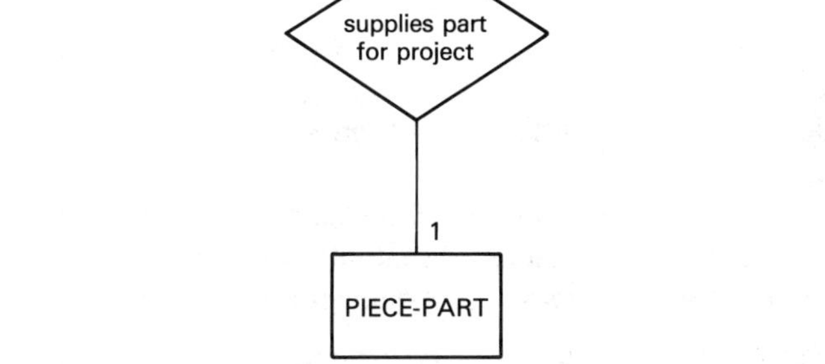

Figure 7.9

Another E-R diagram that I wish to discuss has an attribute in an object data store whose data values are the same as those for the primary key. Consider the case using the EMPLOYEE object data store, where one of its attributes is REPORTS-TO. The data value of the REPORTS-TO attribute for a given employee is the name of another employee that also has an entry in the EMPLOYEE object data store. The E-R diagram of Fig. 7.10 shows this reflective ability of an object data store onto itself as R7[EMPLOYEE/EMPLOYEE, 1-1].

The relationship shown in Fig. 7.10 is known as a *reflective relationship*. **A REFLECTIVE RELATIONSHIP is one defined on a single object data store.**

REFLECTIVE RELATIONSHIP

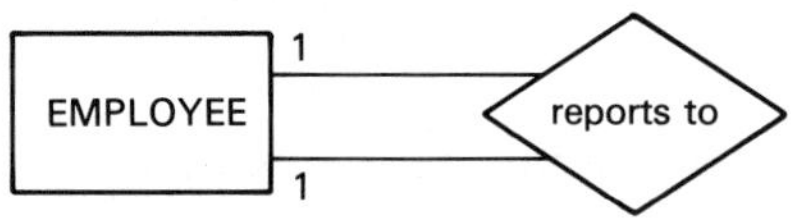

Figure 7.10

There is another special relationship, known as a *dependency relationship*, which is denoted as R8[PARENT/<CHILD>, 1-M]. **A DEPENDENCY RELATIONSHIP exists between two object data stores if an entry from one object data store is dependent on the existence of one (or more) associated entry from another object data store.** A usual example is as follows: an entry in the object data store CHILD only exists if there is an entry in the object data store PARENT, as shown in Fig. 7.11. To illustrate the existence dependency on an E-R diagram, the dependent object is shown in a double rectangle.

DEPENDENCY RELATIONSHIP

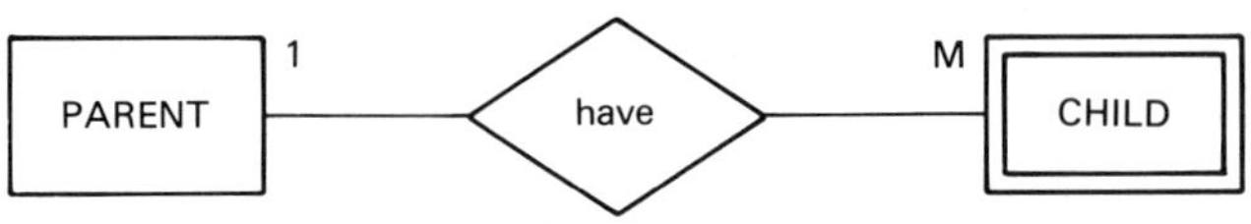

Figure 7.11

Relationships Between Relationships

Within a system there can be relationships between relationships as well as between object data stores. Consider the situation where there is a one-to-many relationship between the following two object data stores COUNSEL and CLIENT:

COUNSEL = {CO-NAME + LAWYER-NAME + CO-ADDR + LAWYERS-PHONE + . . . }

CLIENT = {CLIENT-NAME + CLIENT-ADDR + CLIENT-PHONE + . . . }

COUNSEL/CLIENT RELATIONSHIP

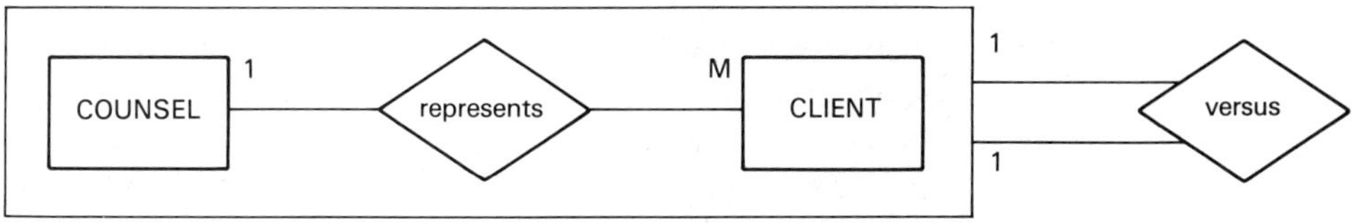

Figure 7.12

When a suit is brought to court, there is a reflective relationship between the objects COUNSEL and CLIENT, as shown in Fig. 7.12, and is denoted by R9[(COUNSEL/CLIENT, 1-M)/(COUNSEL/CLIENT, 1-M), 1-1]. In Figs. 7.13a and 7.13b, we see the relationship to another object data store CASE. The data store CASE is associated to the *relationship* between COUNSEL and CLIENT. In the example in Figs. 7.13a and 7.13b, the relationship R10[(COUNSEL/CLIENT, 1-M)/ CASE, M-M] is a many-to-many relationship between COUNSEL and CLIENT (as a *subrelation*) and the object data store CASE. **A SUBRELATION is a relationship that is used as an object in another relationship.** There are two ways to illustrate subrelations, as shown in the E-R diagrams of Figs. 7.13a and 7.13b.

E-R DIAGRAM INVOLVING SUBRELATIONS

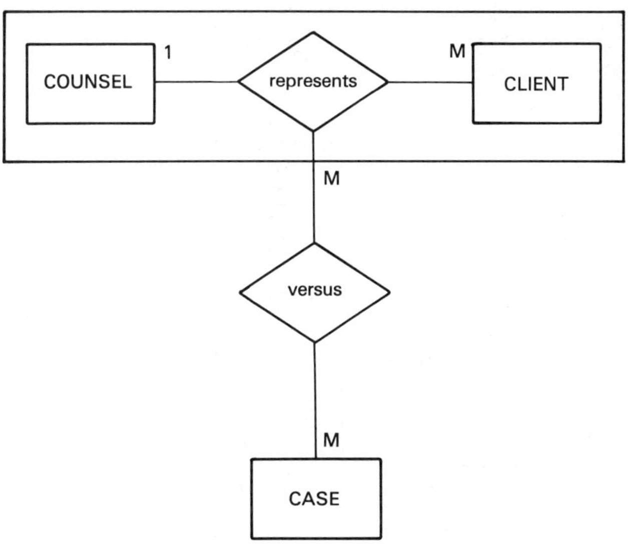

Figure 7.13a

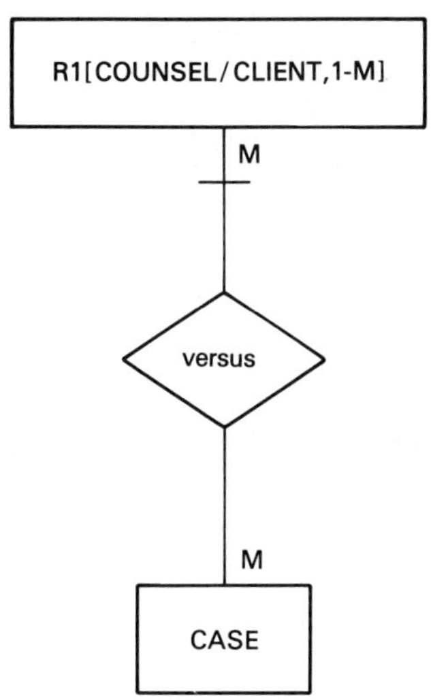

Figure 7.13b

Data Attributes in E-R Diagrams

Data attributes describe relationships as well as objects. Each attribute obtains its values from a specific *domain*. **A DOMAIN is the set of all possible values for a given attribute.** Moreover, several attributes can be defined on the same domain. The concept of domains is useful in two ways. First, it provides a means of confirming the values of a data element and second, it is useful when we explore the relationships between object data stores. There is a special kind of domain called a *primary domain*. **A DOMAIN is said to be PRIMARY if there is at least one data store in the system whose primary key is a single attribute defined on the domain.** By the way, in the case where there is more than one attribute that comprise the primary key, there is no requirement that every attribute of the primary key be defined on a primary domain. A primary domain exists if there is a *single attribute* that is a primary key for some data store in the system.

From an object data store EMPLOYEE, the *domain* of OCCUPATION is TYPES-OF-OCC = ["Manager" | "Accountant" | "Lawyer" | "Salesperson" | "Engineer" | "Programmer" | "Secretary" | "Clerk" | "Systems Analyst"].

OBJECT, ATTRIBUTE, DOMAIN AND DATA VALUE

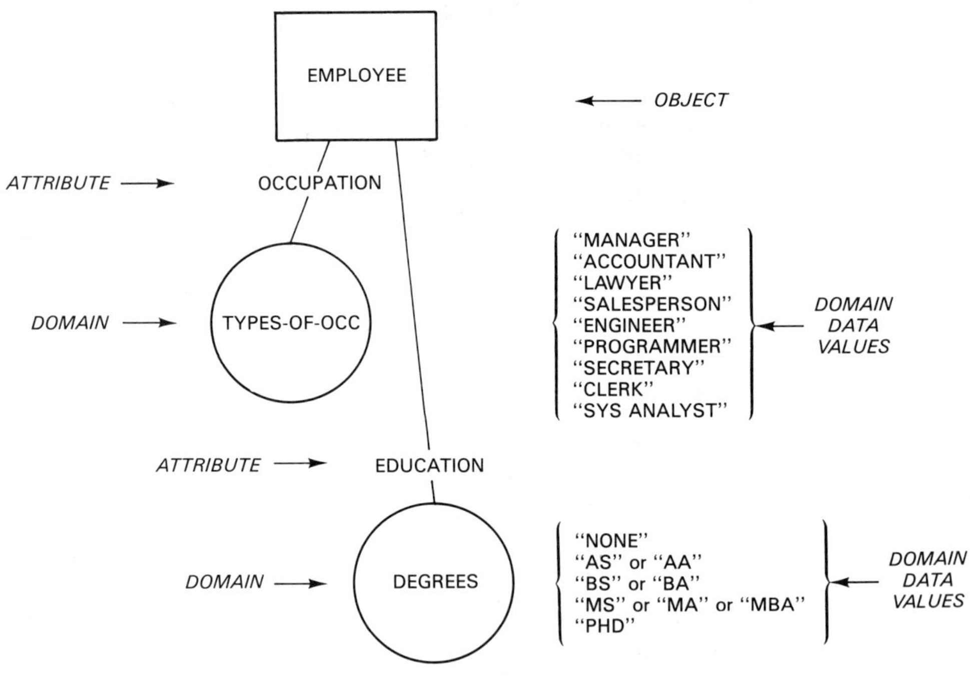

Figure 7.14

If there is a data store in the system defined as follows:

OCCUPATIONS = {JOB-TITLE + SALARY-RANGES + . . . }

and the domain for OCCUPATION is the same domain as that for JOB-TITLE above. Because JOB-TITLE is a single attribute that is the primary key of the data store OCCUPATIONS, the domain for JOB-TITLE (or OCCUPATION) is a primary domain. Figure 7.14 illustrates the relationship between objects, attributes, and domains for a specific example.

GENERAL E-R DIAGRAM

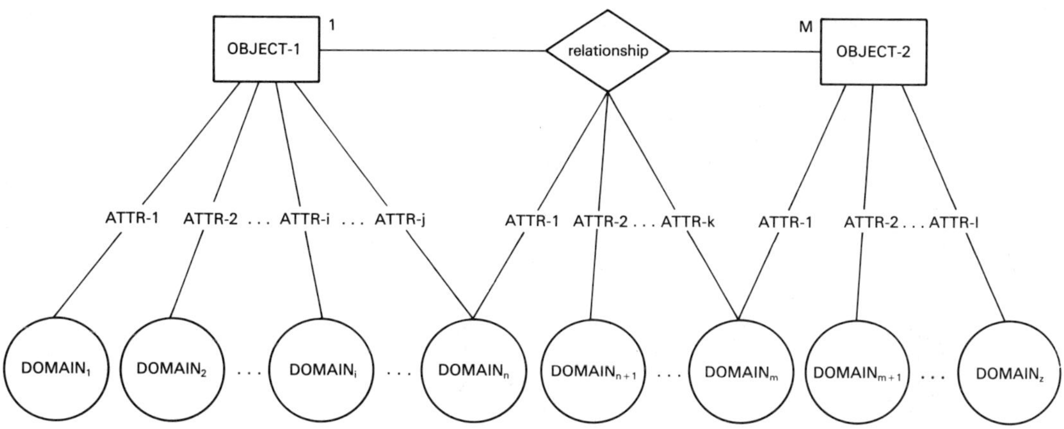

Figure 7.15

The overall general format for an E-R Diagram illustrating the relationships between objects and the relationship of attributes and domains is shown in Fig. 7.15. This graphical notation provides a means of communicating between the users and the data-base designers with regard to the data stored in the system. Although it is not necessary to include all data elements in the diagram, there are many occasions when it is useful to illustrate the relationships among objects, attributes, and domains. One example is when attributes of different objects share the same domain. We will see later that this information is helpful to the data-base designer in defining special keys when using certain types of data-base management systems.

From the illustration, we can tell that *one* entry in OBJECT-DATA-STORE-1 is related to *many* entries from OBJECT-DATA-STORE-2 and is denoted as

R#[object-data-store-1/object-data-store-2, 1-M]

Figure 7.16 is a specific example using the general form of E-R diagram that illustrates the relationship R11[STUDENT/PARENT, M-1].

E-R DIAGRAM

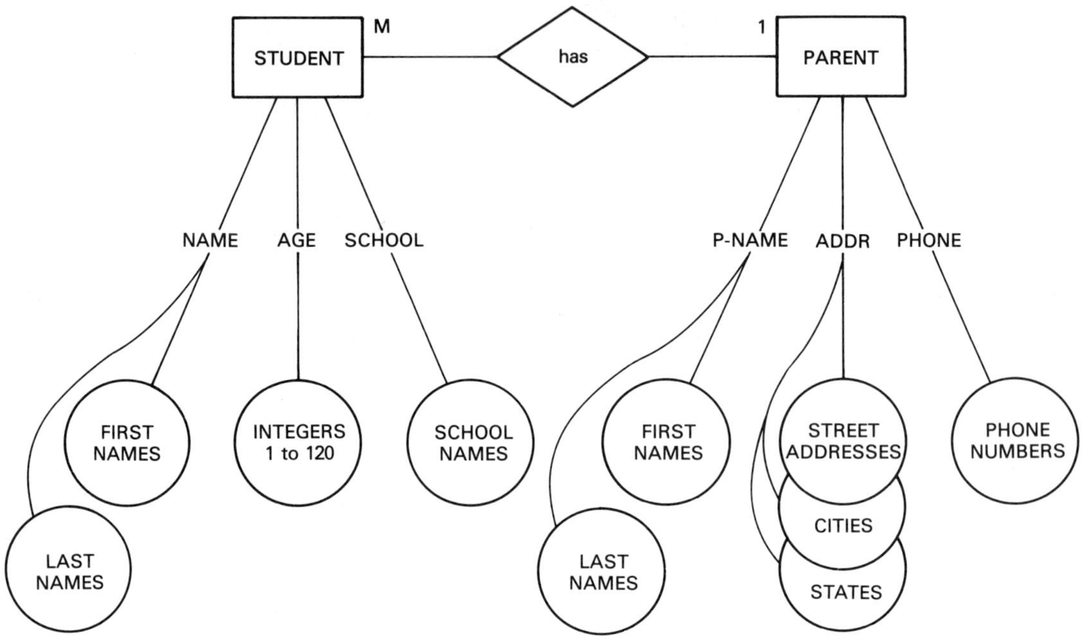

Figure 7.16

OBJECTS AND SUBTYPES
. . . some attributes relate to only a certain type of object

Once the objects of a system have been defined, we can ask how to specify requirements that pertain to just a particular subset of an object. For example, consider an object data store EMPLOYEE back in Fig. 7.14 with the attribute OCCUPATION. The domain of OCCUPATION is TYPES-OF-OCC = ["Manager" | "Accountant" | "Lawyer" | "Salesperson" | "Engineer" | "Programmer" | "Secretary" | "Clerk" | "Systems Analyst"]. There can be a set of attributes that pertain to just one or two of the particular OCCUPATIONs. An example would be professional certification. The professional certification for accountants is a CPA (Certified Public Accountant); for a systems analyst, it is a CDP (Certified Data Processor). Likewise, certain engineering professionals, as well as lawyers, are licensed by the state, and programmers can be certified as CCP (Certified Computer Programmer), etc. Assume that the system requirements specify that the system contain information about professional certification/license, together with fees and qualifications. However, only the employees with the occupations of "Accountant" (acct), "Lawyer" (lawyr), "Engineers" (engr), "Programmer" (pgmr), and "Systems Analyst" (sys anlst), are associated with the information about each type of certification or license for their respective professions. How would one represent this in the data model for a system?

First of all, an object can be partitioned into *subtypes*. **An OBJECT SUBTYPE is a distinguishable category or class of an object that has its own attributes, characteristics, and/or relationships.** An object can contain one or more subtypes. The subtypes can be mutually exclusive, that is, an object can be of *only one* subtype. On the other hand, a subtype cannot be mutually exclusive, that is, an object can be considered a member of more than one subtype. In our earlier example of the object data store EMPLOYEE, the subtypes are the various occupations of the employees. Some occupations may not have a certification process, such as managers (mgr), secretaries (secy), clerks (clk), etc. Therefore, to illustrate that only certain occupations are associated to the object data store CERTIFICATION, we can use the illustration in Fig. 7.17.

All object types can contain both mutually exclusive and nonmutually exclusive subtypes. In Fig. 7.17, an employee can be either male or female, not both, and there are no other alternatives. Thus, we use a mutually exclusive notation for that object subtype. On the other hand, an employee can be a licensed engineer or a lawyer *and* a manager. Also, there are other valid occupation types not specified in the illustration. Thus, we use the nonmutually exclusive notation for the object subtype OCCUPATION. Because an employee can have a license in more than one state or hold multiple certifications and/or licenses, there is a one-to-many relationship between EMPLOYEE and CERTIFICATION.

RELATIONSHIPS INVOLVING SUBTYPES

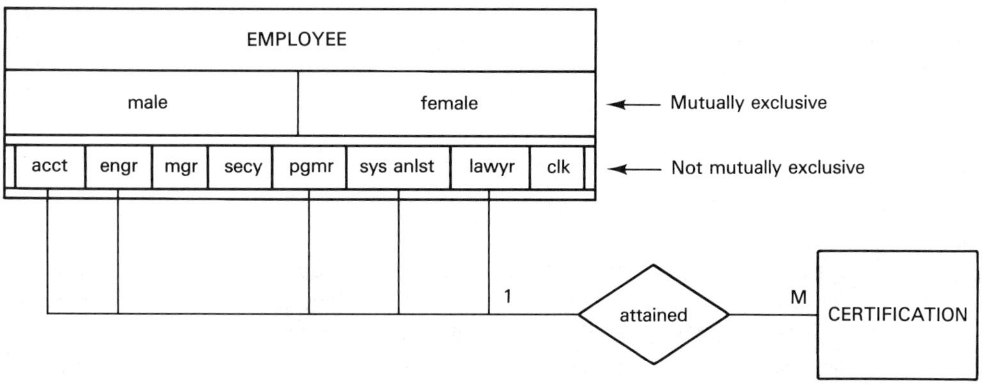

Figure 7.17

Object subtypes behave in every way as objects; they have their own attributes, characteristics, and relationships to other objects. Subtypes can be divided themselves into subtypes. The object type that is subdivided into subtypes is known as the *supertype*. **An OBJECT SUPERTYPE is an object or object subtype that contains distinguishable categories or classes each having its own attributes, characteristics, and/or relationships.** Subtypes of objects can form hierarchies in which certain subtypes can have a relationship to another object or object subtype, whereas the other subtypes do not have any other relationships, as shown in Fig. 7.18.

SUBTYPE HIERARCHIES

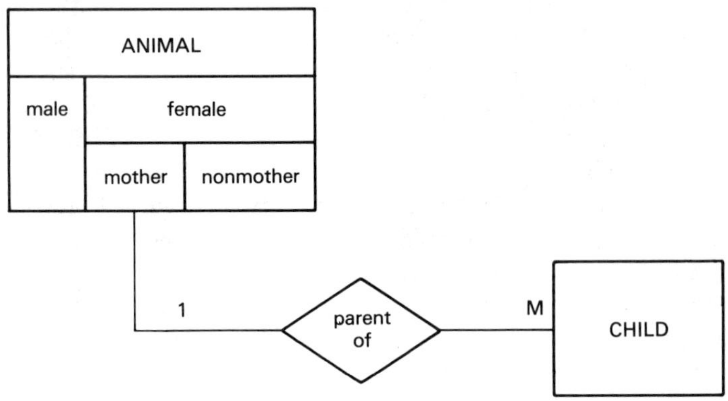

Figure 7.18

Care should be taken not to confuse the concepts of objects, subtypes, attributes, and relationships. The systems analyst can use these concepts and the corresponding tools of E-R diagramming to represent the system's requirements. There are tools that permit much greater details of structuring data that are in the jurisdiction of the systems designers and/or data-base designers, and are beyond the scope of this text.

CHARACTERISTICS OF DATA STORES
. . . the size and shape of stored data

To complete the discussion of specifying data requirements, there are two remaining characteristics of data stores to discuss: *cardinality*, and *degree*. **CARDINALITY is the number of unique entries in a data store.** Cardinality is used during the data-base design stage. The analyst can provide the cardinality of a data store as an estimate of the number of entries in each of the normalized object data stores. Cardinality can also be used as a means of specifying an upper bound (maximum number) of entries for a data store. Thus, it is used to specify the required capacity of data stores. **DEGREE is the number of key and nonkey attributes in a data store.** The degree of each normalized object data store can be easily computed; just count all of the unique *elementary data elements* for any given entry in the data store.

Example

CLIENTS = $_{1}${NAME + ADDR + OCCUPATION + BIRTH-DATE + OFFICE-ADDR
+ {CHILD-NAME}}^{1000}

The *degree* of the CLIENTS data store is 6. If the requirement for the data store is to hold 1000 unique entries, the *cardinality* would be 1000.

The last item I want to discuss in this section is linking elements of object data stores. Consider the following object data stores that we placed in Boyce/Codd normal form earlier in this chapter and the relationships we identified.

BOYCE/CODD NORMAL FORM

CLIENT = {NAME + ADDR + OCCUPATION + BIRTH-DATE + CHILD-NAME
+ OFFICE-ADDR}

RESIDENCE = {ADDR + [HOUSE | APT | CONDO] + DATE-CONSTRUCTED
+ HOUSE-STYLE}

CHILD = {CHILD-NAME + CHILD-BIRTH-DATE + SCHOOL-NAME + ROOM#}

HOME-ROOM = {SCHOOL-NAME + ROOM# + TEACHER}

OFFICE = {OFFICE-ADDR + OFFICE-PHONE + OFFICE-SIZE}

***RELATIONSHIPS**

R1[CLIENT/<CHILD>, 1-M]

R2[CLIENT/RESIDENCE, 1-1]

R3[CLIENT/OFFICE, 1-1]

R4[CHILD/HOME-ROOM, M-1]

The E-R diagram of Fig. 7.19 illustrates the relationships between the objects.

E-R DIAGRAM

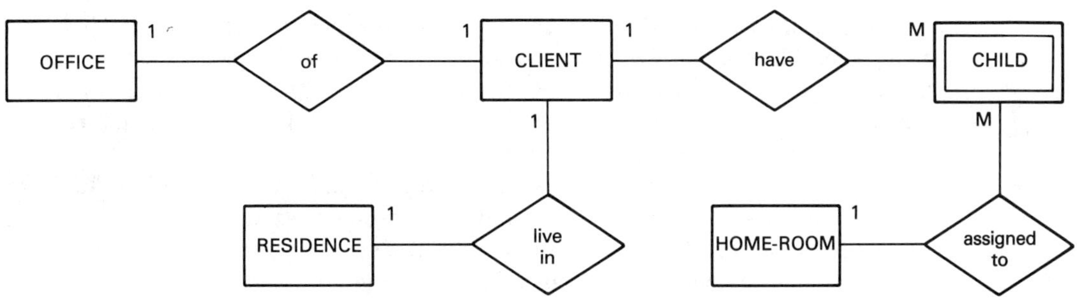

Figure 7.19

In the previous object data store definitions, there are four instances where there is a common data element(s) in two or more object data stores:

1. The data element that is common in the object data stores CLIENT and RESIDENCE is the data element ADDR. ADDR is only a part of the primary key in the CLIENT object data store and ADDR can be used to reference data entries in the RESIDENCE object data stores where it is the primary key.
2. The data element CHILD-NAME is a nonkey data element in the CLIENT object data store and can be combined with ADDR to form the primary key for the object data store CHILD.
3. The attributes SCHOOL-NAME + ROOM# in the CHILD object data store are used as the primary key in the HOME-ROOM object data store.
4. The data element OFFICE-ADDR is a nonkey attribute in the CLIENT object data store and is the primary key in the OFFICE object data store.

The nonkey attributes described in these four cases are *foreign keys*. **A FOREIGN KEY is a data element that is a nonkey attribute in one data store and is used to identify a related entry in another data store where the attribute is the primary key.** In the case where the foreign key is a single attribute, we could say that a foreign key is a nonkey attribute that is either null or derives its values from a primary domain. However, foreign keys can also be composite, that is, consisting of more than one data element. Likewise, the composite data value of the entire foreign key is permitted to be null in the data store where it is a *nonkey element*. When the foreign key is not null, its value matches the value of a primary key for some other data store in the system.

Not all data-base management systems (DBMS) use a foreign key to link data entries between two data stores. Relational DBMSs *do* require the presence of foreign keys. On the other hand, hierarchical DBMSs usually *do not* require duplicate data elements to link entries from two data stores. However, hierarchical DBMSs employ linkages that must be predefined. To design the structure of a hierarchical data base, we must know how the data is going to be used. As a result, hierarchical data bases have a limited ability to service unanticipated queries. **A QUERY is a request for data made of a data base.** The data-base designer constructs the data structures of a hierarchical data base according to the types of queries the users anticipate making. Performance problems can occur if the data structures do not support the most frequently used queries.

We may not necessarily find a foreign key in a data store definition that links the entries of one data store to those in another. Foreign keys only have to be present for certain types of DBMSs, such as relational DBMSs. When we are designing a system, we should not be concerned about defining foreign keys because by doing so we may inadvertently prejudice the structured specification toward using relational DBMSs.

When modeling existing systems, we simply specify what is actually present in the system.

Data Store Integrity Rule #2 (Referential Integrity)

Consider a data store P that has a primary key made up of a single attribute whose values are taken from domain D. If there is another data store F, which has an attribute A, that is also defined on the same domain D, then at any given time, each value of A in the data store F must be null or equal to some value V, where V is the primary key value of some entry in data store P in the system. Such an attribute A is a foreign key, as defined earlier. The set of all normalized object data stores using foreign keys is equivalent to the set of all physical data stores in the physical model. Thus, we can replace the set of physical data stores with the set of normalized object data stores in the logical model and not lose any data or access capability present in the physical model.

Let's look at referential integrity once again. Basically, referential integrity states that the foreign key is able to be a cross-reference element between two data stores. One of those data stores has the cross-reference data element as a nonkey attribute (A) and the other has it as the primary key (PK).

P = {PK + X + Y + Z}   A is either NULL or equal to V, where V is a value of PK in P,

↕

F = {K + A + B + C + D}   then A in data store F is a foreign key.

A foreign key either has a null value or has a value that is equal to that of a primary key in another data store. This permits an entry from data store F to reference or link to an entry in data store P.

Loosely speaking, when the value of an attribute in a data store is also a primary key value in some other data store, we say that the nonkey attribute is a foreign key. We should also note that the two data stores do not necessarily have to be distinct. Foreign keys can be used to link two entries from the same data store.

Let's look at a specific example.

EMPLOYEE = {F-NAME + L-NAME + ADDR + JOB-TYPE + BIRTH-DATE + HT + WT + SEX + DEPT + DATE-OF-EMPLOYMENT}

OCC-STATISTICS = {OCCUPATION + SALARY-RANGE + MEDIAN-SALARY + MODE-SALARY + DATE-LAST-SALARY-SCALE-CHANGE + NUMBER-OF-EMPLOYEES + PERCENT-OF-SALARY-CHANGE}

JOB-TYPE = ["Manager" | "Accountant" | "Lawyer" | "Salesperson" | "Engineer" | "Programmer" | "Secretary" | "Clerk"].

OCCUPATION = ["Manager" | "Accountant" | "Lawyer" | "Salesperson" | "Engineer" | "Programmer" | "Secretary" | "Clerk"].

In this example with the domains of OCCUPATION and JOB-TYPE as shown, JOB-TYPE is a foreign key. OCCUPATION is a single-attribute primary key of the data store OCC-STATISTICS. The values of the data element OCCUPATION are the same values of the data element JOB-TYPE. Once we locate an entry in the data store EMPLOYEE that is of interest, the attribute value is the primary key value to select the entry in the OCC-STATISTICS data store. Thus, a process in the system has access to the data from the EMPLOYEE data store and also the statistical information about all of the employees that have the same JOB-TYPE or OCCUPATION. OCCUPATION is the primary key in the data store OCC-STATISTICS and is said to be drawn from the same domain as JOB-TYPE.

In Fig. 7.20, the attribute JOB-TYPE is a foreign key in the first data store. Also, the data element OCCUPATION is the primary key for the second object data store. The types of OCCUPATION can be found in the primary domain for OCCUPATION. Furthermore, OCCUPATION as an object can be considered to have the object subtypes as listed in the primary domain of OCCUPATION. We would only need to show the object OCCUPATION as a supertype if there were relationships that existed for some of the types or classes of occupations and not others.

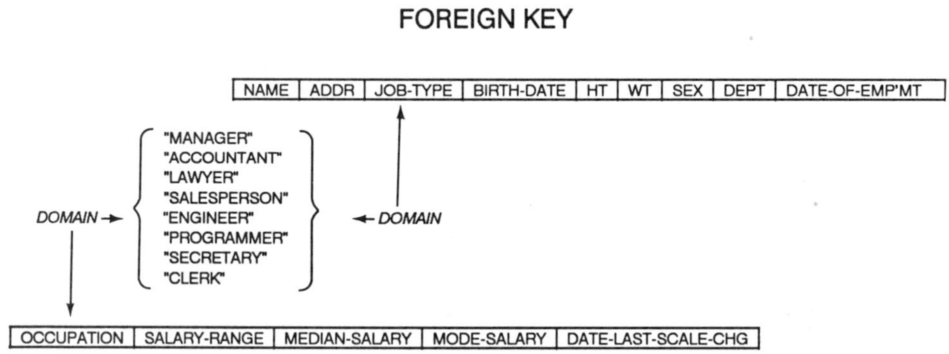

Figure 7.20

Let's consider another example, in figure 7.21, where the object OCCUPATION is shown as a single object and is related to the object EMPLOYEE as a one-to-one relationship. The object OCCUPATION is shown as a subtype, because there are only certain subtypes related to the object UNION-OFFICIAL. In the example in Fig. 7.21 only programmers (pgmr), secretaries (secy), and clerks (clk) are members of a union and eligible to be union officials.

Remember that in all of the notation provided in this text, the intention is to give the systems analyst a means of specifying the *system's requirements*. It is not the purpose of this text to provide a full repertoire of tools for data-base or systems design. However, the tools presented can be useful in documenting the requirements

RELATIONSHIP BETWEEN OBJECT SUBTYPE AND OTHER OBJECTS

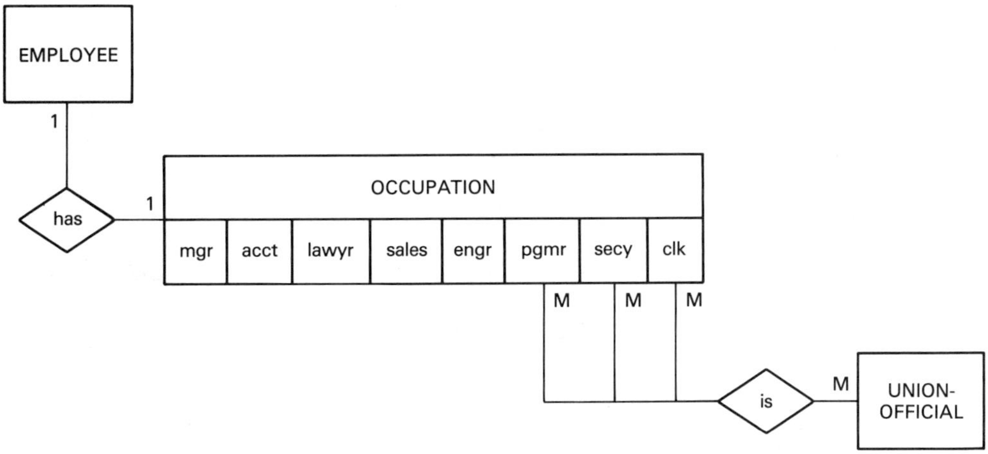

Figure 7.21

specified by the user and understood by the analyst. Once the requirements are verified by the user, they are passed on to system and data-base designer who will implement the system requirements.

IDENTIFYING RELATIONSHIPS
. . . finding relationships in existing systems

Now that we have a means of expressing relationships between objects or other relationships, how do we know what relationships exist in a system?

First of all, in modeling an existing system, we should not assume that a relationship exists just because there is a natural or inherent relationship between the objects themselves. For example, consider a system used by a college to register and maintain a data base of students. It is possible that a parent and a son or daughter could both register and enroll in classes at the college. There is an inherent one-to-many relationship between parents and their children. Although there is this natural relationship, the college registration system does not use or require that the data from the parent's registration be associated with the data for the student's registration, or vice versa. (Unless the college offered a "family" discount on tuition, which is unlikely.) Therefore, even though there is an intuitive or natural association between objects or relationships, the structured specification does not include such relationships unless they are required in the processing of data performed by the system. So, how can an analyst determine which relationships exist between the object data stores?

The following are some indications of relationships that are used in an existing system:

1. **Analyze the unnormalized data stores.**

   If an essential memory contains information on more than one object, examine the data store structure carefully to see if there is an implied relationship that is present in the system.

Example

$$\text{FILE-A} = \{\underbrace{\text{EMPLOYEE + AGE + HEIGHT + } \ldots}_{\text{object EMPLOYEE}} + \underbrace{\text{WIDGET + WIG-SIZE + } \ldots}_{\text{object WIDGET}}\}$$

In FILE-A, there is an implied one-to-one relationship between an EMPLOYEE and a WIDGET. Likewise, in FILE-B that follows, there is an implied one-to-many relationship between EMPLOYEE and WIDGET:

$$\text{FILE-B} = \{\underbrace{\text{EMPLOYEE + AGE + HEIGHT + } \ldots}_{\text{object EMPLOYEE}} + \underbrace{\{\text{WIDGET + WIG-SIZE + } \ldots\}}_{\text{object WIDGET}}\}$$

2. **Find all foreign keys in all of the essential memories in the existing system.**

CAUTION: remember for a foreign key to be valid, the attribute names do not have to be the same. The only requirement is that the nonkey attribute and the single primary-key attribute in the other data store must be defined on the same domain.

Example

EMPLOYEES = {<u>EMPLOYEE-NAME</u> + . . . + WIDGET#}

WIDGETS = {<u>WIDGET-NUMBER</u> + . . . }

If the essential memories contain a foreign key as shown above (WIDGET# and WIDGET-NUMBER are defined on the same domain), then there may be a one-to-one relationship between the objects EMPLOYEE and WIDGET. To confirm the existence and the type of relationship, we need to consult the mini-specs that process the data.

3. **Find all functional primitive processes that input data from more than one essential memory (or object data store). Examine the mini-spec for the functional primitive to determine if there is any relationship between the data stores.**

Example

PARTIAL DATA FLOW DIAGRAM

Figure 7.22

In the example Fig. 7.22, there may be a relationship (one-to-one, one-to-many, many-to-one, or many-to-many) between the objects EMPLOYEE and WIDGET, or no relationship at all. To determine if any relationship exists examine the logic in the mini-spec for process 7.8*. Just because a process inputs data from two separate data stores does not mean that there is a relationship between the entries of those data stores.

4. **Examine the essential memories or object data stores for common data elements (or nonkey attributes that are defined on the same domain), that use two data stores during processing.**

Example

EMPLOYEES = {<u>EMPLOYEE-NAME</u> + . . . + ORDER# + . . . }

WIDGETS = {<u>WIDGET-NUMBER</u> + . . . + ORDER# + . . . }

If there is a process that obtains the above data and has access to one of the other data stores, there can be a relationship between the object EMPLOYEE and WIDGET. A relationship can exist even if the WIDGETS data store were to contain the nonkey attribute ORDER-NUMBER instead of ORDER#, but defined on the same domain as ORDER#. The name of an attribute does not matter. For two attributes to be the same, all that is required is that they be defined on the same domain.

Before we specify a relationship in an E-R diagram, there must be a process that requires the relationship to exist in order to obtain all of the necessary data to perform its job. As stated earlier, an analyst must take great care not to model his or her perception of what a system does, but rather the structured specification of an existing system should be a statement of fact, specifying the requirements of the system under study.

OBJECT ANALYSIS EXAMPLE
. . . Trad'em Often brokerage firm

The following object analysis example takes a structured specification (excluding the detailed data flow diagrams and mini-specs), identifies all of the objects, normalizes the data to Boyce/Codd normal form, and constructs an E-R diagram.

The receptionist of a brokerage firm assigns new customers interested in buying or selling stocks to an available secretary and also forwards calls from established customers who want to buy or sell stocks to their broker's secretary.

The secretary of each stockbroker screens *all* calls. For customers that only ask for a price quote for a particular stock or for information on commission rates, the secretary provides the caller with the information. For new customers, the secretary begins by first entering the caller's name, address, social security number, telephone number, customer's bank name, and bank account number into a data base via a computer terminal. The secretary also inquires of the new customer's credit rating by means of an on-line access link to the National Credit Bureau. The bureau will respond that the new customer's credit rating is a "Good Risk," "Bad Risk," or "Unknown." If the credit is a "Good Risk," the secretary will forward the call to a stockbroker. If the credit rating is "Bad Risk," the secretary will politely end the conversation. If the credit is "Unknown," the secretary will explain to the person that they must first fill out a credit reference form, which will be mailed to them. When the customer returns the form, a credit agent will process it outside this system.

The stockbroker basically gets two type of calls. The first is for information about a specific stock. For this type of call, the stockbroker will mail the appropriate "prospectus" or "Standard & Poor's Stock Information" to the client. The second type of call is to trade stock. Clients can buy or sell stock only. (This system does not handle bonds, "short" positions, "options," etc.) The stockbroker has a computer terminal whereby the name of the company can be entered and the broker can obtain two pieces of information. The computer provides the stock symbol (SYM, up to four letters that uniquely identifies the stock listings for a particular company), and the stock market exchange (EXCHG) that trades the stock, such as New York (NYX), American (AMX), National (NASDAQ) or Over-The-Counter (OTC).

To enter a trade, the customer must provide the name of the stock, whether the transaction is a "buy" or "sell," the number of shares, and either the price used to transact the stock, or whether to complete the transaction at "market price" (the current price of the stock when the transaction order reaches the exchange).

The stockbroker can use the terminal to inquire about the current "market price" of the stock by entering the stock symbol and the exchange that trades the stock. The terminal also indicates the LAST-TICK ("UP-TICK" if the last change in price was an increase; "DOWN-TICK" if the last change was a decrease in price). Also, the terminal indicates the overall change in price (PRICE-CHG) from the current price (CURR$) to the price of the first sale made during that day (OPEN$).

After the transaction is completed at the exchange, the broker is informed on the results of the transaction: stock symbol, quantity of stock, "Bought" or "Sold," cost of stock, broker's fee, taxes, date of transaction, settlement date (usually five business days from the date of transaction). The settlement date is the date by which the brokerage firm must receive the money or shares from the customer to complete the transaction. Failure to complete the transaction by this date will affect the customer's credit standing and the brokerage firm will terminate the customer's account. Each day the secretary checks to see if all customers have settled their accounts by the settlement date. When the secretary receives the appropriate payment or stock certificates, the sale is completed. The secretary mails the purchased stock certificate or a check to the customer. If an account is not cleared (the customer did not send in the money for purchased stock, or the stock certificate for the sold stock), the secretary cancels the purchase order for the stock and anyone else is free to buy it. The customer receives a termination notice indicating the consequences of such an event.

Figure 7.23 shows the context diagram and Fig. 7.24 shows the data dictionary for this example. Detailed data flow diagrams and mini-specs are not required for this example. Material contained in this chapter pertains to object analysis only and, as a result, several sections of the example structured specification are not included. The chapter examines the data in a system and presents a method to organize the data according to the objects of the system. Finally, this chapter includes a means to express the relationships that must exist between those objects to satisfy the requirements of the system.

CONTEXT DIAGRAM

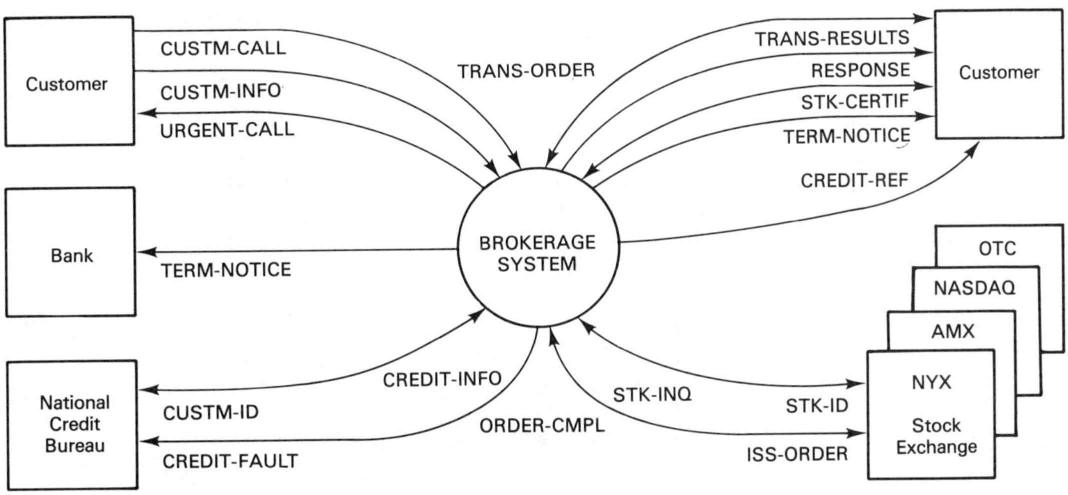

Figure 7.23

DATA DICTIONARY

AMT-OWED	= * Total amount owed to the brokerage firm by the customer. *
BANK-ACCT#	= * Customer's personal bank account number. *
BANK-ADDR	= * Address of the customer's BANK-NAME. *
BANK-NAME	= * Name of the customer's bank used as a credit reference. *
BANK-PHONE#	= * Telephone number of customer's BANK-NAME. *
BROKER-NAME	= * Name of the broker assigned to the customer's account. *
BROKER-OFFICE	= * Office location number of BROKER-NAME. *
BROKER-PHONE#	= * Telephone number of BROKER-NAME. *
CERTIF#	= * Unique serial numbers of the stock certificates. *
CLIENTS	= {CUSTM-INFO + {TRANS# + SYM + QTY + ["BOUGHT"|"SOLD"] + TRANS-DATE + SETTLE-DATE + TRANS-AMT + ["TRANS-COMPLETE" |"TRANS-OPEN"|"TRANS-CLEARED"]+ TRANS-PRICE} + BROKER-NAME}
	= * CLIENTS is organized by SS# but can also be accessed by CUSTM-NAME & CUSTM-ADDR. *
CLOSE$	= *The last price paid for a stock at the close of the trading day. *
COMM-RATE	= * $15 plus 1% of the PRICE times QTY for all QTY in increments of 100, for QTY less than 100 shares an additional flat rate of $10 is charged. *
CO-ADDR	= * Address of the company that issued the stock certificates. *
CO-NAME	= * Name of the company that issued the stock certificates. *
CREDIT-FAULT	= CUSTM-NAME + SS# + CUSTM-ADDR + AMT-OWED + SETTLE-DATE
CREDIT-INFO	= CUSTM-ID + (CREDIT-RATING)
CREDIT-RATING	= ["GOOD"|"BAD"|"UNKNOWN"]
CREDIT-REF	= * Form mailed to a potential customer to ascertain their credit risk. *
CURR$	= * The current price of the stock. *
CUSTM-ADDR	= * Address of the customer. *
CUSTM-BANK	= BANK-NAME + BANK-ADDR + BANK-PHONE#
	= * Name of the customer's bank. *
CUSTM-CALL	= CUSTM-NAME + ["NEW CUSTOMER" | BROKER-NAME] + INFO-INQUIRY
CUSTM-ID	= CUSTM-NAME + CUSTM-ADDR + SS#
CUSTM-INFO	= CUSTM-NAME + CUSTM-ADDR + SS# + CUSTM-PHONE# + CUSTM-BANK + BANK-ACCT# + (CREDIT-REF)
CUSTM-NAME	= * Name of the customer. *
CUSTM-PHONE#	= * The customer's phone number. *
DIRECTORY	= {BROKER-NAME + BROKER-OFFICE + BROKER-PHONE#}
EXCHG	= ["NYX" | "AMX" | "NASDAQ" | "OTC"]
HIGH$	= * The highest price the stock sold for during the day. *
INFO-INQUIRY	= ["PRICE OF STOCK?" + CO-NAME | "COMMISSION RATES?" | "OPEN ACCOUNT" | BROKER-NAME]
ISS-ORDER	= TRANS# + STK-ID + QTY + [PRICE | "MARKET PRICE"] + ORDER-DATE + BROKER-NAME
LAST-TICK	= * The last change in the price of the stock, either an "UP-TICK" or a "DOWN-TICK". *
LOW$	= * The lowest price of the stock during the day. *
OPEN$	= * The price paid for the stock during the first trade of the day. *

Figure 7.24

DATA DICTIONARY (Cont.)

ORDER-CMPL	= TRANS# + STK-ID + QTY + PRICE + ORDER-DATE + TRANS-DATE + BROKER-NAME
ORDER-DATE	= * Calendar data the transaction is placed by the customer. *
PRICE	= * Price in dollars per share. *
PRICE-CHG	= * The overall change in the price from the previous day closing price to the current price. *
QTY	= * Quantity of shares. *
RATES	= {PRICE + QTY + COMM-RATE}
RESPONSE	= [RATES | STK-INQ | TRANS-RESULTS | "other market information or advice"]
SETTLE-DATE	= * Usually five working days after TRANS-DATE. *
SS#	= * Customer's social security number. *
STK-CERTIF	= * STOCK CERTIFICATE with CO-NAME + CO-ADDR + QTY + TRANS-DATE + CERTIF#. *
STK-ID	= SYM + EXCHG
STK-INQ	= SYM + OPEN$ + HIGH$ + LOW$ + CURR$ + VOL + YR-HI$ + YR-LO$ + PRICE-CHG + LAST-TICK
SYM	= * The unique symbol used on the exchange to identify a company. *
SYM-LIST	= {CO-NAME + CO-ADDR + SYM + EXCHG}
TAXES	= * Taxes that apply to trading stock. *
TERM-NOTICE	= * Notice explaining the default on a stock trade, and termination of an account. *
TRANS#	= * A sequence number of integers starting from 000-0001 that uniquely identifies a transaction. *
TRANS-AMT	= PRICE * QTY + (.01 * QTY/100 + 15) + TAXES *For QTY ≥ 100. * = PRICE * QTY + (.01 * QTY/100 + 25) + TAXES *For QTY < 100. *
TRANS-DATE	= * Date the stock transaction is made at the exchange may not always be the same as ORDER-DATE if customer's PRICE is not agreeable to other buyers or sellers. *
TRANS-ORDER	= TRANS# + STK-ID + QTY + ORDER-DATE + [PRICE | "MARKET PRICE"]
TRANS-PRICE	= * The total amount in dollars for the transaction including TRANS-AMT. *
TRANS-RESULTS	= TRANS# + STK-ID + QTY + ORDER-DATE + TRANS-DATE + PRICE
URGENT-CALL	= * Call to client seeking immediate payment or stock certificate to complete transaction.*
VOL	= * The accumulated number of shares traded during the day. *
YR-HI$	= * The highest price the stock sold for during the past twelve months. *
YR-LO$	= * The lowest price the stock sold for during the past twelve months. *

Figure 7.24 (cont.)

Data Stores with All Elementary Elements

The following object analysis results in identifying all the objects from the corresponding data stores normalized to Boyce/Codd normal form, and provides the entity-relationship (E-R) diagram illustrating all of the relationships between the objects.

UNNORMALIZED FORM

CLIENTS = {CUSTM-NAME + CUSTM-ADDR + SS# + CUSTM-PHONE#
+ CREDIT-RATING + CUSTM-BANK + BANK-ACCT# + {TRANS# + SYM
+ QTY + ["BOUGHT" | "SOLD"] + TRANS-DATE + SETTLE-DATE
+ TRANS-AMT + ["TRANS-COMPLETE" | "TRANS-OPEN"
| "TRANS-CLEARED"]} + TRANS-PRICE + BROKER-NAME}

RATES = {PRICE + QTY + COMM-RATE}

SYM-LIST = {CO-NAME + CO-ADDR + SYM + EXCHG}

DIRECTORY = {BROKER-NAME + BROKER-OFFICE + BROKER-PHONE#}

To put the above data stores into first normal form, remove all repeating groups. Transactions within the data store CLIENTS is a repeating group of several data elements. Remove the repeating group and expand the primary key to include TRANS# so that we can identify each entry uniquely in the 1NF data store. Notice here that there is a 1-M relationship between a customer and a transaction. After we put all of the data stores into BCNF, we will construct an E-R diagram, so let's keep track of the relationships as we identify them. To begin with, the first relationship we can identify is **R1[CLIENTS/TRANSACTIONS, 1-M]**.

FIRST NORMAL FORM

CLIENTS = {CUSTM-NAME + CUSTM-ADDR + SS# + CUSTM-PHONE#
+ CUSTM-RATING + CUSTM-BANK + BANK-ACCT# + TRANS# + SYM
+ QTY + ["BOUGHT" | "SOLD"] + TRANS-DATE + SETTLE-DATE
+ TRANS-AMT + ["TRANS-COMPLETE" | "TRANS-OPEN"
| "TRANS-CLEARED"] + TRANS-PRICE + BROKER-NAME}

RATES = {PRICE + QTY + COMM-RATE}

SYM-LIST = {CO-NAME + CO-ADDR + SYM + EXCHG}

DIRECTORY = {BROKER-NAME + BROKER-OFFICE + BROKER-PHONE#}

RELATIONSHIPS

R1[CLIENT/TRANSACTION, 1-M]

To put the previous data stores into second normal form, make sure that each nonkey attribute describes the key, the whole key, and not just part of the key. The RATES data store has a compound primary key, however, upon inspection, we find that the commission rate (COMM-RATE) is truly related to both the price of the stock (PRICE) and the quantity of stock (QTY). Because CLIENTS is the only other data store with a compound key, check to make sure that no data element describes just the CUSTM-NAME or just the CUSTM-ADDR or just TRANS#. In this case, all of the data elements from TRANS# to TRANS-PRICE seem to describe the transaction and not CUSTM-NAME + CUSTM-ADDR (CUSTM-NAME is assumed not to be unique). Thus, all of the attributes that describe TRANS# have to be separated from the CLIENTS data store. Create a new data store with TRANS# as the primary key and all of the attributes that describe TRANS#. The relationship R1 describes the relationship between these two data stores.

BROKER-NAME in the CLIENTS data store seems to be an attribute of both the customer and the transaction. A data element can only be related to two objects if it is an attribute of the relationship between the two objects, or if it is itself an object that has attributes. In this case, BROKER-NAME is an object because it has attributes BROKER-OFFICE and BROKER-PHONE# found in the data store DIRECTORY. As an object, BROKER-NAME is an intersecting data object of the objects CLIENT and TRANSACTION. This is the second relationship we have found to exist among the data stores for this system, and it is denoted by **R2[(CLIENT/TRANSACTION)/BROKER, (1-M)-1]**. A third relationship exists between CLIENT and BROKER because in the data store CLIENTS, we find BROKER-NAME as an attribute with one BROKER able to have many CLIENTs. Thus, there is a 1-M relationship between the two objects BROKER and CLIENT, **R3[BROKER/CLIENT, 1-M]**. The following data stores are in second normal form:

SECOND NORMAL FORM

CLIENTS = {CUSTM-NAME + CUSTM-ADDR + SS# + CUSTM-PHONE# + CREDIT-RATING + CUSTM-BANK + BANK-ACCT# + TRANS#}

TRANSACTIONS = {TRANS# + SYM + QTY + ["BOUGHT" | "SOLD"] + TRANS-DATE + SETTLE-DATE + TRANS-AMT + ["TRANS-COMPLETE" | "TRANS-OPEN" | "TRANS-CLEARED"] + TRANS-PRICE}

RATES = {PRICE + QTY + COMM-RATE}

SYM-LIST = {CO-NAME + CO-ADDR + SYM + EXCHG}

DIRECTORY = {BROKER-NAME + BROKER-OFFICE + BROKER-PHONE#}

RELATIONSHIPS

R1[CLIENT/TRANSACTION, 1-M]

R2[(CLIENT/TRANSACTION)/BROKER, (1-M)-1]

R3[BROKER/CLIENT, 1-M]

To derive the objects of a system, we put the data stores into Boyce/Codd third normal form and then examine each data store to identify the associated object. To put the previous data stores into third normal form, make sure that each nonkey attribute describes the key and not another nonkey attribute. Let's take each data store and analyze it in detail.

Examine the CLIENTS data store to determine if each nonkey attribute only describes the object client. (The client object is identified by the primary key CUSTOM-NAME + CUSTM-ADDR.)

CLIENTS = {CUSTM-NAME + CUSTM-ADDR + SS# + CUSTM-PHONE#
+ CREDIT-RATING + CUSTM-BANK + BANK-ACCT# + TRANS#}

SS# is the social security number of the client. CUSTM-PHONE# is the phone number of the client. CREDIT-RATING is the credit rating of the client provided by the National Credit Bureau. CUSTM-BANK is the name of the bank reference used by the client to establish his or her account with the brokerage firm.

What about the attribute BANK-ACCT#?

Is BANK-ACCT# an attribute of CUSTM-NAME & CUSTM-ADDR? No, the account number alone is not unique; many banks can use the same number.

Is BANK-ACCT# an attribute of CUSTM-BANK? No, it is not a description or fact about the bank itself.

BANK-ACCT# is an attribute of the relationship between a customer and a bank. BANK is a separate object because it has attributes of its own, namely BANK-ADDR and BANK-PHONE#. For the purposes of this system, the brokerage firm only needs one bank reference for a client, but many clients can be serviced by one bank, thus the relationship between CLIENTS and BANKS is a M-1. The fourth relationship is **R4[CLIENT/BANK, M-1]**. BANK-ACCT# is an attribute of R4.

TRANS# is the transaction number of a purchase or sale of stock for the client. Therefore, after examining the CLIENTS data store, we find the following data store to be in third normal form with the respective objects identified.

BANKS = {BANK-NAME + BANK-ADDR + BANK-PHONE#}

object is BANK

CLIENTS = {CUSTM-NAME + CUSTM-ADDR + SS# + CUSTM-PHONE#
+ CREDIT-RATING + TRANS#}

object is CLIENT

Now examine the object data store TRANSACTIONS:

TRANSACTIONS = {TRANS# + SYM + QTY + ["BOUGHT" | "SOLD"] + TRANS-DATE
+ SETTLE-DATE + TRANS-AMT + ["TRANS-COMPLETE"
| "TRANS-OPEN" | "TRANS-CLEARED"] + TRANS-PRICE}

SYM represents the name of the stock bought or sold in the transaction. QTY is the number of shares traded during the transaction. ["BOUGHT" | "SOLD"] is the type of transaction. TRANS-DATE is the date of the transaction. SETTLE-DATE is the deadline date for completing the transaction with the client. TRANS-AMT is the total amount of the commission charged by the brokerage firm for conducting the transaction. ["TRANS-COMPLETE" | "TRANS-OPEN" | "TRANS-CLOSED"] is the status of the transaction. And, finally, TRANS-PRICE is the total amount of the transaction in dollars. All of these attributes describe the primary key and are independent of each other.

Next, we examine the data store RATES:

RATES = {TRANS-PRICE + QTY + COMM-RATE}

What does the key attribute TRANS-PRICE describe? TRANS-PRICE, by definition, is the dollar cost per share of stock for a particular company at a particular date and time. TRANS-PRICE must therefore be an attribute of TRANSACTION. QTY is an attribute of TRANSACTION and indicates the number of shares purchased.

What about the attribute COMM-RATE? COMM-RATE is a function of the TRANS-PRICE and QTY of a purchased stock. The owners of the brokerage firm set the COMM-RATE and it is the same for all trades made by the brokers who work for the firm. COMM-RATE is then an attribute of the object RATES, which has a primary key of TRANS-PRICE + QTY.

RATES = {TRANS-PRICE + QTY + COMM-RATE}

object is RATE

Next examine the SYM-LIST data store:

SYM-LIST = {CO-NAME + CO-ADDR + SYM + EXCHG}

The SYM-LIST data store contains all the information about a company that offers stock for trade at a particular trading exchange (EXCHG). CO-ADDR is the address of the company. SYM is the abbreviation used by the exchange to identify the particular company. EXCHG is the name of the stock exchange that trades stock for the company. Thus, all of the attributes describe the primary key CO-NAME and the object company.

SYM-LIST = {CO-NAME + CO-ADDR + SYM + EXCHG}

object is COMPANY

Finally, let's examine the DIRECTORY data store:

DIRECTORY = {BROKER-NAME + BROKER-OFFICE + BROKER-PHONE#}

Both BROKER-OFFICE and BROKER-PHONE relate to the stockbroker and do not relate to each another. It is assumed that if the broker were to change offices, he or she would keep the same phone number. Otherwise, the phone number would be related to the office and we would have to create another data store with office as the primary key and phone number as the attribute. All of the attributes in the data store are related to the object broker.

DIRECTORY = {BROKER-NAME + BROKER-OFFICE + BROKER-PHONE#}

object is BROKER

After placing the data stores into third normal form, we have to make sure that the preceding set of object data stores is also in Boyce/Codd normal form. To do this, we have to examine each data store and see if there are any determinants that are made up of more than one attribute and that have at least one attribute in common. If so, all such determinants must be valid candidate keys. In the CLIENTS data store SS# (which is not necessarily unique, by the way) + CUSTM-NAME is a determinant of CUSTM-NAME + CUSTM-ADDR. However, the determinant SS# + CUSTM-NAME is also a legitimate candidate key for the data store CLIENTS.

Because all determinants are candidate keys in these data stores, the resulting objects and corresponding object data stores are said to be in Boyce/Codd normal form, and are defined as follows:

BOYCE/CODD NORMAL FORM

BANKS = {BANK-NAME + BANK-ADDR + BANK-PHONE#}

object is BANK

DIRECTORY = {BROKER-NAME + BROKER-OFFICE + BROKER-PHONE#}

object is BROKER

CLIENTS = {CUSTM-NAME + CUSTM-ADDR + SS# + CUSTM-PHONE#}

object is CLIENT

RATES = {TRANS-PRICE + QTY + COMM-RATE}

object is RATE

SYM-LIST = {CO-NAME + CO-ADDR + SYM + EXCHG}

object is COMPANY

TRANSACTIONS = {TRANS# + SYM + QTY + ["BOUGHT" | "SOLD"] + TRANS-DATE + SETTLE-DATE + TRANS-AMT + ["TRANS-COMPLETE" | "TRANS-OPEN" | "TRANS-CLEARED"] + TRANS-PRICE + CUSTM-NAME + CUSTM-ADDR}

object is TRANSACTION

RELATIONSHIPS

R1[CLIENT/TRANSACTION, 1-M]

R2[(CLIENT/TRANSACTION)/BROKER, (1-M)-1]

R3[BROKER/CLIENT, 1-M]

R4[CLIENT/BANK, M-1]

At this point in the analysis, we inspect each of the data stores in Boyce/Codd normal form, and decide what object the primary key identifies such that all of the attributes describe or are facts about that object. Therefore, the objects assigned to each of the object data stores are the following six:

BANK, BROKER, CLIENT, RATE, COMPANY, and a TRANSACTION

E-R Diagram

To construct the resulting E-R Diagram, we have to recall all of the relationships identified as we were normalizing the data stores. So far, we have identified four relationships:

R1[CLIENT/TRANSACTION, 1-M]

R2[(CLIENT/TRANSACTION)/BROKER, (1-M)-1]

R3[BROKER/CLIENT, 1-M]

R4[CLIENT/BANK, M-1]

In examining the object data stores, we find there are other relationships because there are data elements that are common among the object data stores. For example, SYM is common in both TRANSACTION and COMPANY objects. Thus, there is a relationship between the object data stores SYM-LIST and TRANSACTIONS. Because there can be many TRANSACTIONs for each COMPANY, this is an M-1 relationship, **R5[TRANSACTION/COMPANY, M-1]**. Also, consider PRICE and QTY. Both of these data elements are present in the object data stores TRANSACTIONS and RATES. Thus, there is a sixth relationship between the objects TRANSACTION and RATE. Because only one rate applies to a transaction, the relationship between transaction and rate is a 1-1 relationship and is denoted by **R6[TRANSACTION/RATE, 1-1]**.

Figure 7.25 illustrates the six relationships identified in this example:

R1[CLIENT/COMPANY, 1-M]

R2[(CLIENT/BROKER)/TRANSACTION, (1-1)-M]

R3[BROKER/CLIENT, 1-M]

R4[CLIENT/BANK, M-1]

R5[TRANSACTION/COMPANY, M-1]

R6[TRANSACTION/RATE, 1-1]

There could be a seventh relationship, one in which many brokers represent many companies, R7[BROKER/COMPANY, M-M]. However, the relationship R7 is not explicitly mentioned in the system description; nor are there any common data elements between the data stores BROKER and COMPANY to permit us to implicitly define such a relationship. If there is a mini-spec that has access to both data stores, examine the policy statement of the mini-spec to see if such a relationship exists. Although it may exist in reality, for the purposes of this system, the relationship R7[BROKER/COMPANY, M-M] *does not exist* in the requirements for the system being modeled.

E-R DIAGRAM

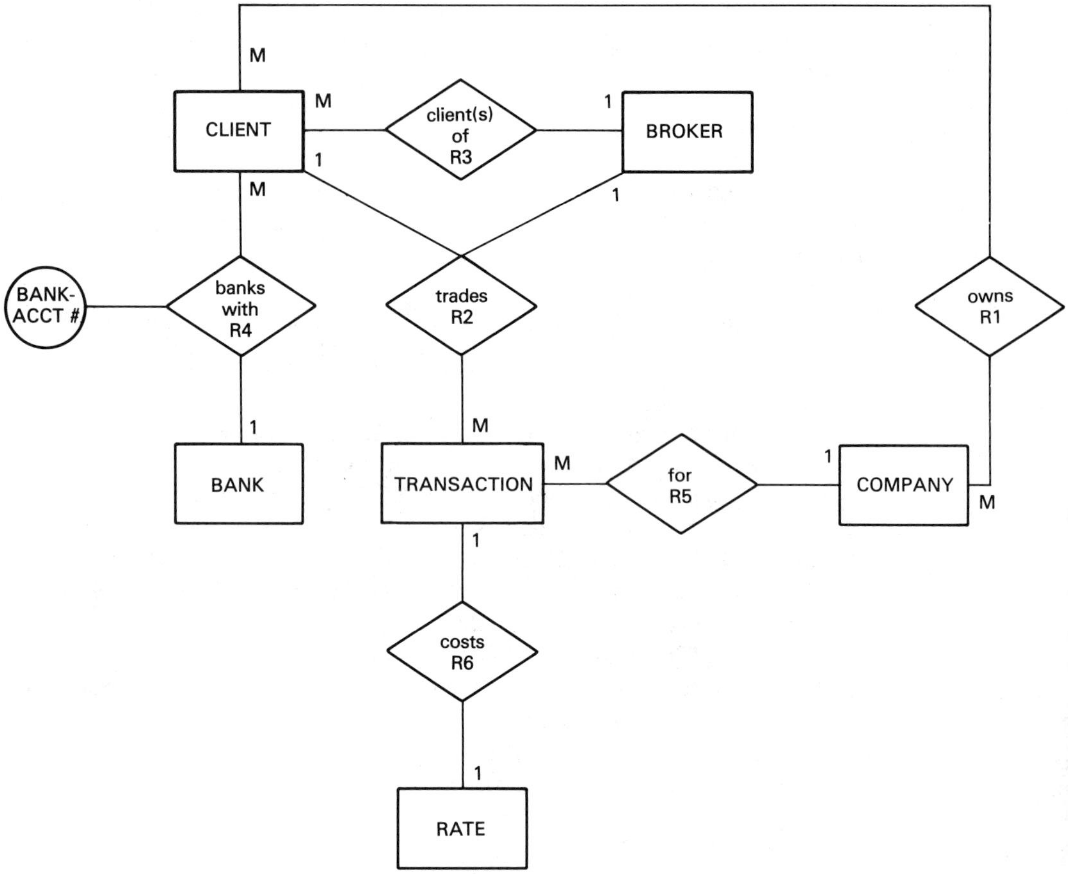

Figure 7.25

SUMMARY
. . . understanding the data in a system

In this chapter, we presented the theory and techniques to analyze the data in existing systems by performing *object analysis*. **OBJECT ANALYSIS can be summarized as the process that reorganizes the essential memories of a system into the objects, attributes, and relationships between the objects and attributes of a system.** There are those analysts that feel object analysis (or the study of data as we outlined in this chapter) should precede process definition. Others place it in the traditional structured analysis phase. Either way, a detailed study of data is essential for good system specifications.

We have separated the processes of a system into those that serve the system and those that serve the users (Chapter 6). In this chapter, we have reorganized the data into objects in order to think about the data requirements of a system instead of how data is to be structured to accomplish the results of the system. With the essential processes and objects, we are now ready to enter *event analysis*, where we will again reorganize the processes and data of a system, but this time into segments, each of which fulfills one of the users' expectations from the system.

QUESTIONS

1. What are the normal forms of data? Give an example of an unnormalized data store.
2. What is an alternate key?
3. What are the steps to derive the objects from the physical data stores in a system model?
4. Define the characteristics of a data store.
5. Why should data be put in normal form?
6. What are the four basic relationships that can occur among entries in one data store and entries in another data store?
7. Define the term candidate key.
8. What is a primary key?
9. What is intersecting data?
10. What is a primary domain?
11. What is an object?
12. a. What is an object subtype?

    b. What is a supertype?
13. What are attributes?
14. Is the primary key of a data store an attribute?
15. Give the definition of foreign key. Why are foreign keys necessary?
16. What are the first and second rules of data store integrity?
17. What is the purpose of Data Store Integrity Rule #1 (Object Integrity)?
18. What is the purpose of Data Store Integrity Rule #2 (Referential Integrity)?

19. Explain how Integrity Rule #1 applies to foreign keys?

20. Provide in relationship notation, R#[data-store-a/data-store-b, n-n], the following relationships:

    a. Entries from object-data-store-A have a one-to-one relationship to entries in object-data-store-B

    b. Entries from object-data-store-C have a one-to-one relationship as an intersecting object of a one-to-many relationship between object-data-store-A and object-data-store-B

    c. Entries from object-data-store-B are dependent on the existence of entries in object-data-store-A and there is a many-to-many relationship between them when it exists

    d. Entries from object-data-store-C have a one-to-many relationship to the one-to-one relationship of entries from object-data-store-A and entries from object-data-store-B

    e. Entries from object-data-store-A have a one-to-many relationship with entries from the same object data store.

PROBLEMS

1. Place the input data form in Fig. 7.26 into third normal form and identify the objects.

2. Construct an E-R diagram illustrating the relationships among the objects found in problem 1.

3. How would the relationships among the objects PROFESSOR, TEXT-BOOK, and COURSE be illustrated using an E-R diagram under the following assumptions?

    a. Assume each professor who taught the same course was required to use the same text book.

    b. Assume each professor who taught the same course was permitted to select any text book he or she wished.

4. a. When a foreign key does not have a null value, the value of the foreign key **must be** from a *primary domain*. (True or False?) If True give an example, if False explain why.

    b. A foreign key in a data store **must** also be a candidate key for that data store (True or False?) If True give an example, if False explain why.

INPUT DATA FORM

ACME WHOLESALERS
123 Main Street
Hometown, USA 99123

FIRM NAME			DATE
STREET ADDRESS			
CITY	STATE		ZIP
TELEPHONE	TYPE OF ACCOUNT (Check ONE): Wholesale ___ Retail ___		

BANK REFERENCES

NAME	BRANCH		PHONE
ADDRESS	CITY	STATE	ZIP
NAME	BRANCH		PHONE
ADDRESS	CITY	STATE	ZIP

OTHER REFERENCES

NAME			PHONE
ADDRESS	CITY	STATE	ZIP
NAME			PHONE
ADDRESS	CITY	STATE	ZIP

OFFICE USE ONLY

ACCT NUMBER	EFFECTIVE DATE	CREDIT CODE Circle ONE: G M P	DISCOUNT CODE Circle ONE: A B C

ACCOUNT: APPROVED ___ DISAPPROVED ___
REASON (if disapproved):______________________________
____

____
MANAGER SIGNATURE

Credit codes are: Good Credit (G), Marginal Credit (M), and Poor Credit - prepayment only (P).
Discount codes are: 50 percent discount (A), 40 percent discount (B), and 33 percent discount (C).

Figure 7.26

8

EVENT ANALYSIS

When we did the logical analysis he current physical model, we identified the essential components (essential fundamental activities, essential custodial activities, and essential memories) of the system. Object analysis focused on the essential memories and reorganized them into data stores of distinguishable objects and their associated attributes. Now we turn our attention to event analysis, to reorganize all of the essential activities found in our logical analysis and all of the object data stores found in our object analysis into *event segments*. **An EVENT SEGMENT is composed of all the processes, data flows, and data stores that are activated when a particular event occurs.** I will define events more formally later in this chapter, but for now let's consider an event simply as an external happening that requires a response from the system. Why do we want to partition systems by events?

Reorganizing the processes and data by events provides a perspective of the system that is closer to the users' view of the system. The users' perspective is important, particularly when we want to understand their needs when they ask for modifications and enhancements to the existing system. A nontechnical view of a system (typically the users' view) is one that considers the system either as a stimulus-response mechanism or as an unsolicited periodic-response mechanism. In both cases, the productivity of the users is typically a function of the ability of the system to respond. Systems respond to either an incoming stimulus or an occurrence of time. There is a third cause of response from a system that pertains mainly to real-time systems discussed later in the chapter. When we organize the components of a system according to the events to which the system responds, both the users and the analyst can effectively discuss the need for change from a common perspective. The complexity of the required change is reduced by the natural partitioning of the system into subsets of activities and data that serve each event to which the system responds (event segments). An additional utility of organizing systems by a stimulus-response event or an occurrence-of-time event is to effectively communicate the value added, purpose, or need for change in the system to the owners of the system. Before we begin analyzing events in more detail, let's summarize what we've done so far in creating a current physical model and developing a current logical view of the system.

PHYSICAL MODEL
. . . the physical model of an existing system

When we model a current system as it physically exists, we begin with the *context diagram*, the highest possible level. We then decompose the system into preferably seven plus or minus two major areas of activity. Repeating the process for each of the major activities, we decompose each major activity into usually seven plus or minus two lower-level functions. This decomposition process continues, creating lower and lower levels of *data flow diagrams*, until we reach the processes known as functional primitives. For each functional primitive, we create a *mini-specification* that specifies the policy used to transform the input data into the output data. At least one additional document is necessary, the *data dictionary*. It contains the definitions of all the data flows, data stores, and common processing routines or functions used in the system. If the physical model uses a data base, or if there is some other type of relationship that exists among the entries in one data store with those in another, we prepare an *E-R (entity-relationship) diagram*. With these modeling tools, we can specify the following:

The scope, bounds, and objectives of the system

The sources and destinations of all data

The path that data flows through the system

The detailed explanations of all data transformations used by the processes

The composition of all data and data stores

The relationship between the data stores (if any)

Figures 8.1a and 8.1b illustrate all of the documentation for a structured specification used to define a current physical model of an existing business system. From the current physical structured specification, we can come to a better understanding of how the current system works. We then wish to examine the system from a more abstract or logical view.

Logically, we want to know three things: (1) What processes and data are associated with the technology used to implement the system (infrastructure), and which processes and data are associated with the essential functions of the system (essence)? (2) What are the events to which the system responds? (3) What are the "things" (objects) about which the system stores data? The essence, events, and objects are the components of a system that are necessary and independent from the implementation of the system.

STRUCTURED SPECIFICATION
(Current Physical Model)

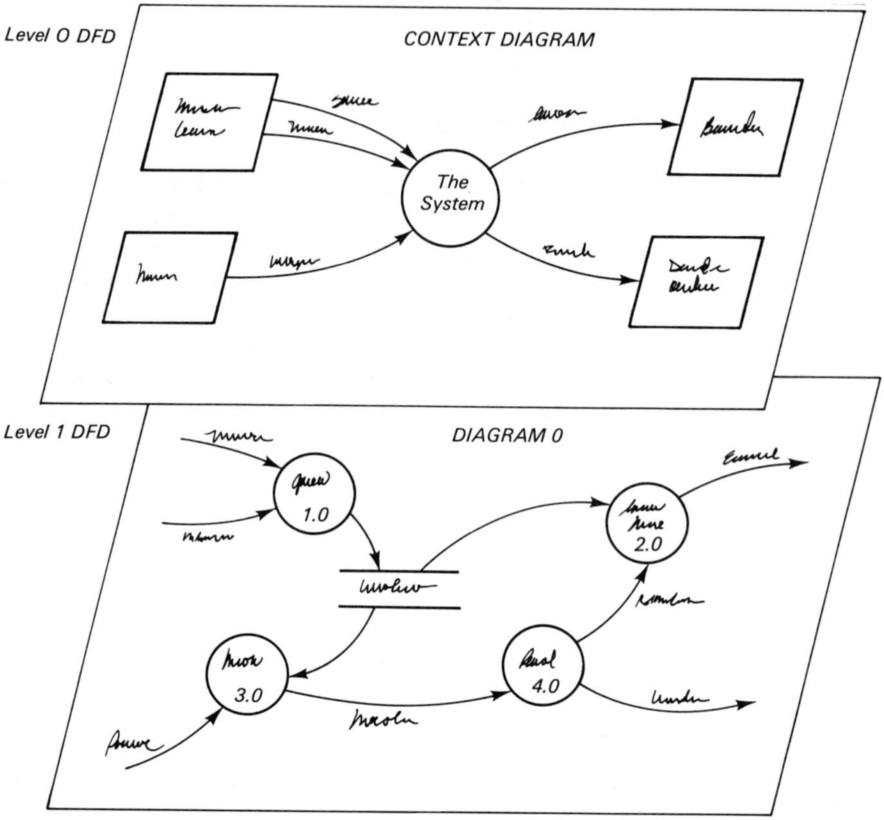

Figure 8.1a

Level 2 DFD:

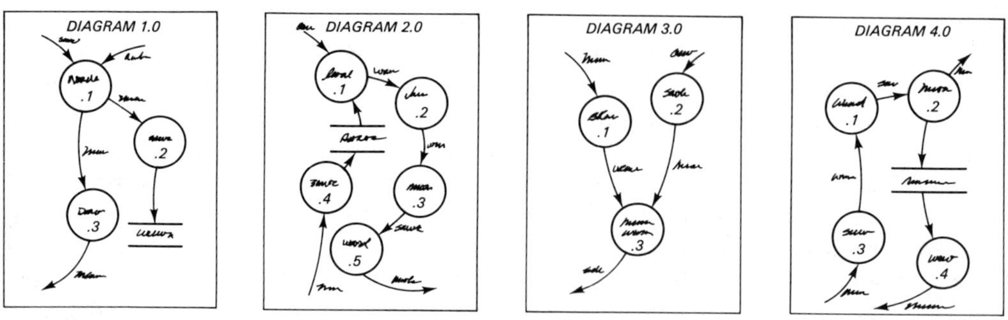

Figure 8.1b

Level 3 DFD:

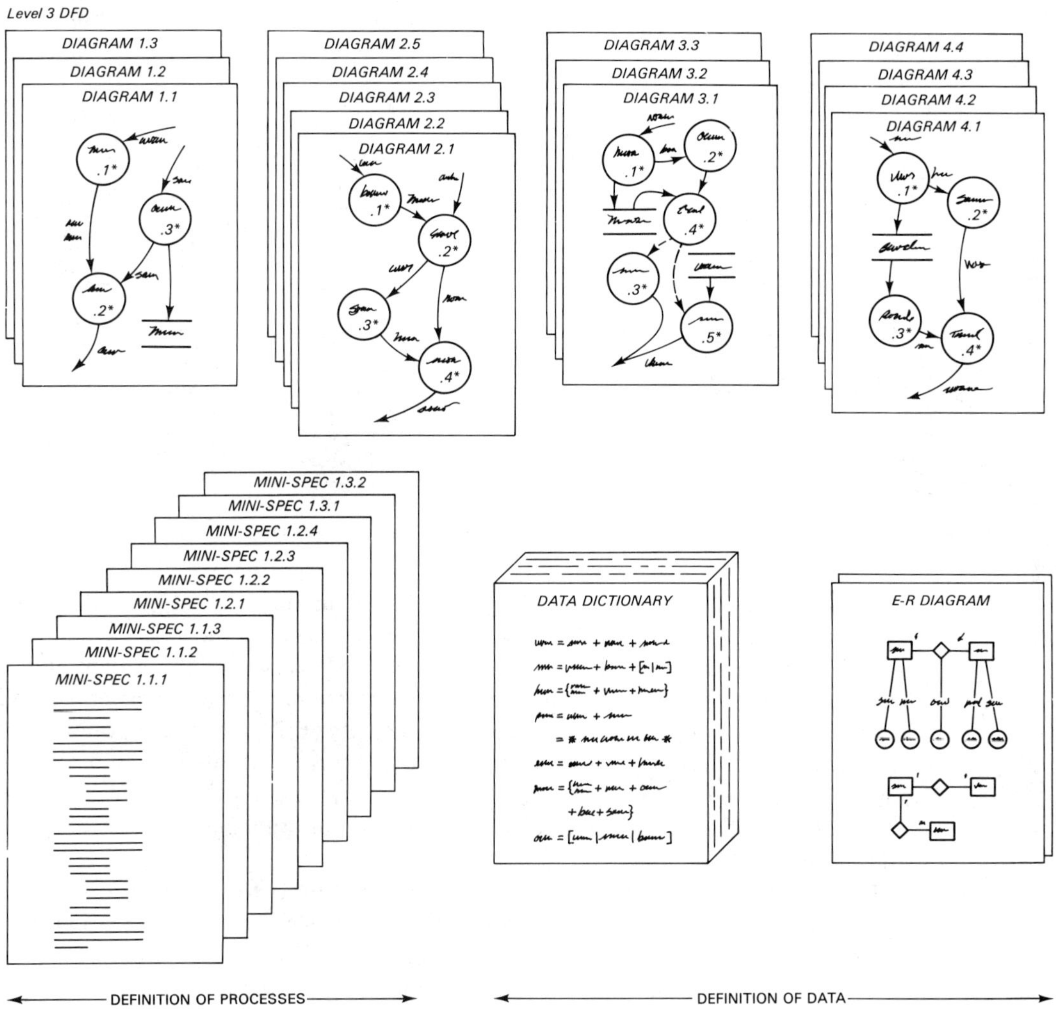

Figure 8.1b (cont.)

LOGICAL MODEL
. . . the logical model of an existing system

When we create a current logical model, we analyze a system in a manner similar to the way we created the current physical model but in reverse. We start with the *expanded data flow diagram.* **The EXPANDED DATA FLOW DIAGRAM (DFD) consists of all of the functional primitives linked with the normal data flows shown**

EXPANDED DATA FLOW DIAGRAM

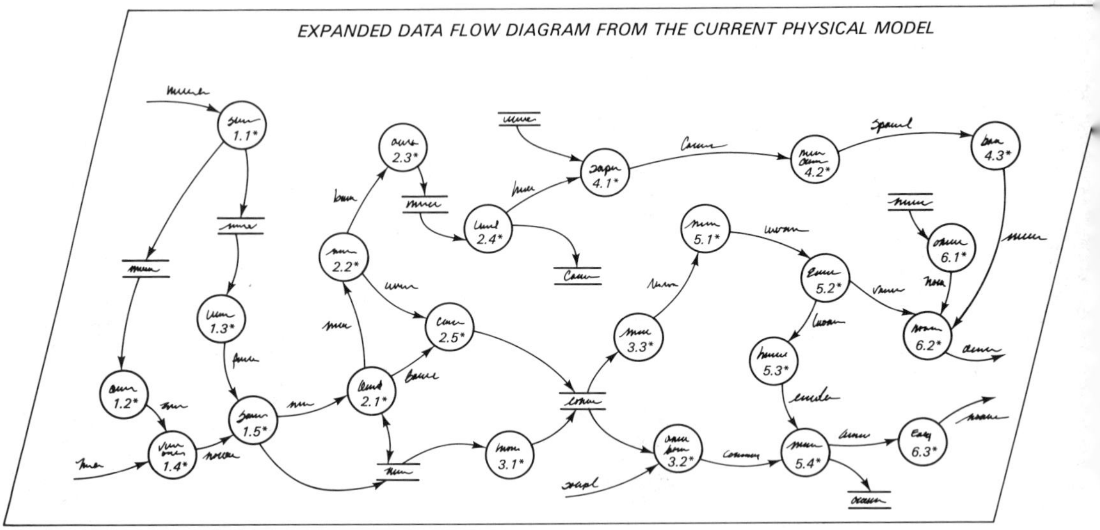

Figure 8.2

on the various levels of data flow diagrams found in the current physical model. The expanded DFD can be a rather large diagram, as shown in Fig. 8.2.

From the lowest-level data flow diagrams in the current physical model, we build an expanded DFD and classify each of the functional primitives as an essential fundamental activity (EFA), essential custodial activity (ECA), or a component of the infrastructure. The processes of the infrastructure consist of internal administrative activities (IAA), and the internal communications activities (ICA). Likewise, each data store is classified as an essential memory (EM), internal administrative memory (IAM), or internal communications memory (ICM).

By classifying all of the processes and memories on the expanded DFD, we can easily identify the infrastructure and ultimately remove it from the model. The result is the essence of the system. When the infrastructure (ICA, IAA, ICM, and IAM) is separated from the expanded DFD, the remaining elements are known as the *essential fragments* (namely, EFAs, ECAs, and EMs). They comprise the lowest level of the current logical data flow diagram. Figure 8.3 illustrates the removal of the infrastructure found after we have completed the logical analysis on Fig. 8.2.

The original mini-specs for each of the corresponding essential activities still apply. The difference in the logical model after the infrastructure is removed is that the essential system uses "*perfect*" technology to transport data in, around, and out of the system. The remaining activities perform the necessary data transformations to accomplish the objectives of the system. Everywhere the infrastructure is removed, data flows must be extrapolated to reconnect the essential activities and the essential memories. The resulting diagram represents the business functions that *must be performed* regardless of the type of technology used in the system.

CURRENT LOGICAL DATA FLOW DIAGRAM

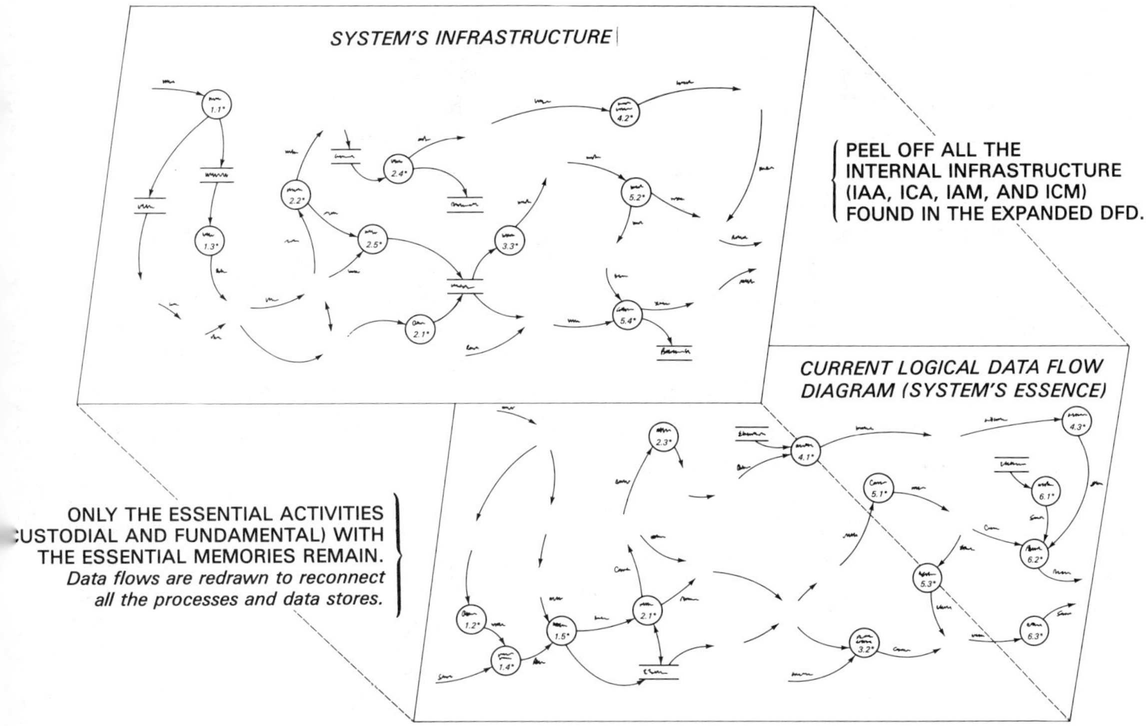

Figure 8.3

REMOVING SUPERFLUOUS DATA
. . . starve the bubbles

Now that we have limited the system to a relatively small set of activities and memories, we have to look at the data that flows between these remaining components. From the mini-specs of each essential fundamental activity (EFA) and each essential custodial activity (ECA), we can determine the minimal amount of data that must flow through the system. In cases where the input data flow is a group element containing more than the minimal set of *required data*, replace it with only the names of the elementary or group data elements that are absolutely necessary for each activity. This action is known as *starving the bubbles*, and it eliminates the data that are passed to an activity either because of convenience due to the technology used, or by coincidence. For example, when a customer orders merchandise, the form can be filled out using a triple-copy carbon form. The three copies go off in three different directions. The shipping dock gets a copy to be sure that the shipment contains what the customer ordered. The shipping dock also needs the SHIP-TO-ADDRESS. The shipping dock, however, receives all of the other information contained on the form, such as BILL-TO-ADDRESS, PERCENT-OF-DISCOUNT, etc. This data reaches the shipping dock only because of the technology of carbon copy forms. If different technology were used, such as on-line terminals, data such as the billing information would not be transmitted to the shipping dock. Therefore, in our logical view of the current system, we wish to eliminate all superfluous data elements and get down to the essentials.

When we starve the bubbles, we also need to identify what data elements are absolutely necessary for each physical data store. Once we complete the action of minimizing the flow of data elements, we have two remaining steps to complete the current logical model:

1. Replace the physical data stores with normalized object data stores.
2. Regroup the essential activities and object data stores by the events that invoke responses from the system.

Chapter 7 addressed the normalization process that places the essential memories into Boyce/Codd normal form. After normalization we can identify the objects of the system and if there are any necessary relationships among the objects, we construct an E-R diagram. The remainder of this chapter will discuss *events*.

EVENTS
. . . a system is built to respond to events

The purpose of a system is to provide the necessary response whenever an input stimulus, a time event, or an anomalous condition is recognized by the system. To provide the necessary response, many activities and data memories of the system are called into service. We want to associate those activities and data memories and identify them as processing segments (event segments) of the external event that causes data to enter the system, the time event that occurs, or the anomalous condition requiring special processing.

The first level *logical view* of a system, as the users view the system, is a logical decomposition of the context diagram with a bubble for each event to which the system responds. The graphical illustration of this level 1 logical view is a current, logical data flow diagram known as the *event diagram*. **The EVENT DIAGRAM is the highest-level, logical data flow diagram illustrating the association of the object data stores to each of the events to which the system has a planned response.** In the event diagram, we have regrouped the essential fragments into high-level "parent" nodes. The criteria used to group the essential fragments will be the *events* to which the activities and memories respond. **An EVENT is an independent occurrence outside**

the system, an occurrence of time, or (in real-time systems) an occurrence detected inside the system that causes activity in, or results from, the system. (In the case of real-time systems, see Chapter 10, anomalous conditions are unusual conditions that jeopardize the continued operation of the system.) To regroup processes and data around events, we must analyze all of the inputs and outputs of the system and determine which processes and data memories are activated by which events. It is permissible for a given process (functional primitive) to be in more than one event.

In preparing to create a level 1 event diagram, we must determine the data needed to service each event. Because the data is now organized in object data stores, we can associate each event to the appropriate object data stores that contain the data needed to produce the desired response. A resulting level 1 event diagram usually takes the form shown in Fig. 8.4.

TYPICAL FORM OF AN EVENT DIAGRAM

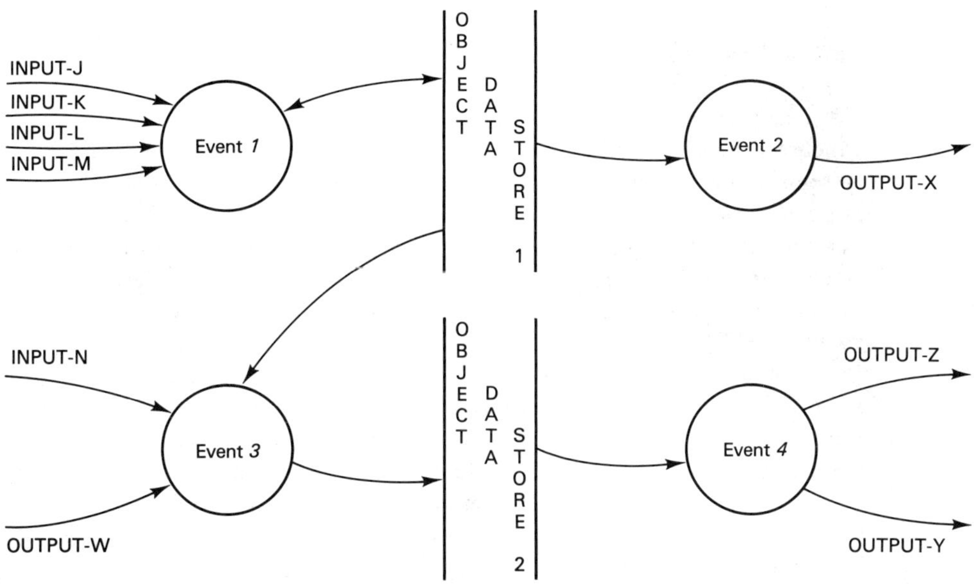

Figure 8.4

In event diagrams, it is convenient to position the object data stores vertically down the center of the diagram. Also, notice that data flows do not exist between event bubbles. If data did flow directly from an event 1 to an activity in another event, say event 2, then some activity in event 2 would be called into service when event 1 occurs. Because a process can exist in more than one event segment, the activity used in event 2 would also be part of event 1, and thus no data flow is necessary between event 1 and event 2. By definition, *all* activities needed to completely service an event are associated in the event bubble on the event diagram. Thus, a *functional primitive process can be part of several event bubbles* if the particular process is used when different events occur. The functional primitive process is identified by the process number and mini-spec from the physical structured specification. The event diagram represents the whole system and is in balance with the context diagram. The event diagram includes the following:

All of the input and output data flows

An event bubble for each event for which the system provides a response

All of the object data stores

The data flows linking the necessary object data stores to the event bubble

We must *balance* the level-1 event diagram with the original context diagram, and with any lower-level data flow diagrams that illustrate in detail the processes, data flows, and data stores used to provide the necessary response(s) for a particular event.

External Events

Now, let's consider the events themselves. There are three types of events, all of which will be discussed in this chapter. The third type, however, applies to real-time systems and will also be discussed in Chapter 10 as well. The first type of event is the *external event*. **EXTERNAL EVENTS are independent occurrences that take place outside the system and cause activity to take place inside the system when a stimulus is presented to the system.** Only when a source presents a stimulus to the system does the system know the external event has occurred and attempts to respond to the event. **A STIMULUS is an input data flow from a source and is the consequence of the occurrence of an external event.** The arrival of a stimulus to the system is an indication that an event has occurred and the system then initiates a planned activity that produces the expected response. **A RESPONSE is an output data flow from the system to a sink, or some internal activity in the system, and is the consequence of the system recognizing that an event occurred.** The relationships among an external event, a source, a stimulus, and a response is that *an event motivates or causes the source to send the stimulus to the system. The system, upon receiving the stimulus, realizes that the event has occurred and activates the process(es) necessary to produce the response that is sent to the sink.* Remember, the external event is the *happening* in the universe that *motivates or causes* the source to present stimulus to the system. The stimulus is always presented to the system when an external event takes place, whereas, an external response by the system is optional.

External events are the most common type in a system. Users most often treat or interact with a system as a stimulus/response mechanism. The response the system provides may be immediate or may come days later depending on the nature of the system. Either way, if a system is dependent on the arrival of an input stimulus to recognize that it must initiate some action or provide some response, the system is being driven by an external event.

Examples of External Events

1. Person decides to open an account.
2. Person decides to make a purchase.
3. Customer needs to withdraw money from his or her account.
4. Person needs information on the cost of a product.
5. Employee expends money on a business trip.

Each of the above external events are outside the boundaries of the system and are independent of the system. It is important to remember that the external event is the motivating force that causes the source(s) to present one or more stimuli to the system. The event is *not* the action of the source presenting the stimulus to the system. This may seem to be an unimportant distinction, however, it is critical for two reasons: when a single event is responsible for causing two or more sources to send their stimuli into the system indicating that a single event occurred as shown later in Fig. 8.7d, and when the system has multiple ways of performing the same service depending on the type of stimulus received. An example of the second reason is when a person needs a loan and uses either stock, or some other asset such as a house or boat as collateral. Upon receiving the input data flow the system performs different activities, but produces in the same response.

Temporal Events

A second type of event is a *temporal event*. **A TEMPORAL EVENT is an occurrence of time that causes activity within the system and usually results in a system response.** In the case of a temporal event, the system does not receive an input stimulus from outside the system. The system must keep track of the day and time, and, at the appointed moment, activate the necessary process(es) that produce the response. On rare occasions, a temporal event can initiate activity in a system that does not result in an external output sent to a sink. Nevertheless, the activity within the system does accomplish the intended task at the appointed time.

Examples of Temporal Events

1. END OF THE DAY, when activities summarize the day's transactions.
2. END OF THE DAY, when a process gathers all currency from a data store and sends a deposit to the bank.
3. BEGINNING OF THE DAY, when a process interrogates a master file to see if any customer's subscription has expired.

4. BEGINNING OF THE DAY, when a process examines all accounts and generates an overdue notice for accounts having outstanding amounts of more than 30 days.
5. END OF THE WEEK, when a process generates a weekly status report.
6. NOON, when processes collect and post receipts in the account ledger.
7. AT 3:00 P.M., when a process delivers all outgoing mail to the U.S. Postal Service.

For a temporal event to exist, the system must be responsible for keeping track of time. A temporal event occurs upon the completion of the passing of some interval of time. The system may or may not provide an external response to a temporal event. A temporal event could cause a system to just move data from one data store to another with no direct output leaving the system. An example would be a system that determines the cost of telephone calls made from a hotel room. If the hotel has the following policy:

1. Charge the customer the customary time and distance (TD) charges using the standard time and distance table (STD-TD-TABLE), plus a fixed surcharge for all calls initiated between 8:00 A.M. and 5:00 P.M.
2. Apply a discount of 15 percent to all charges computed as specified for calls initiated between 5:00 P.M. and 8:00 P.M.
3. Because the late evening is the busiest calling period in a hotel, compute the time and distance charges from a specially modified TD table (PREMIUM-TD-TABLE) for all calls made between 8:00 P.M. and midnight.
4. Since the hotel's telephone system is used least in the very late hours, compute the customer's charges using a special nighttime TD table (NIGHT-TD-TABLE) and a reduced surcharge for all calls made between midnight and 8:00 A.M.

Figure 8.5 shows a data flow diagram for the portion of the system that determines the customer's telephone charges. In this example, a temporal event occurs at 8:00 A.M., 5:00 P.M., 8:00 P.M. and midnight that causes activity in the system with no input or output data flows. These activities and events are difficult to identify. The analyst needs to take great care to capture all *hidden events* of this kind.

Event classification can be deceiving. There are cases where the response from a system appears to be activated by an *external event*, when in fact it was a *temporal event* that caused the system to recognize that a response was necessary. For example, "a customer failing to make an installment payment" has the appearance of an external event. However, there is no input stimulus by which the system can recognize that the event occurred. Instead, there is some routine activated on some defined time interval that verifies that the system has received all of the necessary

TEMPORAL EVENT WITH NO RESULTING OUTPUT DATA FLOW

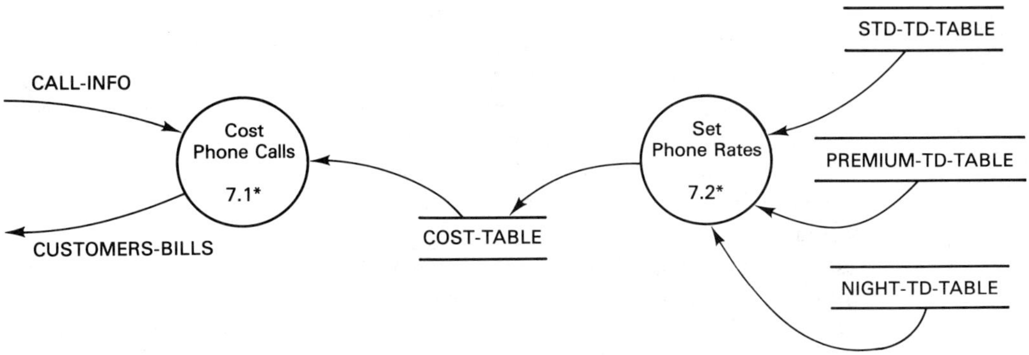

Figure 8.5

input payments. In such cases, the actual event to which the system responds is a temporal event.

Furthermore, one must be careful not to confuse *temporal events* with *external events* that are triggered by the arrival of a stimulus. Consider a payroll system that issues payroll checks at the end of the week.

> An EXTERNAL EVENT would exist and be responsible for the system's output if upon the arrival of all the timecards into the system, the payroll processes activate and produce the employees' checks.
>
> A TEMPORAL EVENT would exist and be responsible for the system's output if the system reads in timecard information continuously or at some regular interval from an on-line time clock, and, at an appointed time, processes the collected data and produces the payroll checks.

The distinction between external and temporal events is determined by whether an input data flow initiates the activities in the system (*external event*) or whether the system must keep track of time via some internal calendar or clock (*temporal event*).

Internal (Anomalous) Events

The third type of event is an *internal* or *anomalous event.* **An INTERNAL EVENT (or ANOMALOUS EVENT) is an occurrence of an anomaly that is recognized by the system, and that requires action to preserve or restore the system to a normal condition.** Such events occur in real-time systems or in operating systems, and the system responds to them to preserve normal processing activities.

Example

A paging operating system is a multiuser environment where the operating system can start more tasks than there are resources available to handle the work. **PAGING is a memory management technique in which the memory address space is split into fixed-length blocks called "page frames" and the software is also split into the same-size blocks called "pages."**[T-1] During execution, the operating system continuously loads and reloads into main memory only those pages being referenced for one of the active jobs. As a result, a situation can occur whereby the operating system is doing more work in setting up and allocating the system resources than it is doing productive work by executing the application software. Such a condition is known as *thrashing*. When thrashing is recognized by the system, several processes are activated that literally deactivate or suspend the execution of lower-priority tasks until thrashing subsides. In this example, no input or output data flows enter or leave the system. The system maintains sanity and ultimately restarts the tasks that were temporarily suspended from executing. The thrashing event is an anomalous event whereby the operating system has a planned set of activities waiting to be called into service. Those activities may or may not produce any direct output data flow to a sink. Other systems that encounter similar situations are usually real-time systems.

Real-time systems often have self-monitoring or auditing functions that run for the specific purpose of detecting anomalies. The processes involved in such activities may or may not necessarily be associated with any external or temporal event. However, it is an anomalous event that causes activities within the system. The processes and data involved in *internal events service the system rather than service the user*. In real-time systems, as you will see in Chapter 10, it is helpful to be able to associate such processes and data to specific types of events. This is most helpful when specifying requirements for new real-time systems.

IDENTIFYING EVENTS
. . . what motivates a source to seek a response from the system?

Many students have difficulty identifying the events to which a system responds. In identifying external events that drive the system, we should *neither* identify the stimulus *nor* the source as the event. Nor is the source's presentation of the stimulus to the system the external event. Rather, an external event is some *happening* in the universe that motivates a source to present a stimulus to the system. The illustration of Fig. 8.6 may better show the relationships among events, sources, stimuli as well as the resulting responses and sinks. Figures 8.7a to 8.7f give some specific example.

RELATIONSHIP OF EVENT, SOURCE, STIMULUS, SYSTEM, RESPONSE, AND SINK

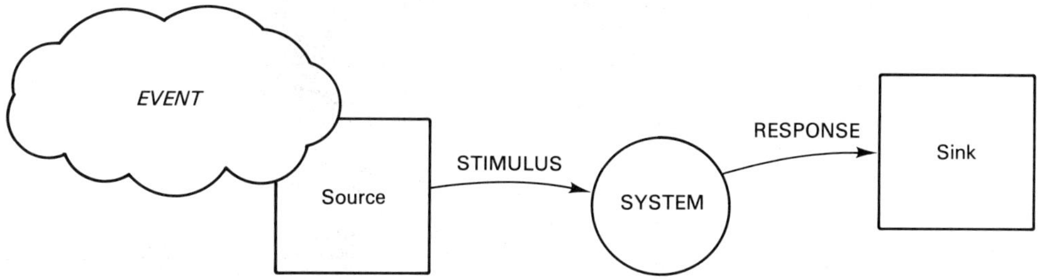

Figure 8.6

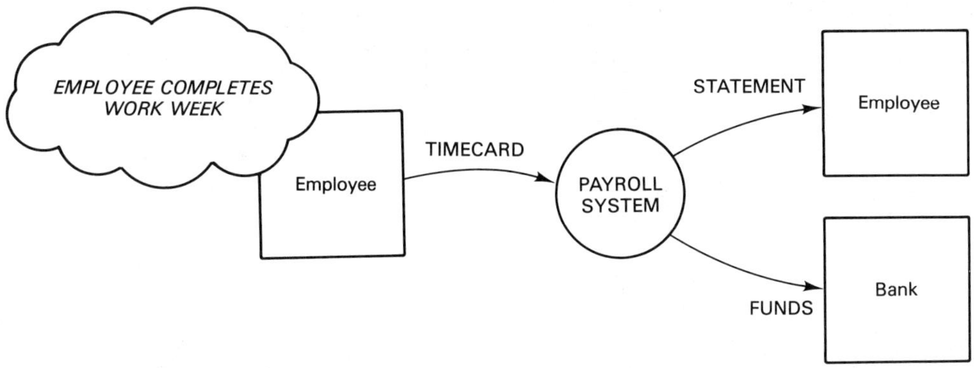

An external event motivates a source to present one stimulus to a system, causing several output data flows as responses.

Figure 8.7a

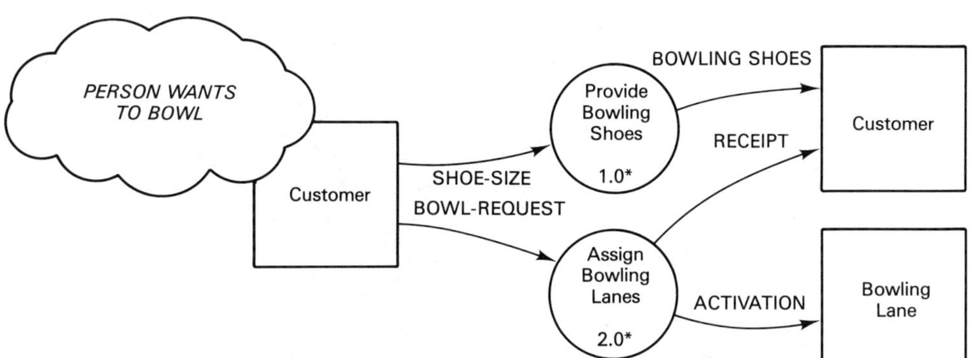

An external event motivates a source to present two stimuli to the system, causing several activities to provide various responses.

Figure 8.7b

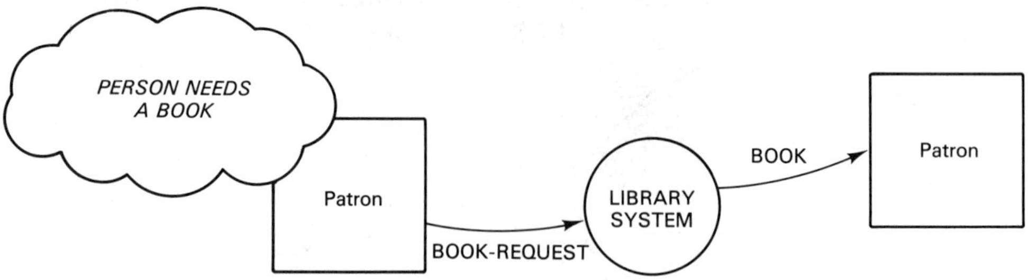

An external event motivates a source to present a stimulus to the system, causing the system to respond with an expected result.

Figure 8.7c

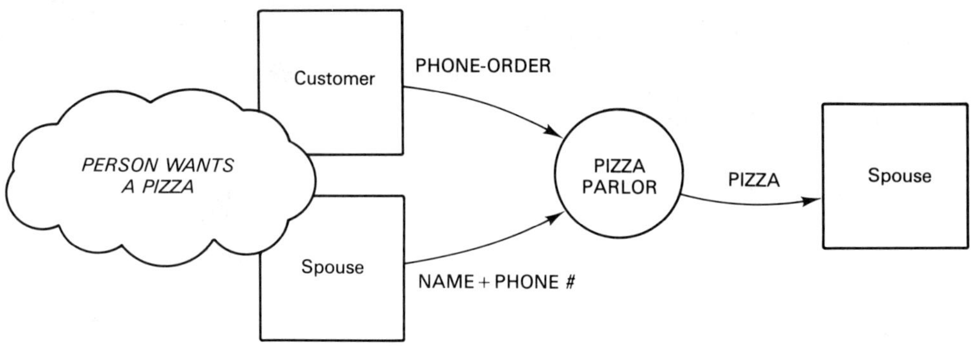

A single external event motivates two separate sources to present two stimuli to the system at different times, causing the system to respond with an expected result.

Figure 8.7d

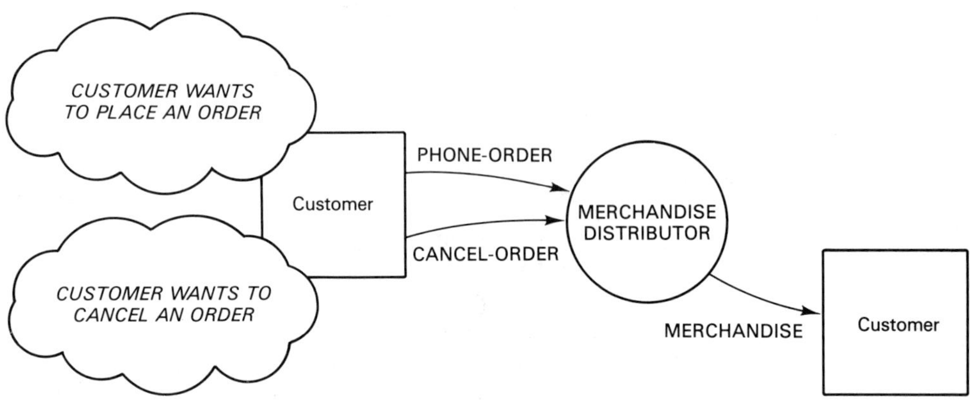

Two external events motivate the same source to present two different stimuli to the system at different times, causing the system to respond or take a planned action.

Figure 8.7e

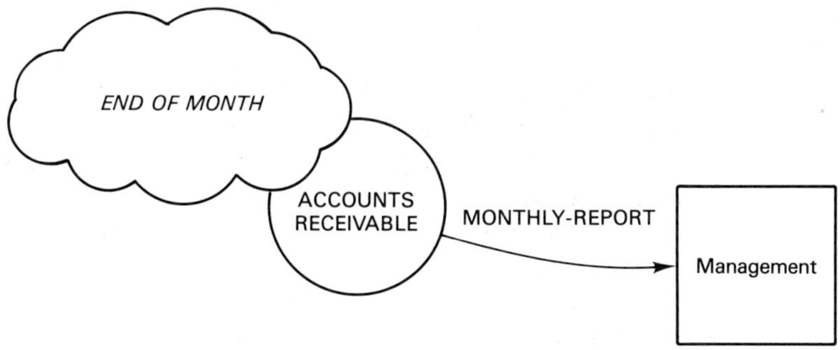

A time event occurs, causing activity within the system to respond with an expected result.

Figure 8.7f

In Figs. 8.7a – 8.7e, the receiving of a BOOK-REQUEST, BOWL-REQUEST, TIMECARD, PHONE-ORDER, or a CANCEL-ORDER is *not* the event. The event is an independent *happening* outside the system that causes or motivates the source to present the stimulus to the system. In some cases, we can have a single event that causes two or more sources to present several stimuli to the system, as shown in Fig. 8.7d. In Fig. 8.7f, the system receives no input stimulus; the system responds when the appointed time occurs.

Examples of statements that are **not** the external events are *the employee submits a time card,* or *the customer asks to bowl,* or *the patron requests a book.* The event must be the immediate cause or motivating factor of the SOURCE to present the STIMULUS to the system, such as *employee completes work week* (Fig. 8.7a), or *person wants to bowl* (Fig. 8.7b), *student needs a book* (Fig. 8.7c) or *a person wants a pizza* (Fig. 8.7d). We could carry the cause and effect relationship back too far. An example would be if we extrapolated back and proposed the event *the teacher assigned a term paper* for Fig. 8.7c. Although we can argue that this may be the ultimate reason why the student needed the book and presented a BOOK-REQUEST to the system, it is probably carrying things back "a little too far" for the purposes of reorganizing activities and data stores of a system.

SCENARIOS
. . . looking beyond events

Once we have identified the basic events to which the system responds let's focus our attention on each event, one at a time. We may find that there are several ways that the system recognizes that an external or internal event occurred. Let's consider an example whereby the system responds to a single event, but detects it through two different stimuli.

Example

A message system takes messages for clients and records them in a computer system. The clients can obtain their messages in two ways: first, the clients can call the message center and have their messages read to them, or second, the clients can dial the computer system with their personal computers and retrieve their messages by issuing a retrieval query. In both cases, the system responds to the same event, namely, *the clients wishes to receive messages.*

In this example, there are two other events. The first is the external event of *a caller wishes to leave a message.* The second is a temporal event, *time to purge unretrieved messages* from the system after some extended period of time (defined by the user) has elapsed. Figure 8.8 is a context diagram illustrating the three events.

From an object analysis of the data in the system, we find that there are only two objects involved in this system: client and message. The event diagram of the system shows three events and two object data stores, CLIENT and MSG (message), as shown in Fig. 8.9.

For event 2, Client Wants Messages, the system is given the CLIENT-NAME either from the attendant or by a computer terminal and responds by displaying each

MESSAGE SYSTEM CONTEXT DIAGRAM

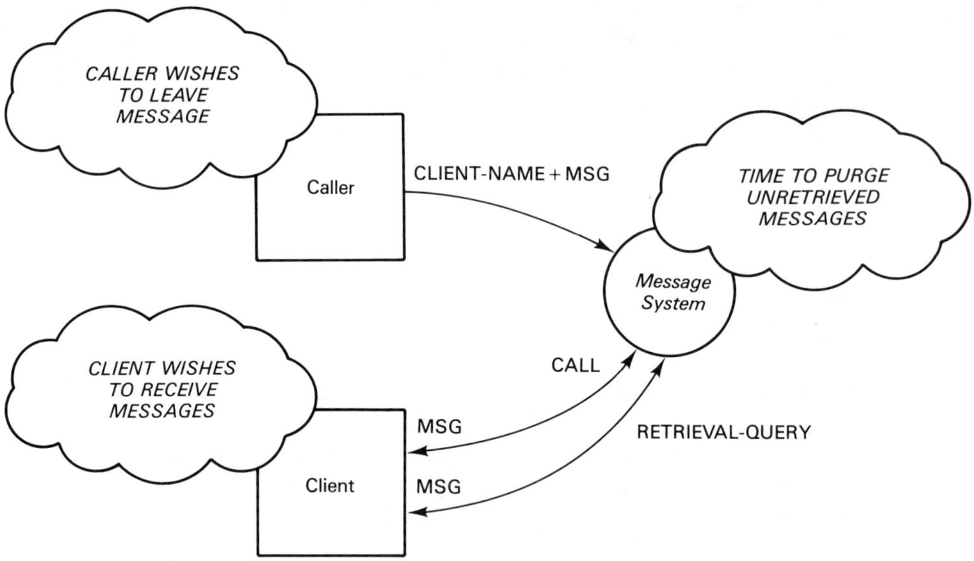

Figure 8.8

message (MSG) on the screen, or obtains a RETRIEVAL-QUERY from the client by means of a personal computer to which the system responds by establishing a session and sending the client his or her messages (MSG). When there are multiple and distinct ways for the system to recognize a single event, we want to associate the processes and data necessary to respond to each distinct *scenario*. **A SCENARIO is a set of input and output data (or control) flows, and behaviors of a system brought about by an instance of an event.** Each scenario can be labeled .1, .2, .3, etc., which is concatenated onto the event number. Each group of input and output data flows on the event diagram that pertains to a specific scenario is numbered according to the scenario to which it applies. The event diagram is balanced with the original context diagram for the system. Event partitioning and scenarios within events allow us to partition the system in order to discuss the users' expectations of the system in very specific circumstances.

MESSAGE SYSTEM EVENT DIAGRAM

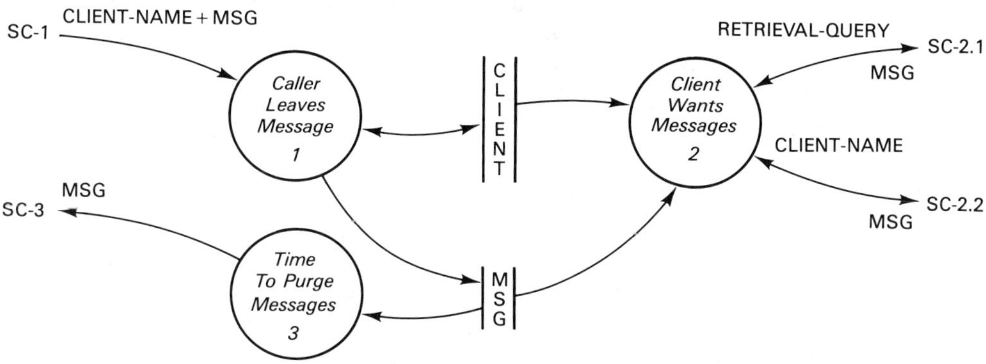

EVENT 1: Caller Leaves Message (EXTERNAL)
Scenario 1: Person Calls Into The Message System To Leave A Message (MSG) For A Client

EVENT 2: Client Wants Messages (EXTERNAL)
Scenario 2.1: Client Calls Into The Message System With A Dial-Up Terminal And Asks To Have His Messages
Scenario 2.2: Client Calls Into The Message System By Phoen And Asks To Have His Messages

EVENT 3: Time To Purge Messages (TEMPORAL)
Scenario 3: At The End Of The Work Week, The System Prints All Messages That Are More Than A Week Old

Figure 8.9

In performing event analysis, we are reorganizing the system's behaviors and data according to the particular scenarios for each event. In this manner, the analyst and other system developers can visualize the system and its behaviors and data *from the user's perspective*. Because the first-level data flow diagram is a decomposition of the context diagram for a system (see Fig. 8.10a), the event diagram is another way of organizing the processes and data of the system (see Fig. 8.10b).

"PHYSICAL" FUNCTIONAL DECOMPOSITION

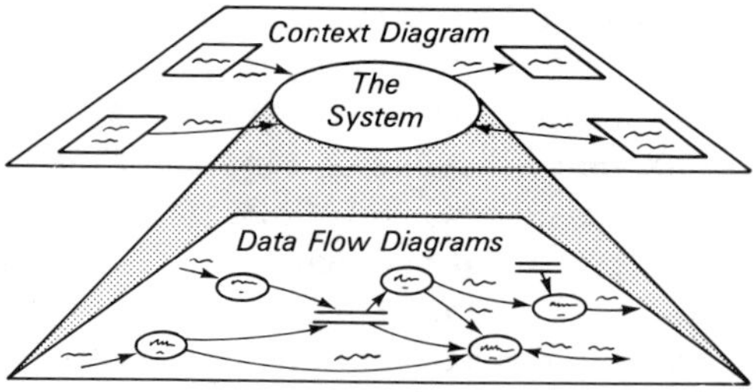

"How It's Done"

Figure 8.10a

"LOGICAL" EVENT DECOMPOSITION

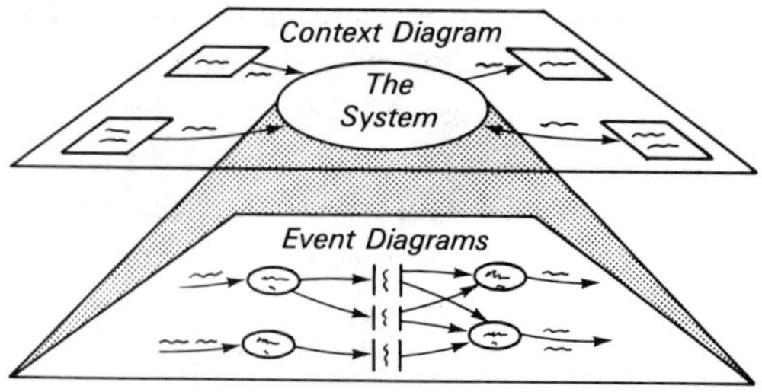

"What Is Done"

Figure 8.10b

As we continue to analyze each scenario, we find that it can be decomposed into *behaviors*. **A BEHAVIOR is a set of data flows, control flows, and processes that performs a specific function or accomplishes a result.** We can think of behaviors as small subsystems within a system. Often behaviors can be *reusable* subroutines that perform specific tasks. Consider the two scenarios in the previous example for event 2, *client wishes to receive messages.*

Scenario 2.1 (*Client retrieves messages by personal computer*):

The set of behaviors that are required to accomplish the scenario may be the following:

1. Establish communications with the client (sign-on).
2. Screen the client for permission to access the system (Security).
3. Retrieve the client's messages.
4. Format and display the client's messages.
5. Purge the client's messages from the data base.
6. Terminate communications with the client (sign-off).

All of the above behaviors service the single scenario for the event *client wishes to receive messages*, whereby the client accesses the message system by means of a personal computer.

Scenario 2.2 (*Attendant retrieves messages for a client*):

The second scenario pertains to the attendant retrieving the client's messages in order to read them to the client who has called in for their messages. Notice that in the set of behaviors for scenario 2.1, some do not apply when the attendant accesses the system as opposed to when the client accesses the system directly.

The set of behaviors that are required to accomplish the scenario may be the following:

1. Retrieve the client's messages.
2. Format and display the client's messages.
3. Purge the client's messages from the data base.
4. Save the client's messages as "old."

Notice in scenario 2.2 that after an attendant retrieves a client's message, behavior 4 stores the message as "old". An "old" message is one that was retrieved by an attendant, but the client wishes to save the message for future access. This feature of the system relates to a separate behavior that can be activated during event 2, *client wishes to receive messages*.

Some analysts may find it useful to decompose event bubbles into behavior bubbles. Behavior bubbles can be further decomposed into process bubbles in keeping with the traditional data flow diagram method. Knowing the events of a system, the number of scenarios the system supports, and the number of behaviors the system performs, we can provide a means of quantifying the work done by a system. We can derive a useful metric that correlates events, scenarios, and behaviors to the amount of development effort very early in the development life cycle. Furthermore, when systems are defined by events, scenarios, and behaviors, the structured specification does not prejudice the developers into any particular high-level design; it merely states the requirements from the users' perspective.

To specify the system requirements from the users' perspectives, the following documentation can be used. First, the *context diagram* defines the boundaries of the system and is the same diagram used in the structured specification, as discussed in the previous chapters. The level-1 decomposition is the *event diagram* that shows the system partitioned into the events (and scenarios) to which the system responds. Each scenario can be further factored into the behaviors (on level 2) that are activated during the scenario and are shown on a scenario data flow diagram.

For real-time systems, it may be noted that the context diagram and event diagram represent data flows only. These diagrams do not indicate any relationship to time. The scenario diagram, however, illustrates the data flows between the behaviors needed to accomplish the appropriate response in a given situation. The scenario can have associated documentation, including a time-dependency chart that specifies the time constraints and requirements for a given scenario.

Each scenario is numbered according to the event to which it is associated. However, behaviors within scenarios are *not* numbered. A behavior can occur in several scenarios and is only given a descriptive name. If we wish to provide more specifications for a complex behavior, the behavior can be viewed as a system in and of itself, and decomposed in the traditional manner of a structured specification, using control diagrams if required, mini-specs, etc., as shown in Fig. 8.11.

The tools and documentation of event models provide system requirements as the users' expectations of system behavior. Users expect the system to respond in a precise manner when presented a stimulus; to respond in an expected manner when a particular time occurs; or in the case of real-time systems, to respond in the appropriate way to maintain the system's integrity and reliability. By specifying the behaviors the system will exhibit in each scenario, there is little room for false expectations or misunderstanding.

There is a fine line between specifying requirements and providing a high-level design. Events, scenarios, and behaviors are intended to specify the users' expectations. The traditional structured specification can be applied to complex business behaviors. However, remember that the documentation is intended to specify *requirements* and not be a high-level design.

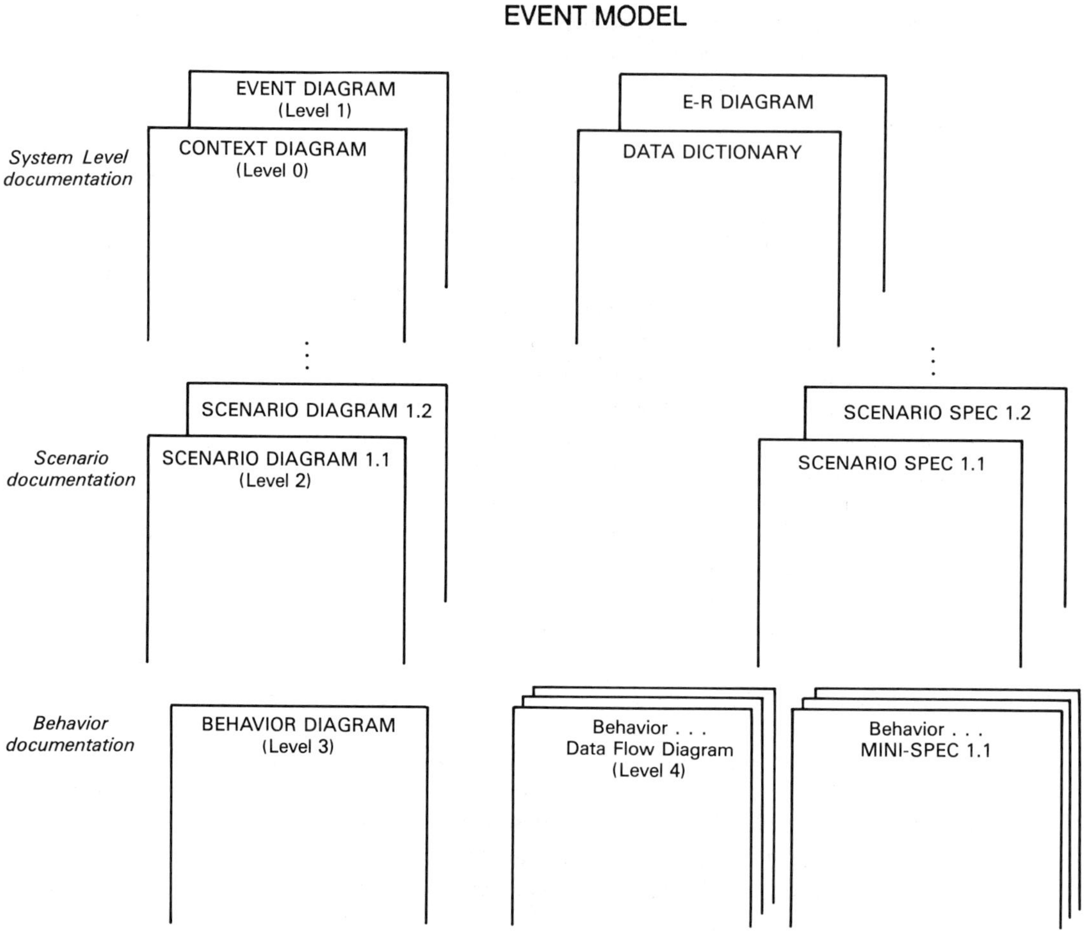

Figure 8.11

EVENTS IN SYSTEM MAINTENANCE
. . . assessing the maintenance efforts

When processes, data stores, and data flows are associated by events, there may be processes and data stores that overlap events, that is, they are common to more than one event. The fact is that certain processes, data stores, and/or data flows can serve different events. One of the benefits of identifying event segments is to be able to identify quickly the areas affected in a system when a change is proposed. Most changes to existing business data-processing systems relate to one or more events. First, there may be a desire to produce additional or different output when an event occurs. This type of change affects the fundamental data processing. The level-1 event diagram indicates the data flows to and from the sources and sink, and the data stores called upon when each event occurs. Thus, we are able to identify quickly not only the area affected by the proposed change, but we can clearly see the impact of the proposed change on the whole system.

Another type of change to a system is one that does not affect the fundamental processing of data but only the activities of the infrastructure. This type of change relates only to the technology used in the system. The infrastructure regulates the flow, volume, capacity, and responsiveness of the system. Most changes to business data-processing systems are in fact related to the infrastructure of the system. Think of how many times companies have built *new* payroll systems, most of which never changed the fundamental processing of data. We have, instead, increased the speed of processing and improved the turn-around time to produce the payroll checks, and/or handle larger volumes, all by collecting the data differently or improving the efficiency of the infrastructure. Collection methods in payroll systems evolved from handwritten time cards, to keypunch cards, to time-clock cards, to automatic punch cards, to magnetic-card readers, electronic palm scanners, and other types of scanners. Technology will continue to offer new techniques and devices to gather, transport, and produce output. Thus, it is helpful to identify and separate the processes and the data stores of the infrastructure from the essential activities, to minimize the efforts of incorporating new technology in the system.

A third type of change to a system occurs when the stimulus for a given event changes. Usually such a change only impacts the initial activity that receives the input data. However, the change may affect the data flows within a system and may, or may not affect the fundamental activities.

The final type of change to a system presented here is the adding or eliminating a whole event. From the level-1 event diagram, we can easily see the consequences of this type of change. However, by factoring an event into several scenarios, users can specify which scenario should be implemented first. In this way the users can begin to obtain some benefits from the system when the new event occurs, while waiting for the other scenarios to be completed. Also, by specifying the behaviors of each scenario, the behaviors that are common among the different scenarios become apparent, and hence encourage reusable code.

EVENT ANALYSIS EXAMPLE

. . . event analysis adds the users' perspective to the logical model

By organizing a system according to the events to which it responds, we can view the system as the users see it. To define a system by events forms a *users' view* of the system requirements.

Identification of external events begins by analyzing each of the input data flows that enters a system. Consider the physical model used as an example in the previous chapter, the brokerage system, and also as shown in Fig. 8.12.

CONTEXT DIAGRAM

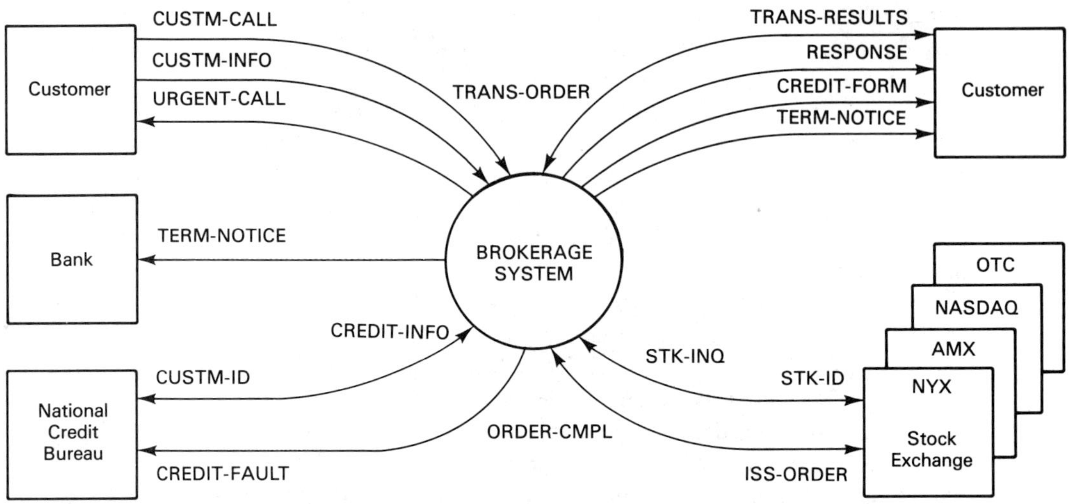

Figure 8.12

We begin event analysis by expanding the input/output processing chart presented in Chapter 6 to include columns for EVENT-# and EVENTS. Then we consider each input data flow shown on the chart, and after examining the data flow diagrams (Fig. 8.13), data dictionary (Fig. 8.14), and mini-specs (Fig. 8.15), we can determine what external events occurred that caused each of the sources to present a stimulus to the system. Because all processes are functional primitives, no lower-level data flow diagrams are needed. The reader is referred to the mini-specs for further details.

After completing the input section of the chart we continue our analysis with the output section. It is in the output section that we can detect temporal events that have caused activity and output in the system, in addition to the output caused by external events. Finally, we must examine those processes that are not used to process inputs or produce outputs to see if any internal events exist.

DATA FLOW DIAGRAM

(Diagram 0, Level 1)

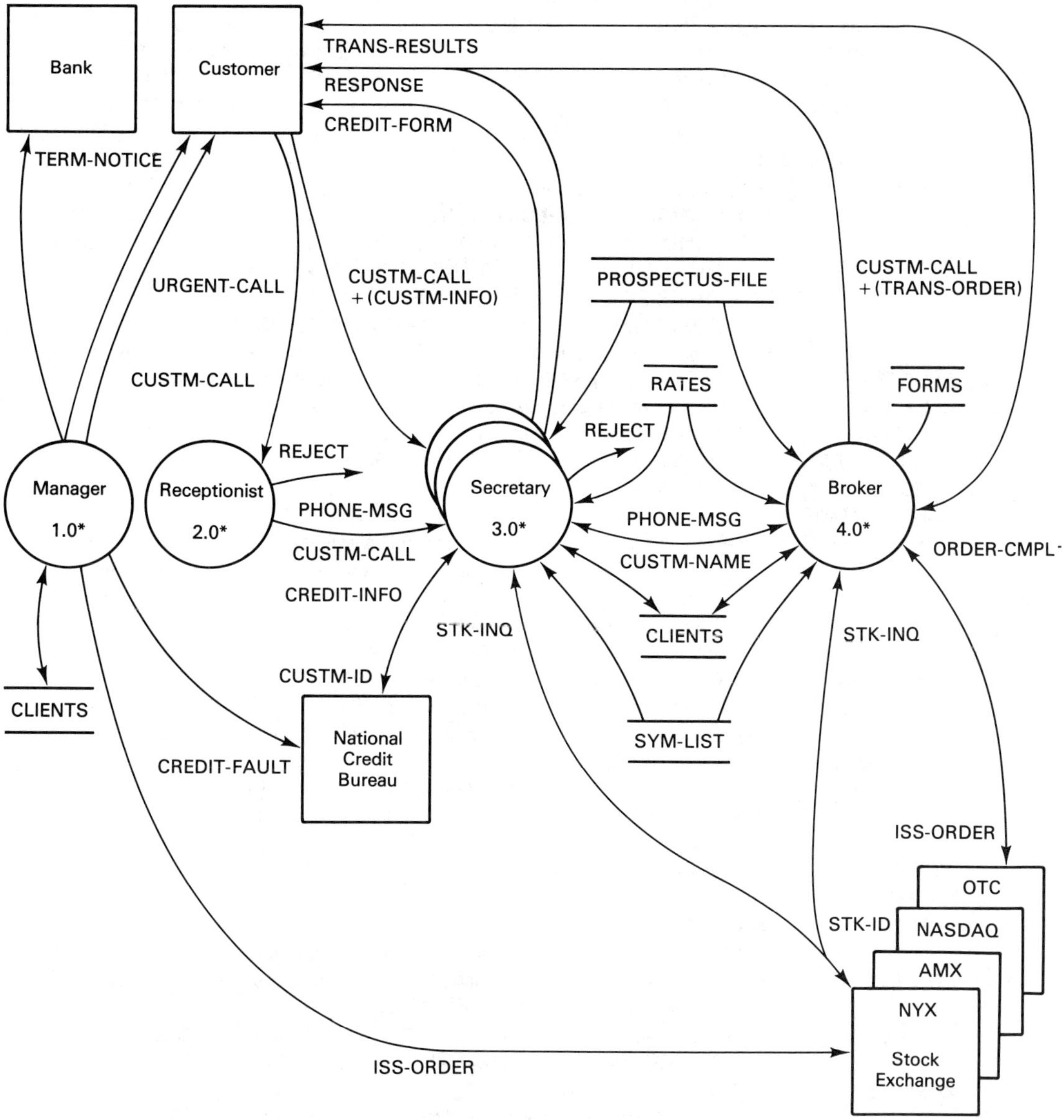

Figure 8.13

DATA DICTIONARY

AMT-OWED	= * Total amount owed by the customer to the brokerage firm. *
BANK-ACCT#	= * Customer's personal bank account number. *
BANK-ADDR	= * Address of the customer's BANK-NAME. *
BANK-NAME	= * Name of the customer's bank they used as a credit reference. *
BANK-PHONE#	= * Telephone number of customer's BANK-NAME. *
BROKER-NAME	= * Name of the broker who handles the customer"s account. *
BROKER-OFFICE	= * Office location of BROKER-NAME. *
BROKER-PHONE#	= * Telephone number of BROKER-NAME. *
CERTIF#	= * Unique serial number of the stock certificate. *
CLIENTS	= {CUSTM-INFO + CREDIT-RATING + {TRANS# + SYM + EXCHG + QTY + ["BOUGHT"|"SOLD"] + TRANS-DATE + SETTLE-DATE + TRANS-AMT + ["TRANS-COMPLETE"|"TRANS-OPEN"|"TRANS-CLEARED"]} + TRANS-PRICE + BROKER-NAME}
	= * CLIENTS is organized by CUSTM-NAME + CUSTM-ADDR + TRANS#. Since TRANS# is unique, CLIENTS can also be accessed by TRANS# only.*
CLOSE$	= * The last price paid for a stock at the close of the trading day. *
CO-ADDR	= * Address of the company that issues stock certificate. *
CO-NAME	= * Company name of the company that issues stock certificate. *
CREDIT-FAULT	= CUSTM-NAME + SS# + CUSTM-ADDR + AMT-OWED + SETTLE-DATE
CREDIT-INFO	= CUSTM-ID + CREDIT-RATING
CREDIT-RATING	= ["GOOD"|"BAD"|"UNKNOWN"]
CREDIT-REF	= * Form mailed to a potential customer to ascertain the customer's credit rating. The credit rating is determined by the credit manager outside this system. *
CURR$	= * The last price paid for a stock. *
CUSTM-ADDR	= * Address of the customer. *
CUSTM-BANK	= * Name of the customer's Bank. *
CUSTM-CALL	= CUSTM-NAME + ["New customer" + CUSTM-INFO | BROKER-NAME | INFO-INQUIRY | "Stock quote" + CO-NAME] + (CUSTM-PHONE# + MSG)
CUSTM-ID	= CUSTM-NAME + CUSTM-ADDR + SS#
CUSTM-INFO	= CUSTM-NAME + CUSTM-ADDR + SS# + CUSTM-PHONE# + CUSTM-BANK + BANK-ACCT# + (INFO-INQUIRY)
CUSTM-NAME	= * Name of the customer. *
CUSTM-PHONE#	= The customer's phone number. *
EXCHG	= ["NYX"|"AMX"|"NASDAQ"|"OTC"]
FORMS	= {TRANS# + ISS-ORDER form}
HIGH$	= * The highest price the stock sold for during the day. *
INFO-INQUIRY	= ["Price of a stock?" + CO-NAME | "Commission rate?" + PRICE + QTY | "Open new account" | BROKER-NAME]
ISS-ORDER	= TRANS# + STK-ID + QTY + [PRICE | "Market price"] + ORDER-DATE + BROKER-NAME + CUSTM-NAME + CUSTM-ADDR
LAST-TICK	= ["UP-TICK"|"DOWN-TICK"]
	= * The last change in the price of the stock, either an Up-tick or a Down-tick depending on the different between the last PRICE paid for the stock and the previous PRICE paid for the stock. *

Figure 8.14

DATA DICTIONARY (Cont.)

LOW$	= * The lowest price of the stock during the day. *
MSG	= * Message left by client. *
OPEN$	= * The price paid for the stock during the first trade of the day. *
ORDER-CMPL	= TRANS# + STK-ID + QTY + PRICE + CURRENT-DATE + TRANS-DATE + BROKER-NAME + CUSTM-NAME + CUSTM-ADDR
ORDER-DATE	= * Calendar date the transaction is placed by the customer. *
PHONE-MSG	= CUSTM-NAME + CUSTM-PHONE# + (MSG) + TIME-OF-CALL
PRICE	= * Price in dollars per share for a stock. *
PRICE-CHG	= * Overall change in the PRICE from the previous day's CLOSE$ to the CURR$. *
PROSPECTUS	= * Information a company supplies about their stock and other information about the company to lure investors. *
PROSPECTUS-FILE	= {{PROSPECTUS} + {STD-POOR}}
	= * The {PROSPECTUS} in the data store is organized by CO-NAME, the {STD-POOR} is a stack of STD-POOR booklets of information. *
QTY	= * Quantity of shares of stock. *
RATES	= {PRICE + QTY + COMM-RATE}
RESPONSE	= [RATE | STK-INQ | TRANS-RESULTS | PROSPECTUS | STD-POOR | "other market information or advice"]
SETTLE-DATE	= * Usually five working days after TRANS-DATE. *
SS#	= * Customer's social security number. *
STD-POOR	= * Standard and Poor's analysis of companies and their stocks. *
STK-ID	= SYM + EXCHG
STK-INQ	= SYM + OPEN$ + HIGH$ + LOW$ + [CURR$ | CLOSE$] + VOL + YR-HI$ + YR-LO$ + PRICE-CHG + LAST-TICK
SYM	= * Unique symbol used by the stock exchange to identify a company's stock. *
SYM-LIST	= {CO-NAME + CO-ADDR + SYM + EXCHG}
TAXES	= * Taxes that apply to trading stock. *
TERM-NOTICE	= * Notice explaining the default on a stock trade by a customer and the termination of the customer's account. *
TRANS#	= * A sequence number of integers starting from 000-0001 that uniquely identifies a transaction preprinted on the ISS-ORDER form. *
TRANS-AMT	= PRICE + QTY + ($0.01 * QTY/100 + $15) + TAXES * Note: FOR QTY ≥ 100.*
	= PRICE + QTY + ($0.01 * QTY/100 + $25) + TAXES * Note: FOR QTY < 100.*
TRANS-DATE	= * Date the stock is traded made at the exchange, which may or may not always be the same as ORDER-DATE. *
TRANS-ORDER	= TRANS# + CO-NAME + QTY + ORDER-DATE + [PRICE | "Market price"] + CUSTM-NAME
TRANS-PRICE	= * The total amount of dollars for a particular stock trade. *
TRANS-RESULTS	= TRANS# + CO-NAME + QTY + ORDER-DATE + PRICE
URGENT-CALL	= CUSTM-NAME + CUSTM-ADDR + "Payment has not been received for TRANS#. The stock will be sold and a default notice is being sent to BANK-NAME and the National Credit Bureau."
VOL	= * The accumulated number of shares traded during the day. *
YR-HI$	= * The highest price the stock sold for during the past twelve months. *
YR-LO$	= * The lowest price the stock sold for during the past twelve months. *

Figure 8.14 (cont.)

MINI-SPECIFICATIONS

Manager MINI-SPECIFICATION 1.0

At the beginning of the day.

While there are CLIENTS entries, Do:
  Search CLIENTS for TRANS# where SETTLE-DATE is greater than CURRENT-DATE and "TRANS-OPEN".
  Input CUSTM-INFO form CLIENTS.
  Place phone call to CUSTM-NAME + CUSTM-PHONE#.
  Output URGENT-CALL to Customer.
  Input TRANS# + SYM + QTY + EXCHG from CLIENTS.
  Set ISS-ORDER to "Market price".
  Set STD-ID to SYM + EXCHG.
  Output ISS-ORDER to EXCHG Stock Exchange.
  Output "TRANS-CLEARED" and CREDIT-RATING = "BAD" for CUSTM-NAME + CUSTM-ADDR + TRANS# in CLIENTS.
  Set TERM-NOTICE to CUSTM-NAME + CUSTM-ADDR + "Bad credit due to default on stock transaction".
  Output TERM-NOTICE to Customer.
  Input BANK-NAME + BANK-ADDR for CUSTM-NAME + CUSTM-ADDR in CLIENTS.
  Output TERM-NOTICE to Bank.
Endwhile.
Stop.

Processor Requirements: . . .

Security Requirements: . . .

Time Requirements: . . .

Processing Priorities: . . .

Performance Requirements: . . .

Safety Requirements: . . .

Other Requirements: . . .

Figure 8.15

MINI-SPECIFICATIONS (Cont.)

Receptionist MINI-SPECIFICATION 2.0

```
Input CUSTM-CALL from Customer.

Case 1 (CUSTM-CALL is "New Customer"):
        Determine which secretary is available.
        Send the CUSTM-CALL to the available Secretary.
        Stop.

Case 2 (CUSTM-CALL is for a specific BROKER-NAME):
        Determine if the BROKER-NAME secretary is available.
        If the secretary is not available,
            Ask the customer to hold.
        If the customer cannot hold,
            Input CUSTM-NAME, CUSTM-PHONE#, and MSG,
            Set TIME-OF-CALL to CURRENT-TIME + CURRENT-DATE,
            Output PHONE-MSG to Secretary,
            Stop.
        Endif.
        Repeat (every two minutes):
            If the secretary is not available,
                Ask the customer to continue to hold,
                If the customer cannot hold,
                    Input CUSTM-NAME, CUSTM-PHONE#, and MSG,
                    Set TIME-OF-CALL to CURRENT-TIME + CURRENT-DATE,
                    Output PHONE-MSG to Secretary,
                    Stop.
                Endif.
            Else, Send the CUSTM-CALL to BROKER-NAME's Secretary.
            Endif.
        Until secretary is available.
        Stop.
Endcase.
Stop.
```

<u>Processor Requirements</u>: . . .

<u>Security Requirements</u>: . . .

<u>Time Requirements</u>: . . .

<u>Processing Priorities</u>: . . .

<u>Performance Requirements</u>: . . .

<u>Safety Requirements</u>: . . .

<u>Other Requirements</u>: . . .

Figure 8.15 (cont.)

MINI-SPECIFICATIONS (Cont.)

Secretary MINI-SPECIFICATION 3.0

```
Input PHONE-MSG from Receptionist.
Place call to Customer.
Input CUSTM-CALL from Customer.
Case 1  (Customer wants a price quote):
        Search SYM-LIST for CO-NAME.
        Input SYM + EXCHG from SYM-LIST.
        Set STK-ID to SYM.
        Output STK-ID to EXCHG Stock Exchange.
        Wait for reply from Stock Exchange . . .
        Input STK-INQ from EXCHG.
        Set RESPONSE to STK-INQ.
        Output RESPONSE to Customer.
        Stop.
Case 2  (Customer wants information or PROSPECTUS on a CO-NAME):
        Search PROSPECTUS-FILE for CO-NAME.
        Input PROSPECTUS from PROSPECTUS-FILE.
        Set RESPONSE to PROSPECTUS.
        Output RESPONSE to Customer.
        Stop.
Case 3  (Customer wants to know Commission Rates):
        Input PRICE + QTY from Customer.
        Search RATES for PRICE + QTY.
        Input COMM-RATE from RATES.
        Set RATE to COMM-RATE.
        Output RESPONSE to Customer.
        Stop.
Case 4  (New customer wants to open an account):
        Input CUSTM-INFO from Customer.
        Output CUSTM-INFO to CLIENTS.
        Set CUSTM-ID to CUSTM-NAME + CUSTM-ADDR + SS#.
        Output CUSTM-ID to National Credit Bureau.
        Wait for reply from National Credit Bureau . . .
        Input CREDIT-INFO from National Credit Bureau.
        If CREDIT-INFO is "Good", Determine next available Broker,
            Output CUSTM-NAME to Broker,
            Set CREDIT-RATING to "Good",
            Output CREDIT-RATING to CLIENTS for SS#.
        Endif.
        If CREDIT-INFO is "Bad", End conversation with Customer,
            Set CREDIT-RATING to "Bad",
            Output CREDIT-RATING to CLIENTS for SS#.
        Endif.
        If CREDIT-INFO is "Unknown", Output "We will send you a credit reference form, fill it out and return to us.
                No transactions will be done until we receive the form and credit has been approved." to Customer.
            Set CREDIT-RATING to "Unknown",
            Output CREDIT-RATING to CLIENTS for SS#,
            Output CREDIT-FORM to Customer.
        Endif.
        Stop.
Case 5  (Customer wants to speak to BROKER-NAME):
        Repeat (every two minutes) while BROKER-NAME is unavailable,
            Output "Please continue to hold." to Customer.
            If the Customer can not continue to hold,
                Set PHONE-MSG to CUSTM-NAME + CUSTM-PHONE + (MSG),
                Output PHONE-MSG to BROKER-NAME.
            Endif.
        Until BROKER-NAME is available.
        Stop.
Endcase.
Stop.
```

Processor Requirements: . . .

⋮

Other Requirements: . . .

Figure 8.15 (cont.)

MINI-SPECIFICATIONS (Cont.)

Broker MINI-SPECIFICATION 4.0

Input PHONE-MSG from Secretary.
Place call to Customer.
Input CUSTM-CALL from Customer.

Case 1 (*Customer wants a price quote*):
Search SYM-LIST for CO-NAME.
Input SYM + EXCHG from SYM-LIST.
Set STK-ID to SYM.
Output STK-ID to EXCHG Stock Exchange.
Wait for reply from Stock Exchange . . .
Input STK-INQ from EXCHG.
Set RESPONSE to STK-INQ.
Output RESPONSE to Customer.
Stop.

Case 2 (*Customer wants information or PROSPECTUS on a CO-NAME*):
Search PROSPECTUS-FILE for CO-NAME.
Input PROSPECTUS from PROSPECTUS-FILE.
Set RESPONSE to PROSPECTUS.
Output RESPONSE to Customer.
Stop.

Case 3 (*Customer wants to know Commission Rates*):
Input PRICE + QTY from Customer.
Search RATES for PRICE + QTY.
Input COMM-RATE from RATES.
Set RATE to COMM-RATE.
Output RESPONSE to Customer.
Stop.

Case 4 (*Customer want to buy or sell stock*):
Input TRANS-ORDER from Customer.
Search SYM-LIST for CO-NAME.
Input SYM + EXCHG from SYM-LIST.
Input TRANS# and ISS-ORDER form from FORMS.
Set ISS-ORDER form to TRANS-ORDER.
Output ISS-ORDER to EXCHG Stock Exchange.
Stop.

Case 5 (*Stock transaction completed at Stock Exchange*):
Input ORDER-CMPL from EXCHG Stock Exchange.
Search SYM-LIST for SYM + EXCHG.
Input CO-NAME from SYM-LIST.
Set TRANS-RESULTS to ORDER-CMPL.
Search CLIENTS for CUSTM-NAME + CUSTM-ADDR.
Input CUSTM-PHONE# from CLIENTS.
Place a phone call to CUSTM-NAME + CUSTM-PHONE#.
Output TRANS-RESULTS to Customer.
Stop.

Processor Requirements: . . .

⋮

Other Requirements: . . .

Figure 8.15 (cont.)

INPUT/OUTPUT PROCESSING CHART WITH EVENTS

(Input Section)

BROKERAGE SYSTEM	← FUNCTIONAL PRIMITIVES → 1.0*	2.0*	3.0*	4.0*	EVENT-#	EVENT
Inputs:						
CREDIT-INFO			ECA		1.	**Person decides to become a client.**

CREDIT-INFO comes into the system from the National Credit Bureau when the system requests a credit check on a new client. However, CREDIT-INFO is part of a dialogue data flow whereby the system sends CUSTM-ID to the National Credit Bureau and awaits the reply CREDIT-INFO. Because CREDIT-INFO is a consequence of the system sending CUSTM-ID to a sink, the event is traced back to the reason why the system sent CUSTM-ID to the sink. The output response of CUSTM-INFO is a consequence of the external event, "A person decides to become a client."

BROKERAGE SYSTEM	1.0*	2.0*	3.0*	4.0*	EVENT-#	EVENT
CUSTM-CALL		ICA	EFA		1.	(Same as event #1)
	ICA	ICA	EFA		2.	**Client decides to buy or sell stock.**
	ICA	EFA			3.	**Client needs to know current price of a stock.**
	ICA	ICA	EFA		4.	**Person wants rate info, stock info, or advice.**
CUSTM-INFO			ECA		1.	(Same as event #1)
ORDER-CMPL				EFA	2.	(Same as event #2)

Dialogue data flows (CUSTM-CALL, CUSTM-INFO, and ORDER-CMPL) are the response to several events. A person may decide to become a client (#1), a client may wish to buy or sell stock (#2), or want to know the current price of a stock (#3), or other information (#4).

BROKERAGE SYSTEM	1.0*	2.0*	3.0*	4.0*	EVENT-#	EVENT
STK-INQ			EFA	EFA	3.	(Same as event #3)

STK-INQ is part of a dialogue data flow initiated when the broker inquires on the price of a stock. Both STK-INQ and STK-ID are results of event #3.

BROKERAGE SYSTEM	1.0*	2.0*	3.0*	4.0*	EVENT-#	EVENT
TRANS-ORDER				EFA	2.	(Same as event #2)

TRANS-ORDER is part of a dialogue data flow with the client that originally initiated event #2.

> *Note: In analyzing events using inputs or outputs that are part of a dialogue data flow, the data flow actions must be traced back to the original independent event that caused the system to initiate the dialogue. In this case, the event of a person deciding to become a client causes activity in the system to initiate the dialogue with the National Credit Bureau. The fact that it may take several hours, or days, for the Bureau to respond is not a factor. The response time of a source or sink during a dialogue is a consequence of the technology employed by the source/sink and is beyond the scope of the system analysis.*

Figure 8.16

After completing the input section of the input/output processing chart, as shown in Fig. 8.16, we turn our attention to the output data flows, as shown in Fig. 8.17. Consider each output data flow and determine whether it is a result of an external event already identified in the input section or if it is the result of a temporal or internal event in the system.

The first output data flow in Fig. 8.17 (CREDIT-FAULT) is produced by process 1.0* at the beginning of each new day. The system examines all of the entries in the CLIENTS data store and determines if any of them are overdue. After the appropriate processing the system issues a CREDIT FAULT. This is a new event from those identified in the input section. Continuing with our analysis you will see that all of the other outputs are a result of previously defined events.

INPUT/OUTPUT PROCESSING CHART WITH EVENTS

BROKERAGE SYSTEM	←-- FUNCTIONAL PRIMITIVES --> 1.0*	2.0*	3.0*	4.0*	EVENT-#	EVENT
Inputs:						
CREDIT-INFO			ECA		**1.**	**Person decides to become a client.**
CUSTM-CALL		ICA	EFA		1.	(Same as event #1)
	ICA	ICA	EFA		**2.**	**Client decides to buy or sell stock.**
	ICA	EFA			**3.**	**Client needs to know current price of a stock.**
	ICA	ICA	EFA		**4.**	**Person wants rate info, stock info or advice.**
CUSTM-INFO			ECA		1.	(Same as event #1)
ORDER-CMPL				EFA	2.	(Same as event #2)
STK-INQ			EFA	EFA	3.	(Same as event #3)
TRANS-ORDER				EFA	2.	(Same as event #2)
Outputs:						
CREDIT-FAULT	EFA				**5.**	**Begin new day (Transaction not completed)**
CREDIT-FORM			EFA		1.	(Same as event #1)
CUSTM-ID			EFA		1.	(Same as event #1)
ISS-ORDER				EFA	2.	(Same as event #2)
RESPONSE			EFA	EFA	3.	(Same as event #3)
				EFA	4.	(Same as event #4)
STK-ID		EFA	EFA		3.	(Same as event #3)
TERM-NOTICE	EFA				5.	(Same as event #5)
TRANS-RESULT				EFA	2.	(Same as event #2)
URGENT-CALL	EFA				5.	(Same as event #5)
PROCESSING SUMMARY:	EFA	EFA	EFA	EFA		
Data Stores:						
CLIENTS	EM		EM	EM = EM		
FORMS				IAM = IAM		
PROSPECTUS-FILE			EM	EM = EM		
RATES			EM	EM = EM		
SYM-LIST			EM	EM = EM		

Figure 8.17

To complete the event analysis, we have to examine the input/output processing chart for any temporal or internal events that do not produce any output data. In this example, we have identified four external events and one temporal event. The results of the analysis can be illustrated on an event diagram showing these five events together with the object data stores that were identified in Chapter 7:

1. Person decides to become a client (Become Client).
2. Client decides to buy or sell stock (Buy or Sell Stock).
3. Client needs to know current price of a stock (Price Inquiry).
4. Person needs rates, stock info or advice (Rates, Stock Info or Advice).
5. Begin new business day, verify transaction completed (Temporal).

To construct an event diagram, we take the events and the object data stores and create a level-1 data flow diagram. The following lists all of the objects and the object data stores for this system. The event diagram is shown in Fig. 8.18.

OBJECTS	OBJECT DATA STORES
CLIENTS	= {CUSTM-NAME + CUSTM-ADDR + SS# + CUSTM-PHONE# + ["GOOD" | "BAD" | "UNKNOWN"]}
TRANSACTIONS	= {TRANS# + SYM + QTY + ["BOUGHT" | "SOLD"] + TRANS-DATE + SETTLE-DATE + TRANS-AMT + ["TRANS-COMPLETE" | "TRANS-OPEN" | "TRANS-CLEARED"] + TRANS-PRICE + CUSTM-NAME + CUSTM-ADDR + BROKER-NAME}
BANKS	= {BANK-NAME + BANK-ADDR + BANK-PHONE#}
RATES	= {PRICE + QTY + COMM-RATE}
COMPANY	= {CO-NAME + CO-ADDR + SYM + EXCHG + PROSPECTUS + STD-POOR}

EVENT DIAGRAM

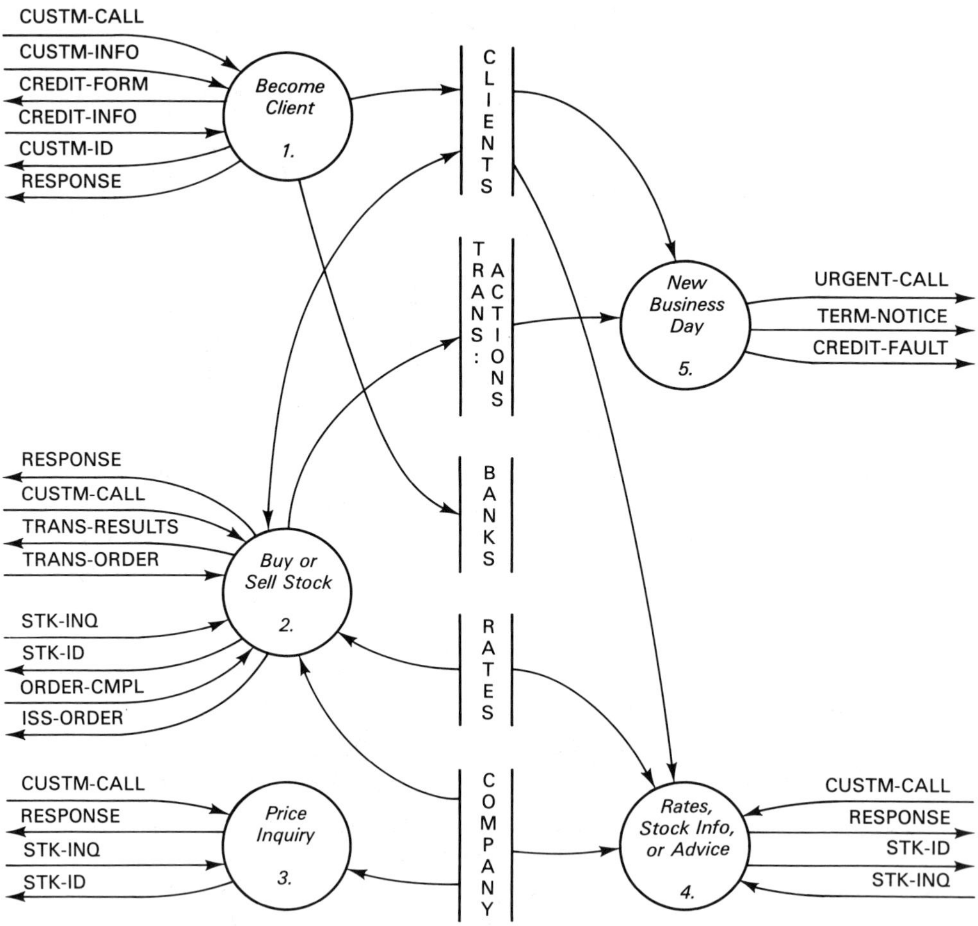

Figure 8.18

In the final step of the event analysis, we need to summarize the functional primitives associated with each corresponding event. There are two ways of accomplishing this: using *event data flow diagrams* or using an *event table*. We can produce a complete set of level-2 data flow diagrams by using the standard means of decomposing each event. We can repeat the process on each successive level until all functional primitives are identified. For those who do not have a convenient means of producing graphic data flow diagrams, the *event table* of Fig. 8.19 may be more practical to construct. **An EVENT TABLE is a list of every event to which the system responds, together with the related set of essential processes and, optionally, the set of infrastructure processes associated to each event.** Figure 8.19 is the event table for the brokerage system.

EVENT TABLE

EVENT	ESSENTIAL ACTV	(Optional) Infrastructure Activities
1. PERSON DECIDES BE BECOME A CLIENT.	2.0*, 3.0*	*List all infrastructure*
2. CLIENT DECIDES TO BUY OR SELL STOCK.	2.0*, 3.0*, 4.0*	*processes here*
3. CLIENT NEEDS TO KNOW CURRENT PRICE OF A STOCK.	2.0*, 3.0*	*accordingly.*
4. PERSON NEEDS RATE OR STOCK INFORMATION OR ADVICE.	2.0*, 3.0*, 4.0*	
5. BEGINNING OF A NEW BUSINESS DAY. (Temporal)	1.0*	

Figure 8.19

The activities of the infrastructure are not necessarily included in event analyses. However, for existing systems, software maintenance professionals may find it useful to include the activities and data stores used in the infrastructure of the existing system in the event table. The benefits of having an event table that includes both the essential components as well as the infrastructure of a system is in assessing the work effort and impact of proposed modifications and enhancements.

From event analysis, we have reorganized the processes and data of the system, according to the events to which the system responds. If a change to the inputs, outputs, or processing requirements occurs, and we can associate the change to one or more events, we know immediately which processes and data stores are affected. To know the scope and impact of a change provides a means by which the system professionals can estimate the time and effort necessary to accommodate the change. Also, after the change is made, the event table is helpful in preparing test plans to verify that the change did not inadvertently affect the system response to any related events. **RELATED EVENTS are two or more events that use the same behavior, process, or data to produce their response. (Sharing the same common processing routine or function is not sufficient to qualify two events as related.)** An event table provides information to those responsible for testing the product to attest to the product's quality. From event analysis, we know what processes and data are used for every event, the type of event, the stimulus for the event, and, optionally, the infrastructure used during each event. If a change were made to one or more processes or data, we can tell every event that uses the changed process or data. In this way, we not only verify that the new change or enhancement works, but also that all related events still produce their expected results.

SYSTEM REQUIREMENTS BY EVENTS
. . . event analysis for new software products

Up to now, we have discussed primarily business information systems. In most all cases, the enterprise has some form of working business system that currently does the basic functions or services. Analysts begin by modeling the existing system, then analyzing the changes or enhancements required. Some modifications involve the infrastructure, others involve the essence of the system, and still others involve both.

There is another area of systems analysis, namely, the specification of requirements for new software products. Unlike general business systems, it is often the case that there is no existing product that performs the functions or services to model using the structured methods. So how do we begin?

First of all, ideas are submitted to the analyst to be worked up as a proposal for a new product. The analyst or product engineer produces a structured specification for the new product, feature, or service. The structured specification serves several purposes:

1. To specify in a complete, well-defined, unambiguous manner the intended functions, features, and services of the new product.
2. To create a document that can be reviewed with potential users to refine the concept of the product and obtain information as to how the product is intended to serve the users.
3. To submit the proposal to a sponsor for approval and funding to develop the new product, feature, or service. A SPONSOR is a person or organization that provides the funds to develop a new product, feature, or service.
4. To continually revise the proposal until one has specified all of the functions, features, and services of the product.
5. To submit the structured specification as the requirements for a new product, feature, or enhancement to the development organization and to an independent test organization to verify that the new product performs as intended.

The objective is to prepare a structured specification that is the *blueprint* that specifies the requirements of the new product, feature, or service. In specifying a new product, we can use event diagrams to specify the logical intent of the system, then endow the structure with the infrastructure to specify an implementation with existing or proposed technology. Consider the sequence of steps as follows.

1. Identify the external events expected to occur in the users' world that will prompt an action or response from the new software.
2. Define the input stimulus that signals the occurrence of the event.
3. Begin constructing a context diagram, an event diagram, and a data dictionary.
4. Define the appropriate behavior the users expect or desire from the product. Create a high-level action diagram in the mini-specs that outlines the major functions for each behavior for each process.
5. Identify the object data stores necessary to support the events to which the system must respond. Define the object data stores in the data dictionary.
6. Identify the temporal events that require action or a response from the system. Add the temporal events to the event diagram, data dictionary with scenarios, behaviors, and the high-level action diagrams for each process as mini-specs.

At this point, the new product can be reviewed with potential users. The system analyst has a context diagram, event diagram, data dictionary, and mini-specs for behaviors/processes. Each mini-spec contains a policy statement or a high-level action diagram to define the major processes by the functional primitives for each event. The structured specification documents the basic inputs, outputs, processes, and object data stores necessary to service the users. When the system boundaries are well-defined and the context diagram and event diagram are complete, we can begin the next phase.

7. Decompose each event by using the functional decomposition technique to prepare data flow diagrams for the behaviors necessary for each scenario.
8. Create lower-level data flow diagrams for each behavior. The level-3 data flow diagram defines behaviors in terms of the processes and data necessary to accomplish their task.
9. Continue functionally decomposing the processes until the functional primitives are reached and we can prepare the mini-specs for each lowest-level process. The mini-specs include the policies that govern the transformations of input data into output data.
10. Update the data dictionary by specifying greater detail that firms up the required input and output data flows as well as internal data flows. Additions are made to the contents of the data stores that are necessary to accomplish the intended results.
11. Create E-R diagrams that specify any required relationships between the object data stores used.

At this point, the new product structured specification can be included in a new *product proposal*. **A PRODUCT PROPOSAL is a document that specifies the intended objective of the new product, the targeted customers, the planned operating environment (type and size of computer, memory, and peripheral options), and the structured specifications, including the events that the product will support.** The new product proposal can be reviewed with the potential users, developers, sponsors, and other management personnel to verify that the intended functions, features, and services are complete, well-defined, and unambiguous. The developers begin to understand the objectives of the proposed new product, its functions, and the expectations the sponsor has for the new product. If approved to continue, the project enters the final analysis phase that completes and publishes the system's requirements.

12. Specify any internal/anomalous events to maintain integrity and provide for continuous operation of the new product (if necessary).
13. Plan and specify the system's architecture. At this point, the realities of current technology begin to *contaminate* the requirements.
14. Add the details of the infrastructure to the structured specifications.
15. Add the detailed requirements as to processors requirements, security requirements, time requirements, processing priority, performance requirements, safety requirements, and other requirements to complete the scenarios, behaviors, and process mini-specs of the structure specification.
16. For real-time systems, include state-transition diagrams and time-dependency charts to specify in detail any known timing requirements in the product, feature, or service.

A third review to complete the system requirements is held to cover in detail the concepts, objectives, architecture, and requirements contained in the structured specification. We include development estimates in the new product proposal together with the additional details and architecture of the new product, feature, or service. Following any required updates and funding approval from the sponsor, the structured specification is sent to the development organization to begin the high-level design. From there, the system will progress in the development life cycle according to the methodology used to develop and implement new products, features, or services.

In the area of developing new products, there is a need to address the product's intended market, expected functional capabilities and services that lead to defining the overall product strategy. The systems analyst or product engineer, as the architect of the product, can effectively communicate with product managers and marketing professionals by means of the event diagrams. Information from an event diagram can be used in a *service description*. **A SERVICE DESCRIPTION is a document that specifies in narrative text the functions, features, and services provided by the product.** The document specifies how and under what conditions the users can benefit from the new product. The service description becomes the sales literature and can include technical specifications for the new product.

It is the event diagram that defines the new product's functions, features, and services in terms of responses to stimuli. Product enhancements, feature contents, enhancements to services, and integration of related products all can be done initially from the event diagram. The most important elements of a product are presented in the event diagram:

The events

The stimulus that causes the system to recognize that the event occurred

The response

The necessary stored data required to make the desired response

A lower levels of detail, the essential, fundamental, and custodial activities and even the infrastructure necessary to service the users in the intended fashion

Management, developers, users, and those responsible for the independent testing of the system or product can all use the event diagram and related documentation to understand the system or product they are involved with in their work.

SUMMARY
. . . communications is the key to good requirements

Many of the tools and techniques discussed in this text can be used in requirements specification for not only understanding an existing system but as an aid during the planning phases of new products and during the maintenance phase of mature systems and products. Based on the type of product, the intended user community, and the functions and capabilities of the product, we can combine the methods and tools in various ways to form a methodology that produces good planning and requirement specifications. Outside of the tools and methods, keep the following precepts in mind when preparing structured specifications as system requirements:

- **Human beings do their best work by iteration.** The requirements specification, as well as the methodology, must provide a convenient means of making iterative changes.
- **Submit your work to a peer review** to obtain feedback to improve the end product.
- **Talk to the users**. Ideally, build prototypes and study how the users will use your system or product.

- **Plan for a high degree of "fault tolerance."** Build systems and products that have the capability to detect and recover from internal faults and are resilient to user-induced faults (erroneous or unexpected inputs).
- And remember, **great engineering is simple engineering.** The most popular software products are very powerful, but *simple*!

QUESTIONS

1. Define event analysis.
2. What is an expanded data flow diagram?
3. What comprises the lowest-level, current, logical data flow diagram?
4. What is an event diagram?
5. For what purpose would we create an event diagram?
6. What are the three types of events and how do they differ?
7. What types of changes can users request?
8. What is the purpose of a system?
9. What is wrong with the event diagram of Fig. 8.20?

EVENT DIAGRAM

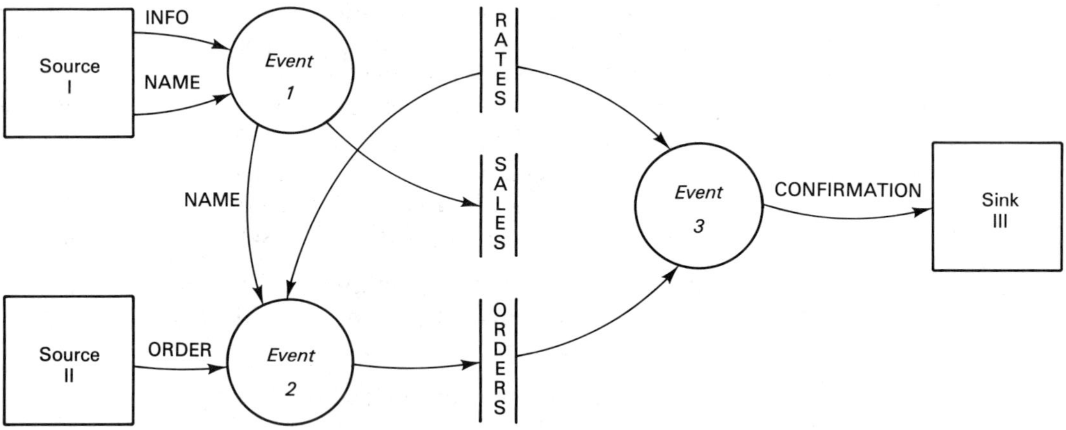

Figure 8.20

10. Should an event table include the processes of the infrastructure? Why or why not?

11. When analyzing events by using an input/output processing chart, we trace back all of the inputs to determine the external events that motivated the source(s) to send the input(s) into the system. All output(s) from the system are traced back into the system to determine if an input triggered the response (external event) or if a temporal or anomalous event occurred. How is the analysis conducted for dialogue data flows?
12. Explain the relationship among sources, sinks, events, stimuli, responses, processes, and object data stores.
13. Are events mutually exclusive? (Explain your answer.)
14. Explain how event analysis can benefit the final testing of a system or product.
15. What type of documentation is suggested for new products? What does it include? What is the primary audience for each document?
16. What is a scenario?
17. What is shown on a scenario diagram?
18. What is the relationship between events and scenarios?
19. What is a behavior?
20. Where are time constraints and requirements shown in an event model?

PROBLEMS

1. From the problem description of the second problem in Chapter 2, determine all of the events for the Purchasing System.

2. The Free Blood Center supplies whole blood to area hospitals upon request and plasma to the Plasma Distribution Center. Hospitals send in requests for whole blood that is needed immediately or at some scheduled date in the future. Plasma is sent out once a day.

   After whole blood has been in inventory for 21 days, it is withdrawn, converted into plasma, and sent to the Plasma Distribution Center.

   When the inventory of types A and O blood fall below 10 pints, an appeal is sent out asking for donations from previous donors who have the type of blood needed. No minimum level of inventory is kept for types B and AB blood. This blood may be in stock; usually, however, the center must ask for donors when such blood is requested.

   A history of each donor is maintained in the system and contains the person's name, address, type of blood, date of last donation, and personal health history. When a donor arrives to donate blood a check is made to insure that: (1) sufficient time has elapsed since his or her last donation, and (2) the donor has a clean health history. If the person is accepted to donate blood, the history file is updated to indicate the date of last donation. The blood is then taken and inventoried with the type and the date of donation.

No money is exchanged between the Free Blood Center and the donors, hospitals, or the Plasma Distribution Center. However, once a week a secretary obtains the names and addresses of all the donors who contributed blood during the previous week and sends them a "Thank You" card. If it was the first time a person donated blood, he or she also receives a "Life-Time" membership card entitling him or her to free blood from the center.

Occasionally a worker will discover that the amount of a particular type of blood reported by the system disagrees with the actual amount of blood in inventory. When this occurs, a supervisor is notified. The supervisor examines the inventory, and if a discrepancy really exists, the supervisor authorizes a special data-base update to bring the information stored in the system up-to-date with the actual amount onhand.

From the above system description, identify all of the events to which this system responds and specify the type of event.

3. Perform an event analysis on problem #2 at the end of Chapter 2 (Purchasing System). It may be helpful to use the input/output processing chart created in problem #3 at the end of Chapter 6.

4. From problem #1 at the end of Chapter 6, perform an event analysis and hypothesize events to which the system responds. It may be helpful to use the input/output processing chart created in problem #1 (Chapter 6).

5. During an event analysis it was learned that all employees fill out time-cards and drop them off with the department clerk. At the end of the day on Friday, the clerk collects all the time-cards and mails them to headquarters unless an employee has failed to submit his or her time-card. All cards must be mailed together. If the clerk does not have all the time-cards, he or she waits until the forgetful employee submits the missing time-card on Monday morning. Upon receiving the card the clerk mails all the cards to headquarters.

   a. Does the system that mails the time-cards to headquarters respond to an external, temporal, or internal event, and why?
   b. Would your answer to 5a be the same if the clerk always mailed whatever time-cards were submitted by 5:00 P.M. on Friday?
   c. Would your answer be different if the clerk were instructed to mail the time-cards whenever all of the time-cards for the week were received?

9

LOGICAL MODELS

Before we can propose alternatives to a system that would either alleviate some problem, bring the system to operate within new constraints, or provide some desired feature or enhancement, we must clearly understand the existing system. To derive that understanding, it is suggested that we accurately create a model representing the system as it physically exists, not just the analyst's perception of the existing system. The model has to be verified by those that operate or work with the system. In this chapter, we continue to apply the tools and methods of structured systems analysis to the example of a physical structured specification presented in Chapter 5.

Although the model of the Rent-em Video system from Chapter 5 details exactly what the system does, there is a need to look deeper into the system. First, as discussed in Chapter 6, we have to analyze the system model to identify the system's infrastructure and essential components. The infrastructure of processes and data keep the system working by using a specific level of technology. On the other hand, the essential processes and data perform the functions that satisfied the system's *reason for being*. The methods in Part II provide a means to identify the essential components, normalize the essential memories, and define the objects of the system. Once we have identified the essential components and objects of the system, we can apply event analysis, as described in Chapter 8 to present a means of reorganizing them around the events to which the system responds.

After completing the described logical analysis, we are in a position to briefly and concisely explain the system under study. We are able to describe the system in terms of its most important activities and tasks, its essential data, and the events to which the system responds.

Once a system undergoes this type of logical analysis, it is relatively easy to convey to the owners of the system, the size, scope, and results of any proposed modifications. In this chapter, we apply the tools and methods of logical analysis to the physical model presented in Chapter 5. There will be examples of how various types of proposed changes can impact the existing system and the role of event analysis in the system's life cycle.

LOGICAL ANALYSIS
. . . identifying the system's essence and infrastructure

To begin logical analysis, we create an input/output processing chart, as shown in Fig. 9.1. To better visualize what is happening in a *small* system, we can create an expanded data flow diagram that contains all of the functional primitives (see Fig. 9.2). Fold out page 280C so that you can follow the expanded data flow diagram while each input and output data flow is analyzed.

In very large systems, the expanded data flow diagram may not be practical to create. Nevertheless, we can always use the input/output processing chart to identify the essential components of a system.

The chart summarizes all of the analyses of the functional primitives and classifies the processes as EFA, ECA, IAA, or ICA. Likewise, the analyses of data stores are summarized and classified as EM, IAM, or ICM. To determine the processing summary line, we examine each process column for the highest-priority classification. The priorities are

highest	ESSENTIAL FUNDAMENTAL ACTIVITY (EFA)
↕	ESSENTIAL CUSTODIAL ACTIVITY (ECA)
	INTERNAL ADMINISTRATIVE ACTIVITY (IAA)
lowest	INTERNAL COMMUNICATIONS ACTIVITY (ICA)

The highest-priority classification found in *each column* is entered in the processing summary row.

Each data store receives its final classification, entered beyond the equal sign at the far right, by examining each row for the highest-priority classification. The priority classifications for data stores are

highest	ESSENTIAL MEMORIES (EM)
↕	INTERNAL ADMINISTRATIVE MEMORY (IAM)
lowest	INTERNAL COMMUNICATIONS MEMORY (ICM)

The highest-priority classification found in *each row* is placed at the end of each row following the equal sign. When completed, the chart provides the classification for each functional primitive process in the system as well as each data store used in the system. Using the chart to do logical analysis provides a systematic way of analyzing every process and data store in the system. In addition, it provides a cross-reference of what processes use, or contribute, in producing certain data flows. If an input data flow was changed, we could read from the chart all of the functional primitive processes and data stores affected.

EXPANDED DATA FLOW DIAGRAM

TURN PAGE AND FOLD OUT

To begin logical analysis set up an input/output processing chart with all of the inputs, outputs, and data stores as shown in Fig. 9.1. Then list all of the functional primitive processes across the top. You are now ready to begin logical analysis.

INPUT/OUTPUT PROCESSING CHART

	←-----FUNCTIONAL PRIMITIVES-----→								
SYSTEM NAME	1.0	2.1	2.2	2.3	2.4	3.1	3.2	4.0	5.0

Inputs:

CALL
CASH
CUSTOMER-INFO
CUSTOMER-NAME
INVOICE
NEW-TAPE
NEW-TAPE-CATALOG
TAPE
TAPE-DEPOSIT
TAPE-REQUEST

Trace each of the system's inputs through the functional primitives and determine from the mini-specs whether the processes are essential or are part of the system's infrastructure.

Outputs:

BANK-DEPOSIT
CALL
LIST-OF-TAPES
NEW-TAPE-PAYMENT
RECEIPT-COPY-2
RENTAL-RECEIPT
TAPE
TAPE-ORDER-FORM
TOTAL-CHARGES

Trace each of the system's outputs back through the functional primitives, and determine from the mini-specs whether the processes are essential or are part of the system's infrastructure.

PROCESSING SUMMARY:

Data Stores:

AVAILABLE-TAPES
CASH-BOX
CHECKBOOK
CUSTOMER-FILE
INVOICE-BOX
NEW-TAPE-LIST-BOX
OLD-INVOICES
PAID-BOX
TAPE-INVENTORY
TAPE-LIBR
TAPE-NUMBERS
TAPE-RECEIPT-BOX

While tracing the above inputs and outputs through the system, determine if the data stores involved in processing each data flow are essential memories or part of the infrastructure.

Figure 9.1

EXPANDED DATA FLOW DIAGRAM - RENT-EM VIDEO RENTAL SYSTEM

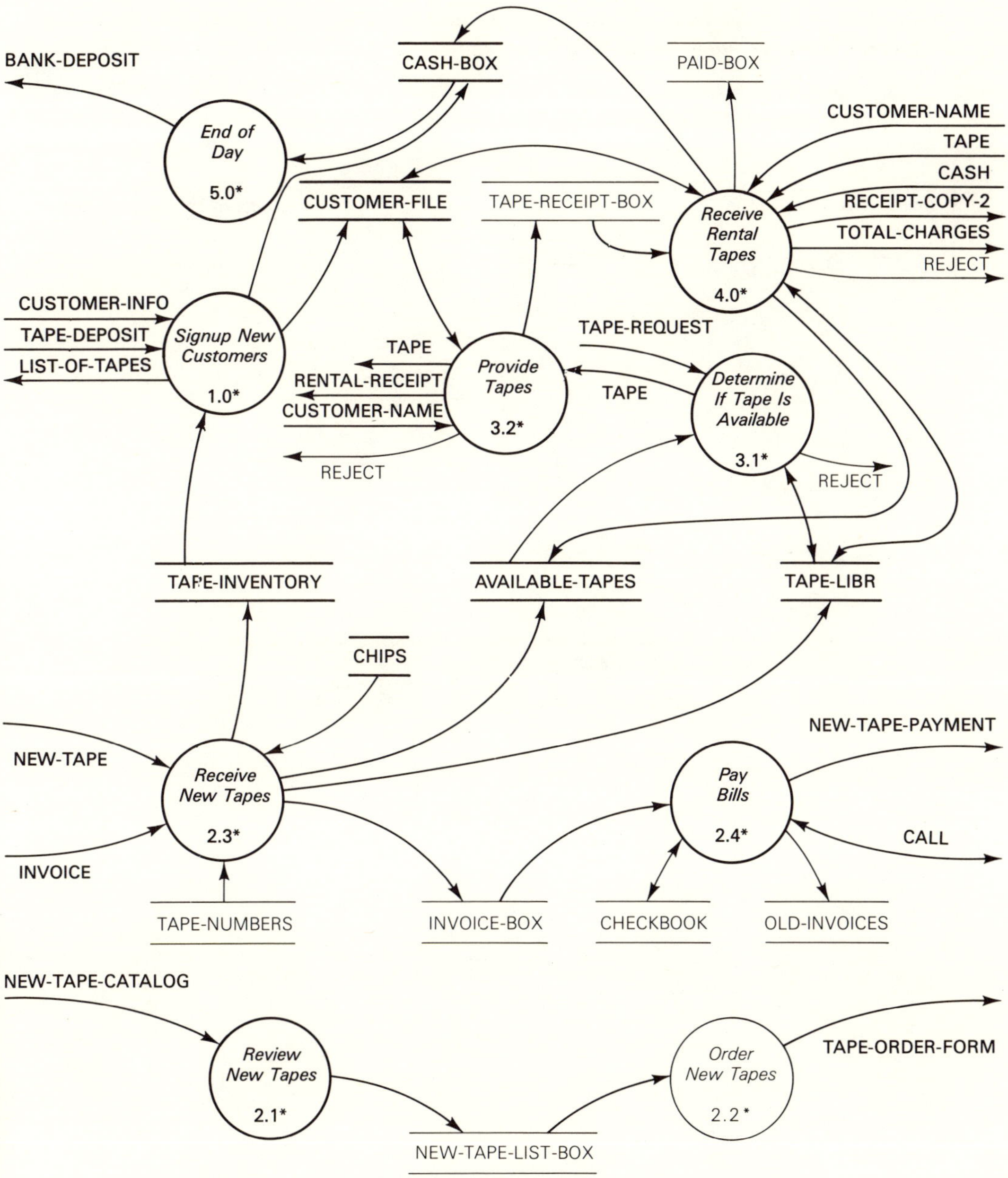

Figure 9.2

For your convenience, we have constructed this chapter so that the expanded data flow diagram for Rent-em Video system (Fig. 9.2) will fold out to the left and be in view while you read through the analysis of each input and output data flow. Likewise, page 286A – D folds out to the right and contains the final input/output processing chart so that both the expanded data flow diagram and the input/output processing chart are in view during the explanation of the logical analysis.

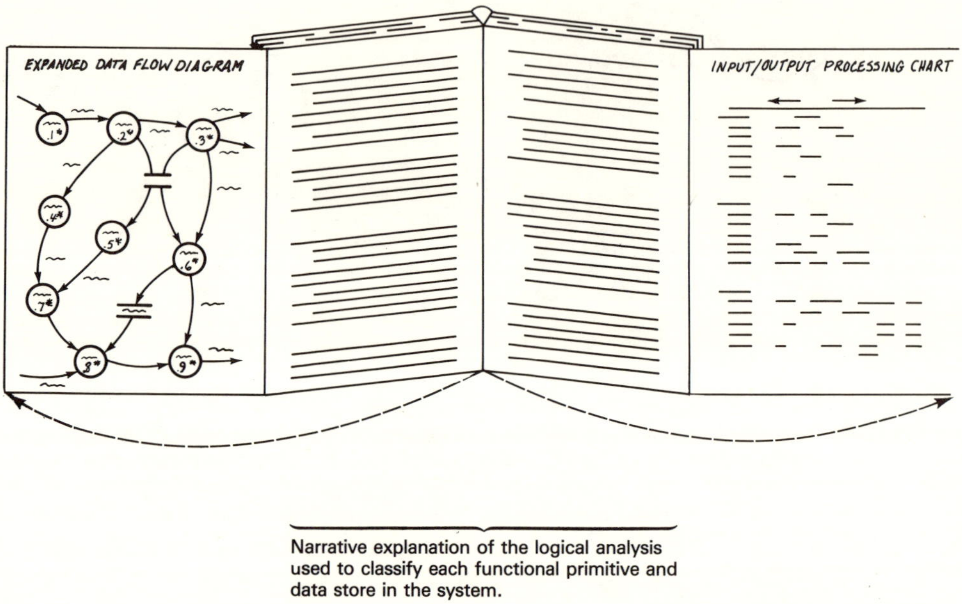

INPUT/OUTPUT PROCESSING CHART
. . . creating an input/output processing chart

To derive the input/output processing chart of Fig. 9.3, consider each input and output data flow as follows.

Input Section

We examine each input of the system and trace it through the system until either an essential fundamental activity (EFA) transforms it into an output data flow or an essential custodial activity (ECA) stores the data in an essential memory (EM). While tracing each input data flow, we will identify any internal communications or administrative activities and/or memories (IAA, ICA, IAM or ICM) that are in the direct data path until we reach an EFA or ECA/EM.

CALL is a dialogue (input/output) data flow of the process Pay Bills 2.4*. It enters process Pay Bills 2.4*, and from the mini-spec in Chapter 5, we can determine that it is an interactive data flow used to resolve any discrepancies found on the INVOICE. INVOICE is retrieved from the data store INVOICE-BOX, where INVOICES are held until process 2.4* is ready to receive them. INVOICE-BOX functions as an ICM. Because process 2.4* is responsible for carrying on the interactive dialogue (CALL) with the Video Tape Wholesaler and resolving any discrepancies found on the INVOICE, it is an EFA. Just before the process completes, it places the INVOICE into the internal administrative memory (IAM) OLD-INVOICES for historical purposes.

So we have:

Process 2.4* (EFA), using INVOICE-BOX (ICM), and
OLD-INVOICE (IAM)

CASH comes into Receive Rental Tapes 4.0* when a customer pays for the TAPE(s) he or she rented. According to mini-spec 4.0* in Chapter 5, CASH is put into the CASH-BOX data store and held there until the End of Day 5.0* process transforms it into a BANK-DEPOSIT and sends it to the Bank. Because one of the system's objectives is to send a BANK-DEPOSIT to the Bank, and process 5.0* is solely responsible for fulfilling that objective, process 5.0* is an essential fundamental activity (EFA). The data store that provides the necessary data for process 5.0* to do its job is the essential memory (EM) CASH-BOX. The process Receive Rental Tapes 4.0*, which updates the essential memory CASH, is an essential custodial activity (ECA) .

So we have:

Process 4.0* (ECA), using CASH-BOX (EM)
Process 5.0* (EFA), using CASH-BOX (EM)

CUSTOMER-INFO is received from the customer and is placed in the CUSTOMER-FILE data store by the process Sign Up New Customers 1.0*. When process 1.0* establishes a CUSTOMER-RECORD in the CUSTOMER-FILE data store, the customer becomes a registered member and may rent tapes. To determine whether process 1.0* is a internal communications activity or an essential custodial activity, we need to look ahead and see how subsequent processes use the data from CUSTOMER-FILE. From mini-spec 3.2* in Chapter 5, CUSTOMER-NAME from CUSTOMER-FILE is used by the process Provide Tapes 3.2* to verify the customer's membership when issuing TAPEs. When TAPEs are issued by process 3.2*, the process updates the data store CUSTOMER-FILE to record the fact that a customer has rented one or more TAPEs. CUSTOMER-FILE serves the system as an essential memory (EM), and thus process 1.0* is an essential custodial activity (ECA).

So we have:

Process 1.0* (ECA), using CUSTOMER-FILE (EM)

CUSTOMER-NAME, as shown in Fig. 9.2, enters the system in two places:

1. *Provide Tapes 3.2**. From the mini-spec in Chapter 5, we find that when a customer wishes to rent a tape, CUSTOMER-NAME is used to retrieve the CUSTOMER-RECORD from CUSTOMER-FILE (EM) to be sure that the customer is eligible to rent tapes and to record TAPE-INFO in the CUSTOMER-FILE. We can consider such an activity an internal data verification function and classify it as an internal administrative activity (IAA). However, because the process also updates an essential memory (CUSTOMER-FILE), it is an essential custodial activity (ECA).

2. *Receive Rental Tapes 4.0**. From the mini-spec 4.0* in Chapter 5, we find that CUSTOMER-NAME enters the system when a customer returns TAPEs. The process uses CUSTOMER-NAME to locate CUSTOMER-RECORD in CUSTOMER-FILE to store the DATE-RECEIVED (an essential custodial activity). Process 4.0* also calculates the AMOUNT-DUE. Determining AMOUNT-DUE is necessary to provide TOTAL-CHARGES as an output. However, we will revisit this data flow again when we consider the output data flows. For now, we are just concerned with the processing associated with the input data flow CUSTOMER-NAME.

We can stop tracing an input data flow when we reach an essential memory or when it is transformed by an essential fundamental activity into an output data flow or a component used to form an output data flow.

So we have:

Process 3.2* (ECA), using CUSTOMER-FILE (EM)
Process 4.0* (ECA), using CUSTOMER-FILE (EM)

INVOICE enters the system in process 2.3* and passes to process 2.4* by means of data store INVOICE-BOX. That data store serves only as a means of communicating data between two processes that are performed at different times and is an internal communications memory (ICM). From the mini-spec 2.3* in Chapter 5, we can determine that process 2.3* compares the information on the INVOICE with the NEW-TAPEs received. The process records any discrepancy on the INVOICE and places it into the data store INVOICE-BOX. The comparison of actual TAPEs received with the INVOICE is one of the processing functions of the system and thus we consider process 2.3* an essential fundamental activity (EFA). Process 2.4*, which determines if a CALL has to be issued to the Video Tape Wholesaler, is responsible for the dialogue data flow (CALL) that resolves any discrepancies found on the INVOICE. However, again we are just concerned about processing the input data flows. All other activities will be studied when we consider the output data flows.

So we have:

Process 2.3* (EFA), using INVOICE-BOX (ICM)

NEW-TAPE is an input to process 2.3* and from mini-spec 2.3* in Chapter 5, one can determine that the process obtains the next available tape number from the data store TAPE-NUMBERS. The data store TAPE-NUMBERS only serves one purpose and that is to keep a list of the next available tape number. This type of data store is an internal administrative memory (IAM). (Keeping track of the next available tape number is a function that is exclusively servicing a specific type of technology used to identify tapes uniquely. It is specifically these types of components that we want to identify as part of the infrastructure and eliminate them from the essential model.) The process also inputs a new blue or yellow TAPE-CHIP from another IAM, CHIPS. When a NEW-TAPE arrives, process 2.3* updates three data stores: TAPE-INVENTORY, AVAILABLE-TAPES, and TAPE-LIBR. Because the system needs to use AVAILABLE-TAPES and TAPE-LIBR to meet one of the system's objectives (issuing TAPEs), these data stores are essential memories (EMs). The data store TAPE-INVENTORY is also used by process 1.0* to provide the output LIST-OF-TAPES to a customer, so it is also an EM. Process 2.3*, which updates these essential memories, is therefore an essential custodial activity (ECA).

So we have:

Process 2.3* (ECA), using TAPE-NUMBERS (IAM),
CHIPS (IAM),
AVAILABLE-TAPES (EM),
TAPE-LIBR (EM), and
TAPE-INVENTORY (EM)

NEW-TAPE-CATALOG is an input to process 2.1*. It indicates the new tapes that are available from the Video Tape Wholesaler. The process of selecting new tapes is an essential fundamental activity (EFA) because it provides the output TAPE-ORDER-FORM to the Video Tape Wholesaler. Process 2.1* outputs the MARKED-NEW-TAPE-LIST into the NEW-TAPE-LIST-BOX. The data is held there until process 2.2* is able to retrieve it and place an order for the new tapes. From the data dictionary in Chapter 5 we determine that the MARKED-NEW-TAPE-LIST is actually a marked-up version of the NEW-TAPE-CATALOG. The data store NEW-TAPE-LIST-BOX merely serves as an internal communications memory (ICM) to pass the marked-up catalog to the store clerk, who writes up the order for the new tapes. Process 2.2* provides the output and will be discussed in the output data flow section.

So we have:

Process 2.1* (EFA), using NEW-TAPE-LIST-BOX (ICM)

TAPE as an input is handled by the Receive Rental Tapes 4.0* process. The process accepts the TAPE, puts it into the TAPE-LIBR, takes the TAPE-CHIP from TAPE-LIBR, and outputs the TAPE-CHIP to the AVAILABLE-TAPES data store to reflect the fact that the TAPE is now available for rent. Because AVAILABLE-TAPES and TAPE-LIBR are essential memories (EMs), process 4.0* performs an essential custodial activity (ECA) task. The process also updates CUSTOMER-FILE, indicating the date the customer returned the TAPE. Because CUSTOMER-FILE is also an essential memory, process 4.0* performs an essential custodial activity (ECA).

So we have:

Process 4.0* (ECA), using TAPE-LIBR (EM),
AVAILABLE-TAPES (EM), and
CUSTOMER-FILE (EM)

TAPE-DEPOSIT is an input to the Signup New Customers 1.0* process. Process 1.0* takes the deposit and places it in the CASH-BOX. Because process 5.0* is the essential fundamental activity that uses the money to make a BANK-DEPOSIT and CASH-BOX is an essential memory (EM), process 1.0* is an essential custodial activity (ECA).

So we have:

Process 1.0* (ECA), using CASH-BOX (EM)

TAPE-REQUEST is an input data flow that is handled by the Determine If Tape Is Available 3.1* process. From mini-spec 3.1* in Chapter 5, we know that the process searches the essential memory (EM) AVAILABLE-TAPES to see if the requested tape is available. If it is, the process selects the TAPE from the essential memory (EM) TAPE-LIBR and passes it to the Provide Tapes 3.2* process. Process 3.1* outputs the TAPE-CHIP obtained from the AVAILABLE-TAPES data store to TAPE-LIBR. The TAPE-CHIP is a replacement for the TAPE, indicating the tape is no longer available to rent. If the TAPE-REQUEST is not available, process 3.1* rejects the TAPE-REQUEST as a REJECT. Process 3.1* essentially transforms the customer's TAPE-REQUEST into a TAPE, and is an essential fundamental activity (EFA).

So we have:

Process 3.1* (ECA), using AVAILABLE-TAPES (EM), and
TAPE-LIBR (EM)

Output Section

We now examine each output from the system and trace it back into the system until we reach one of the following:

1. The essential memories used to produce the output
2. An essential fundamental activity that receives the input data flow and transforms it into the output data flow, or a component used to assemble the final output data flow

While tracing the output back into the system, we will identify any internal communications or administrative processes and/or memories that are in the direct data path that produces the output data flow. We have to go back far enough to meet the input data flow or an essential memory so that we can include in this analysis every process and data store in the direct data path.

BANK-DEPOSIT is an output data flow produced by the End of Day 5.0* process. Tracing back the data flow, we find that it originates from the CASH-BOX data store defined earlier to be an essential memory (EM). CASH-BOX receives CASH from the process Signup New Customers 1.0*, which receives the TAPE-DEPOSIT, and the process Receive Rental Tapes 4.0*, which receives CASH payments for the tape rental. Process 5.0* is the only process that is responsible for outputting the BANK-DEPOSIT and is therefore an essential fundamental activity (EFA).

So we have:

Process 5.0* (EFA), using CASH-BOX (EM)

CALL is an interactive data flow that is used to resolve discrepancies found on the INVOICE retrieved from the internal communications memory (ICM) INVOICE-BOX. Process 2.3* performs the essential fundamental activity that determines if the NEW-TAPEs received are listed properly on the INVOICE from the Video Tape Wholesaler. After process 2.3* indicates any discrepancies, it places the INVOICE into the internal communications memory (ICM) INVOICE-BOX. The Pay Bills 2.4* process obtains data from INVOICE-BOX and is the only process responsible for the dialogue data flow CALL. Thus, process 2.4* is an essential fundamental activity (EFA).

So we have:

Process 2.4* (EFA), using INVOICE-BOX (ICM)

LIST-OF-TAPES is an output data flow produced by the Signup New Customers 1.0* process. From mini-spec 1.0* in Chapter 5, we find that the process 1.0* inputs the LIST-OF-TAPES from the essential memory (EM) TAPE-

INVENTORY. Process 2.3* places the data into TAPE-INVENTORY when the input data NEW-TAPES arrive. Because process 1.0* is the only process that is responsible for the LIST-OF-TAPES output, it is an essential fundamental activity (EFA).

So we have:

Process 1.0* (EFA), using TAPE-INVENTORY (EM)

NEW-TAPE-PAYMENT is an output data flow from the process Pay Bills 2.4*. This data flow is a check (NEW-TAPE-PAYMENT) in payment for the NEW-TAPEs received. The input INVOICE, obtained from the INVOICE-BOX, lists the tapes received. INVOICE-BOX is only a temporary storage area and is an internal communications memory (ICM). Process 2.3*, determined to be an essential fundamental activity earlier in the analysis, receives INVOICE as input and places it in the INVOICE-BOX. Because process 2.4* is the only process responsible for creating and outputting NEW-TAPE-PAYMENT, it is an essential fundamental activity (EFA).

The CHECK-DATE, CHECK#, CHECK-AMT, and CHECK-NAME data are updated in the internal administrative memory (IAM) CHECKBOOK. CHECKBOOK does not provide any data to the process, but rather is a means of making payment for tapes received. Basically, CHECKBOOK is a form, and all forms are a specific type of technology used to control the flow of data in the system. If process 2.4* used electronic funds transfer (EFT), or some other technology, CHECKBOOK would not be necessary. Therefore, CHECKBOOK is an internal administrative memory (IAM). (Note that if this system had the responsibility of balancing the CHECKBOOK each month, CHECKBOOK would be an essential memory.)

After processing the data, process 2.4* stores the INVOICE in the internal administrative memory (IAM) OLD-INVOICES for historical purposes.

So we have:

Process 2.4* (EFA), using INVOICE-BOX (ICM),
CHECKBOOK (IAM), and
OLD-INVOICES (IAM)

RECEIPT-COPY-2 is an output data flow from the process Receive Rental Tapes 4.0*. From mini-spec 4.0* in Chapter 5, the process inputs the receipts from the data store TAPE-RECEIPT-BOX, marks both RECEIPT-COPY-1 and RECEIPT-COPY-2 as "paid," outputs RECEIPT-COPY-2 to the customer, and stores RECEIPT-COPY-1 in the internal administrative memory (IAM), PAID-BOX. The reason PAID-BOX is an internal administrative memory is because it is a write-only data store. The data is for historical purposes.

The data store TAPE-RECEIPT-BOX is an internal communications memory (ICM) that holds the receipts placed there by the Provide Tapes 3.2* process. TAPE-RECEIPT-BOX is an internal communications memory (ICM) between processes 3.2* and 4.0* for the following two reasons: first, process 3.2* places the copies of the receipts into the data store to permit a passage of time until the customer returns the TAPEs, and second, the retrieval of an entry from TAPE-RECEIPT-BOX actually deletes the entry in the data store. Because process 4.0* has the responsibility for marking the receipt "paid," collecting the money for the tape rental, and outputting RECEIPT-COPY-2 to the customer, it is an essential fundamental activity (EFA).

So we have:

Process 4.0* (EFA), using TAPE-RECEIPT-BOX (ICM), and
PAID-BOX (IAM)

RENTAL-RECEIPT and **TAPE** are outputs from Provide Tapes 3.2*. When process 3.2* receives the TAPE from the customer, it creates a RENTAL-RECEIPT in triplicate. The TAPE and RENTAL-RECEIPT are given to the customer and the two copies of the receipt (RECEIPT-COPY-1 and RECEIPT-COPY-2) are placed in the internal communications memory (ICM) TAPE-RECEIPT-BOX. Process 3.2* is responsible for creating the output data flow RENTAL-RECEIPT and is an essential fundamental activity (EFA).

So we have:

Process 3.2* (EFA), using TAPE-RECEIPT-BOX (ICM)

For **TAPE**, see RENTAL-RECEIPT and TAPE. Process 3.2* passes the TAPE received from process 3.1* to the customer and performs an internal communications activity (ICA). Process 3.1* is an essential fundamental activity (EFA) from the earlier analysis of TAPE-REQUEST.

So we have:

Process 3.2* (ICA)
Process 3.1* (EFA)

TAPE-ORDER-FORM is an output data flow that is a list of the new tapes the manager wishes to order from the Video Tape Wholesaler. The process Order New Tapes 2.2* obtains the information from NEW-TAPE-LIST-BOX when it is ready to issue an order. From mini-spec 2.2* in Chapter 5, the process merely copies the information obtained from the NEW-TAPE-LIST-BOX onto the TAPE-ORDER-FORM and outputs it to the Video Tape Wholesaler. The copy function is a type of transcribing associated with the transportation function, and, as a result, we classify the process 2.2* as an internal communications activity (ICA). The data store NEW-TAPE-LIST-BOX holds the information until process 2.2* is ready, and therefore, it is an internal communications memory (ICM). Tracing back the data flow one more step, we find process Review New Tapes 2.1*. In the mini-spec for process 2.1* in Chapter 5, the manager makes the decision as to which tapes to order and stores the information in NEW-TAPE-LIST-BOX. Process 2.2* obtains the information from NEW-TAPE-LIST-BOX, transcribes it onto the proper form, and forwards it to the Video Tape Wholesaler. Process 2.1* is functionally responsible for the output TAPE-ORDER-FORM and is an essential fundamental activity (EFA).

So we have:

Process 2.2* (ICA), using NEW-TAPE-LIST-BOX (ICM)
Process 2.1* (EFA), using NEW-TAPE-LIST-BOX (ICM)

TOTAL-CHARGES is an output data flow from the process Receive Rental Tapes 4.0*. From mini-spec 4.0* in Chapter 5, the process inputs the two copies of the receipt (RECEIPT-COPY-1 and RECEIPT-COPY-2) from the TAPE-RECEIPT-BOX, computes the TOTAL-CHARGES, and outputs the data flow TOTAL-CHARGES to the customer. TAPE-RECEIPT-BOX serves as a temporary holding area until process 4.0* can perform its task. Because process 4.0* is the only process responsible for producing the TOTAL-CHARGES output, it is an essential fundamental activity (EFA).

So we have:

Process 4.0* (EFA), using TAPE-RECEIPT-BOX (ICM)

After completing the logical analysis compare the input/output processing chart (Fig. 9.3) to the expanded data flow diagram on Fig. 9.2. All of the essential components (processes and data store) on the data flow diagram are shown in bold print. From this analysis we have identified the essential data stores and can now move on to object analysis.

NOTES

NOTES

INPUT/OUTPUT PROCESSING CHART

RENT-EM VIDEO RENTAL SYSTEM	←------ FUNCTIONAL PRIMITIVES ------→									
	1.0	2.1	2.2	2.3	2.4	3.1	3.2	4.0	5.0	
Inputs:										
CALL					EFA					
CASH								ECA	EFA	
CUSTOMER-INFO	ECA									
CUSTOMER-NAME							ECA	ECA		
INVOICE				EFA						
NEW-TAPE				ECA						
NEW-TAPE-CATALOG		EFA								
TAPE								ECA		
TAPE-DEPOSIT	ECA									
TAPE-REQUEST						ECA				
Outputs:										
BANK-DEPOSIT									EFA	
CALL					EFA					
LIST-OF-TAPES	EFA									
NEW-TAPE-PAYMENT					EFA					
RECEIPT-COPY-2								EFA		
RENTAL-RECEIPT							EFA			
TAPE						EFA	ICA			
TAPE-ORDER-FORM		EFA	ICA							
TOTAL-CHARGES								EFA		
PROCESSING SUMMARY:	EFA	EFA	ICA	EFA	EFA	EFA	EFA	EFA	EFA	
Data Stores:										
AVAILABLE-TAPES				EM		EM		EM		= EM
CASH-BOX	EM							EM	EM	= EM
CHECKBOOK					IAM					= IAM
CHIPS				IAM						= IAM
CUSTOMER-FILE	EM						EM	EM		= EM
INVOICE-BOX				ICM	ICM					= ICM
NEW-TAPE-LIST-BOX		ICM	ICM							= ICM
OLD-INVOICES					IAM					= IAM
PAID-BOX								IAM		= IAM
TAPE-INVENTORY	EM			EM						= EM
TAPE-LIBR				EM		EM		EM		= EM
TAPE-NUMBERS				IAM						= IAM
TAPE-RECEIPT-BOX							ICM	ICM		= ICM

Figure 9.3

INPUT/OUTPUT PROCESSING CHART

FOLD OUT THIS PAGE

OBJECT ANALYSIS
. . . identifying the system's objects

To begin object analysis, consider all of the essential memories found in the input/output processing chart (see Fig. 9.3 on the foldout pages). The following data stores are the essential memories of the system: AVAILABLE-TAPES, CASH-BOX, CUSTOMER-FILE, TAPE-INVENTORY, and TAPE-LIBR. An abridged version of the data dictionary which has only the essential memories and some related elementary data elements is shown in Fig. 9.4..

DATA DICTIONARY

⋮	
AVAILABLE-TAPES	= {TAPE-BOX + (TAPE-CHIP)} = * Organized by TYPE and TITLE. *
⋮	
CASH-BOX	= {CASH}
⋮	
CUSTOMER-FILE	= {CUSTOMER-RECORD} = * Organized by CUSTOMER-NAME + CUSTOMER-ADDRESS. *
CUSTOMER-INFO	= CUSTOMER-NAME + CUSTOMER-ADDRESS + CUSTOMER-PHONE
⋮	
CUSTOMER-RECORD	= CUSTOMER-INFO + $_{0}\{\text{TAPE-INFO}\}^{3}$
⋮	
LIST-OF-TAPES	= {TYPE + TITLE + TAPE-RATING} = * A list of all the tapes the store has to rent. *
⋮	
TAPE	= TYPE + TITLE + TAPE-RATING = * A video tape of a movie. *
⋮	
TAPE-BOX	= TYPE + TITLE + TAPE-RATING + TAPE-NUMBER + *TAPE-VIEWING-TIME + STUDIO + {STARS} + (PLOT)*
⋮	
TAPE-CHIP	= TAPE-NUMBER + TYPE = * TYPE is indicated by the color of the Chip: Blue = "BETA," Yellow = "VHS." *
⋮	
TAPE-INFO	= TITLE + TYPE + DATE-RENTED + (DATE-RETURNED)
TAPE-INVENTORY	= {LIST-OF-TAPES} = * Stacks of printed lists of all movies available for rent at RENT-EM Video Rentals. *
⋮	
TAPE-LIBR	= {TYPE + TAPE-NUMBER + [TAPE | TAPE-CHIP]}
⋮	
TAPE-RATING	= ["G" | "PG" | "PG-13" | "R" | "X"]
⋮	

Figure 9.4

From the above data dictionary, the essential memory data stores are reconstructed to contain only elementary data elements. As an example, to reconstruct the data store AVAILABLE-TAPES, replace all group data elements with elementary data elements from the data dictionary, as shown in the following example.

Example

AVAILABLE-TAPES = {TAPE-BOX + TAPE-CHIP}
= * Organized by TYPE + TITLE.*

Substituting the elementary data elements for TAPE-BOX and TAPE-CHIP,

AVAILABLE-TAPES = { TAPE-BOX + (TAPE-CHIP) }

AVAILABLE-TAPES = {TYPE + TITLE + TAPE-RATING + TAPE-NUMBER + TYPE}

TAPE-NUMBER and TYPE are data elements in both TAPE-BOX and TAPE-CHIP and only have to be shown once. Also, we customarily show the primary key data element(s) first:

AVAILABLE-TAPES = {TYPE + TITLE + TAPE-RATING + TAPE-NUMBER}

Likewise, we have to substitute elementary data elements in all of the essential memories:

CASH-BOX = {CASH}

CUSTOMER-FILE = {CUSTOMER-NAME + CUSTOMER-ADDRESS + CUSTOMER-PHONE
+ {TITLE + TYPE + DATE-RENTED + DATE-RETURNED}}

TAPE-INVENTORY = {{TYPE + TITLE + TAPE-RATING}}

TAPE-INVENTORY actually contains many copies of the lists that have all of the tapes available for rent. Thus, the data store is shown correctly as an iteration of an iteration of data elements, with the entire set of data elements shown as the primary key.

The following data store is an example of a data store containing a physical item (TAPE or TAPE-CHIP) as well as description data elements (TYPE and TAPE-NUMBER):

TAPE-LIBR = {TAPE-NUMBER + TYPE + [TAPE | TAPE-CHIP]}

The TAPE-LIBR data store contains two physical items: the actual TAPE = TYPE + TITLE + TAPE-RATING, and a cardboard TAPE-CHIP = TAPE-NUMBER + TYPE. The TYPE of tape is indicated by a blue ("VHS") or yellow ("BETA") 1-inch circular cardboard chip with the TAPE-NUMBER stamped on it. For the purposes of normalization, we will use the following data store definition and have eliminated duplicate elementary data elements:

TAPE-LIBR = {TAPE-NUMBER + TYPE + TITLE + TAPE-RATING}

NORMALIZATION
. . . putting the essential data stores in Boyce/Codd normal form

To find the objects of the system, take all of the essential memories defined in terms of elementary data elements and normalize them to Boyce/Codd normal form. Then the objects of the system will be apparent and the primary keys of the normalized data stores will serve as a unique identifier for each of the respective objects.

The following is a list of assumptions that apply to this example:

1. CUSTOMER-NAME is not unique. CUSTOMER-NAME must be combined with CUSTOMER-ADDRESS to be a unique identifier of a customer.
2. TAPE-NUMBER is a unique identifier for a particular TYPE + TITLE of a TAPE and not just the TITLE.
3. CASH is a physical item of currency and TAPE is a physical video movie.

To begin the normalization process consider the following data stores:

UNNORMALIZED DATA

AVAILABLE-TAPES = {<u>TYPE + TITLE</u> + TAPE-RATING + TAPE-NUMBER}

CASH-BOX = {<u>CASH</u>}

CUSTOMER-FILE = {<u>CUSTOMER-NAME + CUSTOMER-ADDRESS</u> + CUSTOMER-PHONE + {TYPE + TITLE + DATE-RENTED + (DATE-RETURNED)}}

TAPE-INVENTORY = {{<u>TYPE + TITLE + TAPE-RATING</u>}}

TAPE-LIBR = {<u>TAPE-NUMBER</u> + TYPE + TITLE + TAPE-RATING}

First Normal Form

To put a data store in first normal form, remove all repeating groups and adjust the primary key to uniquely identify each data entry in the data store. In this example there are two data stores that are in the unnormalized form because they contain repeating groups, CUSTOMER-FILE and TAPE-INVENTORY.

In the first case (CUSTOMER-FILE), each entry contains the name of a customer and lists all of the tapes that he or she has rented. To normalize the data store, remove the embedded repeating group and extend the primary key to include a unique identifier of each video tape the customer has rented.

In the second case the data store simply contains several lists of the video tapes available for rent. In this case we simply remove one of the repeating group's symbols. By doing this we have not lost any implied relationships among the data elements. The resulting data stores are in first normal form.

FIRST NORMAL FORM

AVAILABLE-TAPES = {TYPE + TITLE + TAPE-RATING + TAPE-NUMBER}

CASH-BOX = {CASH}

CUSTOMER-FILE = {CUSTOMER-NAME + CUSTOMER-ADDRESS + CUSTOMER-PHONE + TYPE + TITLE + DATE-RENTED + (DATE-RETURNED)}

TAPE-INVENTORY = {TYPE + TITLE + TAPE-RATING}

TAPE-LIBR = {TAPE-NUMBER + TYPE + TITLE + TAPE-RATING}

To identify an entry uniquely in the TAPE-INVENTORY data store, only the attributes TYPE + TITLE are necessary.

The embedded repeating group of TAPEs in the CUSTOMER-FILE data store implies an M-1 relationship between TAPEs and a CUSTOMER. Thus the relationship R1[AVAILABLE-TAPES/CUSTOMER-FILE, M-1] is specified.

RELATIONSHIP

R1[AVAILABLE-TAPES/CUSTOMER-FILE, M-1]

To illustrate the relationship we can use the following E-R diagram in Fig. 9.5.

E-R DIAGRAM

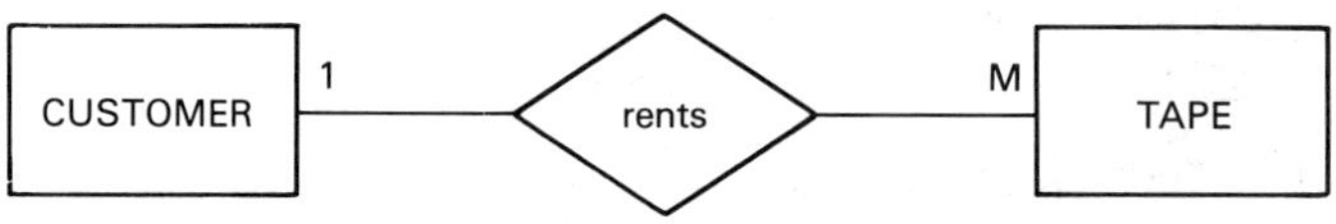

Figure 9.5

Second Normal Form

To achieve second normal form, we start with the data stores in first normal form and examine all of the relationships of the nonkey attributes to the primary key elements. All nonkey attributes must describe or contain a fact about the *whole primary key.* To put a data store in second normal form, examine all of the data stores that have compound primary keys to be sure that each nonkey attribute describes the whole primary key. In cases where the attribute describes or is a fact about only a part of the compound primary key, create two data stores so that each satisfies the condition for second normal form. Let's consider each of the data stores in first normal form that have composite primary keys:

AVAILABLE-TAPES = {TYPE + TITLE + TAPE-RATING + TAPE-NUMBER}

In the data store AVAILABLE-TAPES, TAPE-RATING only describes TITLE and not TYPE + TITLE because it is the specific movie that has the rating and the tape format ("VHS" or "BETA") does not matter. TAPE-NUMBER is unique to TYPE + TITLE, so it remains as an attribute in the AVAILABLE-TAPES data store. The resulting data stores in second normal form are

```
AVAILABLE-TAPES  = {TYPE + TITLE + TAPE-NUMBER}

                 = {TITLE + TAPE-RATING}
```

The next data store that has a compound primary key is CUSTOMER-FILE:

```
CUSTOMER-FILE    = {CUSTOMER-NAME + CUSTOMER-ADDRESS
                   + CUSTOMER-PHONE + TYPE + TITLE + DATE-RENTED
                   + (DATE-RETURNED)}
```

To achieve second normal form, break up the CUSTOMER-FILE data store into two because the attribute CUSTOMER-PHONE is a fact about the customer and not about the tape. The attributes DATE-RENTED and DATE-RETURNED are properly defined as attributes of both customer (CUSTOMER-NAME + CUSTOMER-ADDR) and tape (TYPE + TITLE). When there are attributes associated to two objects, the attribute most often describes the relationship between the two objects. Such attributes were described in Chapter 7 as intersecting data. Therefore, there is a relationship involved here that associates one or more TAPE(s) to one CUSTOMER. As a result, the following data stores and the E-R diagram of Fig. 9.5 are included in the specification as shown in Fig. 9.6.

```
CUSTOMER-FILE    = {CUSTOMER-NAME + CUSTOMER-ADDRESS + TYPE + TITLE
                   + DATE-RENTED + (DATE-RETURNED)}

                 = {CUSTOMER-NAME + CUSTOMER-ADDRESS
                   + CUSTOMER-PHONE}
```

The last data store to consider is TAPE-INVENTORY.

```
TAPE-INVENTORY   = {TYPE + TITLE + TAPE-RATING}
```

As mentioned previously, TAPE-RATING is only a description of TITLE and this data store is already accounted for when we place AVAILABLE-TAPES into second normal form. Also, in considering the data store TAPE-LIBR we find that it is the same as AVAILABLE-TAPES only with the candidate key TAPE-NUMBER selected as the primary key. The following are the resulting data stores in second normal form, together with the relationship we discovered during the normalization through second normal form.

SECOND NORMAL FORM

AVAILABLE-TAPES = {TYPE + TITLE + TAPE-NUMBER}

= {TITLE + TAPE-RATING}

CASH-BOX = {CASH}

CUSTOMER-FILE = {CUSTOMER-NAME + CUSTOMER-ADDRESS + TYPE + TITLE + DATE-RENTED + (DATE-RETURNED)}

= {CUSTOMER-NAME + CUSTOMER-ADDRESS + CUSTOMER-PHONE}

RELATIONSHIP

R1[AVAILABLE-TAPES/CUSTOMER-FILE, M-1]

Third Normal Form (Boyce/Codd Normal Form)

To be in third normal form, all nonkey attributes in the data stores must describe or be a fact about the primary key, the whole primary key (second normal form), and *nothing but the primary key.* This means that a nonkey attribute cannot describe or be a fact about another nonkey attribute. Furthermore, to be in Boyce/Codd (third) normal form, all determinants must be candidate keys. These definitions when applied to data stores form object data stores where a set of related attributes all describe a primary key that is a unique identifier of an object.

Both data stores constructed from AVAILABLE-TAPES satisfy the requirements for third normal form and Boyce/Codd normal form. The same is true for the data store CASH-BOX. Likewise, the data stores formed from CUSTOMER-FILES also satisfy the requirements for Boyce/Codd normal form.

DEFINING OBJECTS
. . . reorganize data attributes by objects

Once the essential data stores are in Boyce/Codd normal form, they relate directly to specific objects in the system. Examine each data store and try to name an object described by the attributes in the data store. The primary key of the data store *is not* necessarily the object; it is a unique identifier of an occurrence of the object. Because there can be many candidate keys that describe the same object, there can be many data stores that relate to the same object. We identify the object associated to each of the data stores in Boyce/Codd normal form . To define the *object data store*, we construct a data store of all the attributes from all the data stores that relate to the same object.

To begin, we build the following table that lists the objects assigned to each of the data stores in Boyce/Codd normal form.

OBJECTS	DATA STORES IN BOYCE/CODD NORMAL FORM	
Tape	AVAILABLE-TAPE	= {TYPE + TITLE + TAPE-NUMBER}
Tape		= {TITLE + TAPE-RATING}
Money	CASH-BOX	= {CASH}
Customer and **Tape**		= {CUSTOMER-NAME + CUSTOMER-ADDRESS + TYPE + TITLE + DATE-RENTED + (DATE-RETURNED)}
Customer	CUSTOMER-FILE	= {CUSTOMER-NAME + CUSTOMER-ADDRESS + CUSTOMER-PHONE}

Basically, there are only three objects for which this system has to store data: CUSTOMER, TAPE, and MONEY. From our analysis, we can combine some of the data stores and construct the following three object data stores:

OBJECTS	OBJECT DATA STORES
CUSTOMER	{CUSTOMER-NAME + CUSTOMER-ADDRESS + CUSTOMER-PHONE}
TAPE	{TYPE + TITLE + TAPE-NUMBER + TAPE-RATING}
MONEY	{CASH}

The primary key for the object data store TAPE can be either TAPE-NUMBER or TYPE + TITLE. In Boyce/Codd normal form, however, TAPE-RATING is an attribute of only TITLE. This raises an important point: *Object data stores do not necessarily have to be in Boyce/Codd normal form.* The process of normalization to third normal form and Boyce/Codd normal form is a means of determining the objects and only aids in defining the object data stores. There are some attributes that describe the relationship between CUSTOMER and TAPE:

RENTS = {CUSTOMER-NAME + CUSTOMER-ADDRESS + TAPE-NUMBER
+ DATE-RENTED + (DATE-RETURNED)}

The attributes DATE-RENTED and DATE-RETURNED relate to the association of CUSTOMER and TAPE. We can better understand this from the E-R diagram of Fig. 9.6.

ENTITY-RELATIONSHIP DIAGRAM
. . . defining relationships among the objects

The E-R diagram illustrates the relationships whereby one CUSTOMER can rent one or more TAPEs, and, when they do, there is one payment of MONEY. DATE-RENTED and DATE-RETURNED are attributes of the relationship (rental agreement) between a CUSTOMER and TAPEs. All the other attributes illustrated in Fig. 9.6 describe either CUSTOMER or TAPE, as indicated.

E-R DIAGRAM

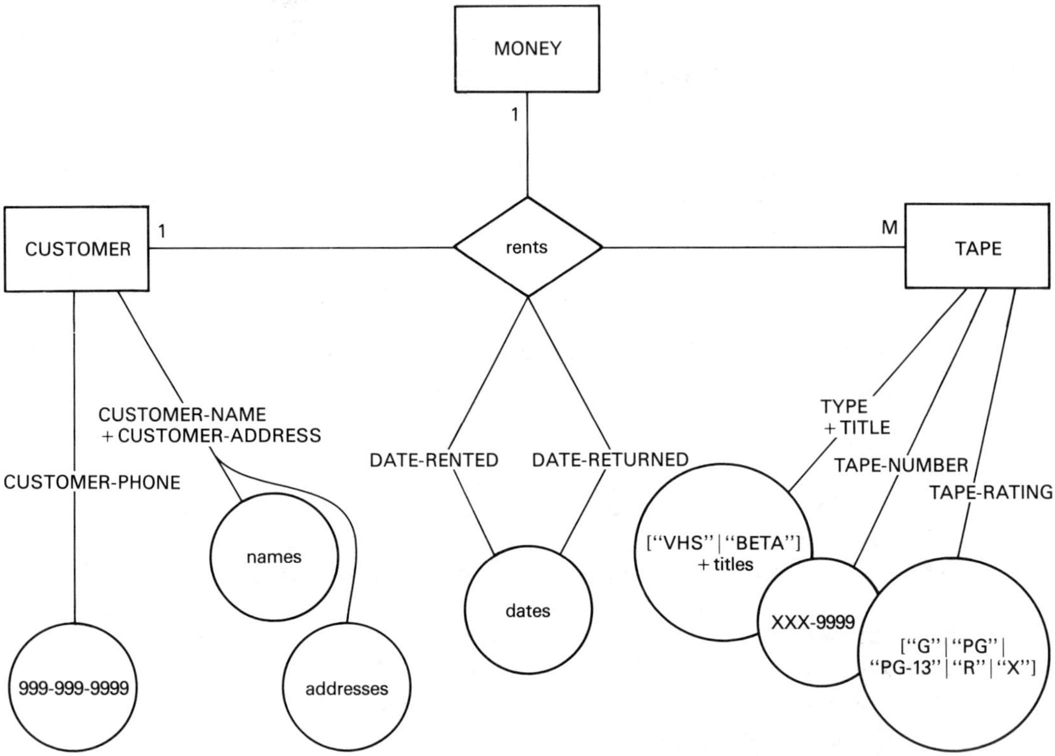

Figure 9.6

EVENT ANALYSIS
. . . the purpose of a system is to respond to events

Event analysis considers each input and output of the system and determines:

1. What independent event occurred that either caused the source to present the input data flow (stimulus) to the system and/or caused the system to perform some activity or send output data out of the system?

2. What kind of event occurred: external (beyond the boundaries of the system), temporal (time-scheduled), or internal (internally generated process(es) called upon to stabilize the system after some unusual condition arises that jeopardizes the continued operation of the system)?
3. What functional primitive processes and data stores contribute to the system's actions and responses when the event occurs?

If we were to expand the input/output processing chart used during logical analysis to contain three additional columns: an event number, an event description, and the type of event, a suitable format exists to do event analysis. With the expanded chart of Fig. 9.7, we can begin to identify the events that caused the input data to enter the system and which events caused the system to produce the indicated output data.

EXPANDED INPUT/OUTPUT PROCESSING CHART

RENT-EM VIDEO RENTAL SYSTEM	1.0	2.1	2.2	2.3	2.4	3.1	3.2	4.0	5.0	EVENT #	DESCRIPTION OF EVENT	TYPE OF EVENT
Inputs:												
CALL					EFA					1	Wholesaler delivers TAPEs & INVOICE.	Ext.
CASH								ECA	EFA	2	Customer finished using the TAPE(s).	Ext.
CUSTOMER-INFO	ECA									3	Person decides to be a customer.	Ext.
CUSTOMER-NAME							ECA	ECA		4	Customer decides to rent TAPE(s).	Ext.
INVOICE				EFA						1	(Same as #1)	
NEW-TAPE				ECA						1	(Same as #1)	
NEW-TAPE-CATALOG		EFA								5	Wholesaler solicits for sales.	Ext.
TAPE								ECA		2	(Same as #2)	
TAPE-DEPOSIT	ECA									3	(Same as #3)	
TAPE-REQUEST						ECA				4	(Same as #4)	
Outputs:												
BANK-DEPOSIT									EFA	6	End of day - Make bank deposit.	Temp.
CALL					EFA					1	(Same as #1)	
LIST-OF-TAPES	EFA									3	(Same as #3)	
NEW-TAPE-PAYMENT					EFA					1	(Same as #1)	
RECEIPT-COPY-2								EFA		2	(Same as #2)	
RENTAL-RECEIPT							EFA			4	(Same as #4)	
TAPE						EFA	ICA			4	(Same as #4)	
TAPE-ORDER-FORM		EFA	ICA							5	(Same as #5)	
TOTAL-CHARGES								EFA		2	(Same as #2)	

Figure 9.7

After completing a detailed analysis of each input and output data flow, we find that the Rent-em Video system responds to six different events. Five are external events and one is a temporal event.

EVENT TABLE AND EVENT DIAGRAM
. . . reorganizing processes and object data stores by events

Event analysis provides a means of partitioning the system by events. In explaining the functions of a complex system, we may find it advantageous to explain what the system does, one event at a time. Consider the following information derived from the event analysis.

EVENT TABLE

EVENT #	TYPE	DATA FLOW	SOURCE/SINK	INPUT/ OUTPUT	FUNC'L PRIMITIVES	ESSENTIAL DATA STORES	OBJECT DATA STORES
1	EXT	CALL	Video Wholesaler	IN/OUT	2.4	None	None
		INVOICE	Video Wholesaler	INPUT	2.3	None	None
		NEW-TAPE	Video Wholesaler	INPUT	2.3	TAPE-INVENTORY AVAILABLE-TAPES TAPE-LIBR	TAPE
		NEW-TAPE-PAYMENT	Video Wholesaler	OUTPUT	2.4	None	None
2	EXT	CASH	Customer	INPUT	4.0	CASH-BOX	MONEY
		TAPE	Customer	INPUT	4.0	AVAILABLE-TAPES TAPE-LIBR	CUSTOMER TAPE
		RECEIPT-COPY-2	Customer	OUTPUT	4.0	None	None
		TOTAL-CHARGES	Customer	OUTPUT	4.0	None	None
3	EXT	CUSTOMER-INFO	Customer	INPUT	1.0	CUSTOMER-FILE	CUSTOMER
		TAPE-DEPOSIT	Customer	INPUT	1.0	CASH-BOX	MONEY
		LIST-OF-TAPES	Customer	OUTPUT	1.0	TAPE-INVENTORY	TAPE
4	EXT	CUSTOMER-NAME	Customer	INPUT	3.2, 4.0	CUSTOMER-FILE	CUSTOMER
		TAPE-REQUEST	Customer	INPUT	3.1	AVAILABLE-TAPES TAPE-LIBR	TAPE
		RENTAL-RECEIPT	Customer	OUTPUT	3.2	None	None
		TAPE	Customer	OUTPUT	3.1, 3.2	None	None
5	EXT	NEW-TAPE-CATALOG	Video Wholesaler	INPUT	2.1	None	None
		TAPE-ORDER-FORM	Video Wholesaler	OUTPUT	2.1, 2.2	None	None
6	TEM	BANK-DEPOSIT	Bank	OUTPUT	5.0	CASH-BOX	MONEY

Figure 9.8

From the event table of Fig. 9.8, we could construct an event data flow diagram, also known as an event diagram. It is a level-1 diagram and balances with the original context diagram from the physical model. For the previous example, the context diagram is found in the beginning of Chapter 5. Level-1 of the event diagram contains one bubble for each event and indicates the interactions between the events and the object data stores, as shown in Fig. 9.9.

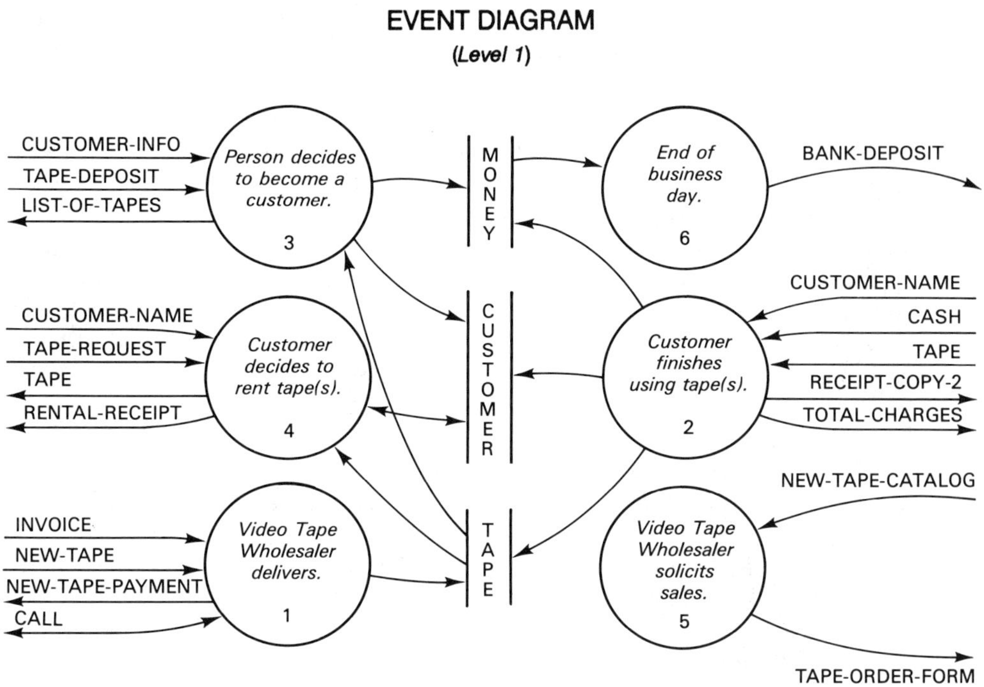

Figure 9.9

In this simple example, each event has only *one* scenario. We could analyze each scenario and determine the system behaviors necessary to accomplish the expected result in each case. In scenario 1.1 (*Video Tape Wholesaler delivers tapes*), the behavior derived from examining the functional primitives and mini-specs involved in servicing this scenario are

> ADD NEW TAPE TO LIBRARY, which consists of the processes that unpackage new tapes, assign a TAPE-NUMBER to the TAPE-CHIP for each new tape, put the TAPE and TAPE-CHIP in their proper places in the TAPE-LIBR and AVAILABLE-TAPE data store, and add the TITLE, TYPE, and TAPE-RATING to the TAPE-INVENTORY data store.
>
> VERIFY INVOICE, which consists of a single process that checks that the tapes received are correctly stated on the INVOICE. All discrepancies are indicated on the INVOICE, which is placed in the INVOICE-BOX.
>
> PAY WHOLESALER, which consists of three processes. The first, examines the INVOICE from the INVOICE-BOX and if any discrepancies are indicated, a CALL is placed to the wholesaler. The

second is the process that actually issues the check and sends it to the wholesaler. The third process stores the "paid" INVOICE in an archive data store.

Likewise, each event is examined and the behaviors that produce the results in response to each scenario are obtained from the physical structured specification. It is possible that two scenarios, in two different events, may perform the exact same behavior. This information is very useful, particularly in automated systems that share the same processing routine. In the event that under one scenario the user requests that the behavior be modified, we could determine under what other conditions the same processing routine is used, and if the requested modification would be acceptable under all of the other scenarios that use the behavior.

EVENTS DURING MAINTENANCE
. . . no system is without modifications or enhancements

Users often request modifications and enhancements for an existing system in terms of *events*. Such changes can be one or more of the following types.

Add/Delete Events

1. *Add an event*: There is a new event to which the system must recognize and provide an appropriate response.

   Example: In the case of the Rent-em Video system, the management may wish to check the files regularly to see if any customer has failed to return the TAPEs rented. Because there is no stimulus presented to the system to announce the event, this activity depends on a time event to initiate the activity. The new event would be a temporal event.

2. *Delete an event*: The system's response to an event is no longer required.

   Example: If the management decided that no new tapes would be purchased from Video Tape Wholesalers, then event 5 would be eliminated and processes 2.1* and 2.2* could be removed from the system.

Recognition/Response to Events

3. *Stimulus change*: If the source changed the input stimulus that signals that an event occurred, then a change in the system is necessary. The system requires a modification, so that when the system receives the new stimulus, it recognizes that the event occurred as if it had received the original stimulus.

Example: When a customer decides to rent tape(s) (event 4), he or she would originally input two stimuli, CUSTOMER-NAME and TAPE-REQUEST. The owner may wish to change the system so that the customer simply inputs their phone number and tape request to the system. A change to the organizational structure (primary key) of CUSTOMER-FILE is necessary so that the system can locate the customer's entry in CUSTOMER-FILE simply by using CUSTOMER-PHONE instead of CUSTOMER-NAME. Notice that the event and the system's response are the same as before. The only change is that CUSTOMER-PHONE announces the event to the system instead of CUSTOMER-NAME.

4. *Response change*: The owner of the system requests that the system respond differently to an event.

Example: In the Rent-em Video system, the management or owner of the business may decide that there is no need to provide the customer with a RENTAL-RECEIPT when they rent the TAPEs. The system could then replace the processing of a three-part receipt with a one-part receipt kept in the TAPE-RECEIPT-BOX. Upon the return of the TAPEs, the process retrieves the receipt, marks it "paid," places it in the PAID-BOX, and gives a polite "thank you" to the customer.

Another example that effects both events 2 and 4 is if the management decides that the customer should pay for one-day rental in advance. Thus, the system would require an additional input CASH with CUSTOMER-NAME and TAPE-REQUEST, and would respond with the TAPE and RENTAL-RECEIPT. With this implementation, the management could install a "drop-off" service, where the customer can simply return the TAPEs after one day without waiting to pay the rental charges.

Enhance Efficiency/Capacity

5. Enhance the system's processing efficiency: We can add, remove, or reorganize the internal processes activated when a particular event occurs. There are no changes to the events or stimulus in this case. The objective here can be to increase processing capability, reduce response time, or reduce the processing costs when a specific event occurs.

Example: In the Rent-em Video system, the system is to increase its ability of service a large number of events 4 and 5. One way to do this is to eliminate the contention for the AVAILABLE-TAPES and TAPE-LIBR accesses by processes 3.1* and 4.0*. To do this, we could create a new process that receives requests from 3.1* and 4.0*, interacts with the two data stores, and responds to the requesting process. The new process is a "chaser" function that retrieves the tapes requested or places tapes back into the data store. This new "chaser" function seems to be a subordinate function to processes 3.0 and 4.0. There are at least two ways of handling such a situation:

a. Create a process 6.0*, which is a functional primitive that takes requests from 3.1 and 4.0 and responds accordingly.

b. Create a common processing routine (CHASER) called by processes 3.1* and 4.0* and document it in the data dictionary.

One purpose for doing event analysis is to organize the information about the system so as to be able to identify quickly the scope and bounds of a change to the system. Users, management, or owners of a system usually give the systems analyst their requirements in terms of the events to which the system responds. The following list enumerates the possible types of changes that can be made to a system:

1. Adding an event and creating a new response from the system.
2. Deleting an event and eliminating the system's response.
3. Recognizing a new or different stimulus for an event.
4. Providing a new or different response for an event.
5. Changing the internal processing for an event.
6. Any combinations of the above.

SUMMARY

. . . a worker is only as good as his or her tools

The following tools have been used to represent the physical model shown in Chapter 5 of an existing system:

Context diagram
Data flow diagrams
Data dictionary
Mini-specifications
E-R diagrams

With the above tools, we can represent in a complete, accurate, unambiguous, and nonredundant way, the detailed operations of an existing system from a physical implementation standpoint. The processes, data, and technology are included in the model from the point of view of the users of the system.

To derive an in-depth understanding of the system, we need tools to look beyond the physical implementation. The following tools are used to represent the logical model of an existing system:

Expanded data flow diagram

Input/output processing chart

Essential data flow diagrams

Normalization to determine object data stores

E-R diagrams of object data stores

Input/output processing chart with events

Event table

Event model
- Event diagram
- Scenario diagrams
- Scenario specs
- Behavior diagrams
- Behavior specs
- Mini-specs

With the above tools, we can obtain a significant amount of information about a system and its operations. By using these tools, we can suggest alternatives to accomplish any new objectives, enhancements, or changes to the existing system. In addition, we have the ability to project the size and scope of the work involved to implement any proposed alternatives.

Many changes to a system do not affect the essence of a system and are merely different ways of implementing the infrastructure of a system. Again, from the above analysis, we have the ability to differentiate between those changes that affect the system's essence and those that only affect the system's infrastructure.

The most valuable aspect of these tools is that they provide a means of communication among the system professionals, administrators, and users of a system. Simple notation is used to encourage user participation, frequent walkthroughs, and iterative changes, whereby the system professionals can review their perceptions, ideas, and proposals in an environment that fosters modifications and iterations. The requirements stage is the least costly place to identify and make corrections to the proposal. It not only provides a means that permits an analyst to become intimately familiar with the operations of an existing system, but it also relates the proposed alternatives and modifications in a way that administrators, users, and owners of systems can understand. These tools provide a means of increasing the effectiveness of the most valued resource of an organization, the *problem solver.*

QUESTIONS

1. Given the following unnormalized data stores from the example in this chapter, assume that the Video Tape Wholesaler assigns the TAPE-NUMBER and that it is unique to the *TITLE only* and not TYPE + TITLE.

   AVAILABLE-TAPES = {TYPE + TITLE + TAPE-RATING + TAPE-NUMBER}
   CUSTOMER-FILE = {CUSTOMER-NAME + CUSTOMER-ADDRESS + CUSTOMER-PHONE + {TYPE + TITLE + DATE-RENTED + DATE-RETURNED}}
   INVOICE = {WHOLESALER-NAME + WHOLESALER-PHONE + INVOICE-DATE + {TYPE + TITLE + TAPE-NUMBER + QUANTITY + PRICE + (NOTE)} + TOTAL-INVOICE-AMT}

   a. Place the set of data stores in Boyce/Codd normal form.
   b. Define the object data stores.
   c. Determine the relationships among the object data stores and provide an E-R diagram.

2. When is the data store CHECKBOOK used as an internal administrative memory and when is it an essential memory?

3. Must the *object data stores* be in third normal form or Boyce/Codd normal form? Explain your answer.

PROBLEMS

1. Update the following to expand the system whereby the owner sells Video Tape Recorders which are obtained from a Video Tape Recorder Wholesaler.

   a. **Data Dictionary** to contain the necessary data and object data store(s).

   b. **Event Diagram** to define the input and output data flows for any new events. Discuss how any existing events need to change and what new events need to be added. (Assume all sales are cash transactions.)

2. What impact on the system's objects and events would there be if the owner also decided to repair video recorders? (Assume all services are paid for in cash.)

3. What changes (objects and events) to the Rent-em Video Rental system would be necessary if the owner decided to offer a cleaning service for VCRs? Customers would bring in their VCR for cleaning and leave their names, address phone number and whether the VCR is used "heavily" (record/playback more than 10 hours a week) or "lightly" (less than 10 hours per week). Upon

completing the cleaning the store employee would call the customer and tell them there VCR is ready.

The store employee would keep a copy of the receipt given to the customer upon receiving payment for the cleaning service and at the time the store employee would fill out a post-card with the customer's name and address. The post-card would be saved in a DATE-FILE under the date it should be mailed. The MAIL-DATE is computed to be six months from the date of the last cleaning if the customer's VCR is used "heavily" or one year from the date of cleaning if the VCR is used "lightly".

Every Monday morning an employee would mail all the post-cards due to be mail during the week to remind customers that it may be time to bring their VCR in for another cleaning.

4. Construct an event diagram and additions to the data dictionary for the Rent-em Video Rental system with the additions stated in problem #3 in this Chapter.

PART III

Real-Time Models and Systems Architecture

This part addresses real-time systems and presents a complete example of a structured specification that includes real-time processes. In the final chapters a basic information-system architecture is proposed. Such an architecture takes advantage of the process-oriented techniques, which transform the data into the desired results, and the data-oriented techniques, which put information retrieval directly into the hands of the end users.

10

REAL-TIME SYSTEMS SPECIFICATIONS

The preceding nine chapters have presented techniques and tools to model most business-information systems. With them, an analyst can describe the elements of a system that are of primary concern, namely:

The sources/sinks, and internal paths along which data flows

The composition of data flows and data stores

The policy governing the transformation of data flows

These elements are all included in a structured specification, consisting of the context diagram, data flow diagrams, data dictionary, and mini-specifications. However, to complete the requirements specification for a system, additional tools are needed to address:

The relationship among data

The relationships among processes

An entity-relationship (E-R) diagram shows the relationships between the data elements in the system. Up to now, we have not had to include any relationships between processes other than normal data flows. In some systems, it is necessary to know when a process is active with relation to time, or how long a process takes to perform its tasks, or whether two or more processes can be active at the same time. In most business-information systems, the relationships between processes have little or no effect on the results of the system. The structured specification is a model of a system that assumes the processes have always existed and are constantly running. It is only in the case where there is a critical relationship between the execution of a process and the element of time that the need to model the relationships between the processes is created.

Systems that have processes that are functionally dependent on real time are known as *real-time* systems. **REAL TIME is defined as the condition to operate concurrently with a regular or predictable time relationship with independent real-world events.** In an attempt to synchronize the execution of the software with independent real-world events, the software may have to respond differently to the same input at different moments in time.

When time becomes a relevant factor regarding the processing of data within a system, other aspects of the processes also become important: the origin of the processes, activation and deactivation of the processes, as well as the control of the internal policies governing the transformation of data as they relate to time.

In this chapter, I will discuss the following three basic ways a process can effect, or have control over, the processing that goes on within a system.

Initiation/Termination of Processes

Feedback and Control of Processes

Finite-State Processes

In defining the requirements for a system that is dependent on time, we must remember the fundamental objective of system analysis is to specify *only* the system requirements. We must resist the temptation of going into the high-level design of *how* the system will accomplish its objectives.

Because of the advances in technology and the advances in software engineering, the use of real-time techniques is more frequently being applied to business problems. Therefore, in modeling real-time systems, we must consider and include in the system model the *flow of control*, which is the distinguishing characteristic of a structured specification for a *real-time system*. **A REAL-TIME SYSTEM is one that contains processes that operate concurrently with a regular or predictable time relationship with independent external or temporal events.** Or, another way, we can recognize real-time systems as those that produce output responses that must be synchronized with independent external events. In most cases, a user sitting at a terminal waiting for an information response *does not* constitute an *independent* external event. The actions of such a user are dependent on the response (or lack of response) from the system.

Users can have more or less patience while interacting with a system. However, the users' ultimate actions are fundamentally *dependent* on the response, or lack of response, by the system. Signaling a valve to open at the precise moment that an independently controlled conveyor has positioned a bottle under the spout does satisfy the definition of a process in a real-time system. The mechanism that controls the conveyor and the mechanism that controls the valve must be synchronized in time so that the bottle begins to be filled the moment it is in proper position. The control mechanism for the valve is independent of the control mechanism of the conveyor, and, thus, their operating characteristics satisfy the conditions for a real-time system.

Before presenting a discussion on modeling the *flow of control* in a structured specification, let's clarify the meaning between **on-line** data-processing systems and **real-time** processing systems.

REAL TIME SYSTEMS VERSUS ON-LINE SYSTEMS
. . . on-line systems are not necessarily real-time systems

Because a computer system employs telecommunications terminals does not necessarily mean that it operates as a real-time system. **An ON-LINE SYSTEM is one that carries on direct data flows that collect input data and/or provide output information while the system is processing data.** There are major distinctions in the operating environments and the operations of on-line systems and real-time systems. Let's discuss each separately.

Operating-System Environment

On-line systems usually operate by conducting a dialogue with the user. The dialogue permits the user to direct the computer in performing tasks. If the operating system is a multiprogramming environment, the scheduling process of the operating system activates or dispatches the tasks according to some predetermined algorithm. The response time the user sees at the terminal is dependent on the amount of time the computer shares with the task that is interacting with the user. As more and more tasks are dispatched and made *ready to run*, the amount of time each task receives diminishes, until the response time increases beyond the tolerance level of the user. In an on-line operating-system environment, the computer takes its direction from the user. As more and more users are added to the system, or as more and more tasks are started by the user, response time is effected until the system appears to come to a halt. The appearance that the computer has come to a *halt* is a result of the operating system spending more of its time managing the work load than running the user's tasks.

In a **real-time system**, the operating system, not the user, controls the amount of central processing unit (CPU) time each task uses according to a predefined set of priorities. As more and more tasks are started, the real-time system will execute the processes that have the highest priority according to a predefined allocation of execution time. In this manner, specific tasks are guaranteed a specific amount of execution time. If response time appears *slow*, it is usually because there are too many high-priority tasks demanding service. In such cases, the operating system takes control and *suspends* the execution of lower-priority tasks until a sufficient amount of computer time is available to guarantee the high-priority tasks the guaranteed minimum amount of execution time. Notice, that in a real-time operating-system environment, the computer takes its direction from the scheduling process and *not the user*. As more and more high-priority tasks are added to the work load, the scheduling process deactivates lower-priority tasks. A real-time system can be overloaded, whereby some of the high-priority tasks must wait an inordinate amount of time due to an insufficient time for the CPU to service all of the high-priority tasks. Often, the operating system detects and reports such overload conditions to the user or operator of the system.

Interrupts (Control)

When an **on-line system** completes a task, such as completing an input retrieval from a data store, the CPU is interrupted and the task waiting for the input is marked *ready-to-run*. When this occurs, the operating system goes through a priority search for the highest-priority task that is ready-to-run. The operating system passes control to that priority task until a predetermined amount of time passes or another interrupt occurs.

In a **real-time system**, when an interrupt occurs, the CPU performs the request immediately if the priority is higher than the task that was currently being executed. Thus, lower-priority tasks do not execute if the machine is constantly being interrupted with high-priority interrupts. The lower-priority tasks are suspended, or held in abeyance, until CPU time is available. The scheduling process can also allocate a portion of time to a particular set of processes. In this type of environment, certain processes are guaranteed to run or have access to a portion of the CPU time regardless of other activities being performed.

Results of Process Failures

In a normal business environment, the business is run by the data generated by computers. Therefore, in a data-processing system (on-line or not), the highest objective is that the **output data be absolutely accurate**. If conditions in the computer exist that jeopardize the accuracy of the output data, it would be more desirable to halt and not output anything, rather than output erroneous data. This philosophy accounts for the fact that most business applications would fail at the point of providing erroneous data, rather than proceed and produce erroneous results. In a typical business program, if the computer attempts to divide by zero, the operating system would take control, stop all activity in the program, and provide the traditional *core dump* (even though most computers do not store anything in ferrite-core memory today). When the application *goes down*, all results being processed by the program at that time are lost. The rule is that it is more desirable to provide no results than to provide erroneous results. Even in a multiuser time-sharing environment with on-line applications running, if a single user activity attempts to produce an erroneous result as detected by the operating system, the application serving the single user will fail and the on-line interaction will cease.

In a real-time system, the highest objective is the **preservation of the system**. Therefore, it is more acceptable to provide erroneous results than to **halt** the application and cause the entire system to *go down*. This philosophy governs such large-scale real-time operating systems as the electronic switching systems that provide the majority of the telephone service in the United States and many other types of monitor and control systems. In the case of the telephone system, it is more desirable for the system to output a call to a wrong number, than to stop all processing of telephone calls in the system. In the case of a heartbeat monitoring system, it is better to give a false alarm for a patient when the input data cannot accurately be interrupted than to abort the application and stop monitoring all of the patients connected to the system.

Difference Among On-Line, Real-Time, and Batch Systems

So what have we said about on-line and real-time systems? First of all, we should consider if the system needs to be continuously running, awaiting input data. If not, perhaps the system should run at some specific time when all of the input data are available. Such systems are the traditional *batch systems* in data processing. **A BATCH SYSTEM is one whose data is collected and is fed into the system only at the specific times when the system is activated.**

Second, if the system must continuously await the arrival of data to serve the user, than the system is required to be, at least, an on-line system. If there is a further need to have the system guarantee that certain processes be allocated for a certain percentage of the processing time in order to be synchronized to some independent external or temporal event, then the system must be a real-time system.

In a real-time system, the time event to which the system is synchronized must be an *independent event* that is external to the system. **A user waiting for the reply from a terminal is dependent on the execution of the process and is usually not an independent event.** Depending on where the boundary of the system lies, the user may or may not be external to the system. The user's waiting tolerance is not a suitable event to qualify the system as a real-time system. The user's actions are dependent on the system's response or lack of response. Examples of external independent events are the movement of freight cars in a railroad yard, releasing product into containers that have been positioned under the outlets by an independently controlled conveyor; and the refreshing of a screen display that shows the current attitude and speed of aircraft as they move about the skies. These are all real-world events that have to be synchronized with the output from some real-time process(es) executing in a computer.

Examples (On-Line Systems and Real-Time Systems)

The following are some examples of interactive system that are either on-line or real-time systems.

1. A pharmacist enters a customer's name into a drug verification system to verify a prescription before filling the customer's order.

The system has to be constantly running, awaiting input from the pharmacist. No other synchronization is required.

On-line system

2. A railroad yardmaster sits in a tower overlooking a railroad yard and enters a command into the freight car routing system to reset the switch tracks to route railroad cars as they travel through the railroad yard.

The system has to be constantly running, awaiting input from the yardmaster. The setting of the switch is related to the independent external event of the moving freight cars to route the freight cars to their proper position in the railroad yard.

Real-time system

3. A point-of-sale terminal in a supermarket, connected to the bar-code scanning and pricing system, is activated by the universal bar code on a product as it passes over the scanner. The system reads the product identification number, looks up the price of the item, calculates a new subtotal, and provides an acknowledgement tone. The tone indicates to the checkout clerk that the system processed the item. If the product identification number does not match that in the system or the system was unable to read the identification number, the system responds with a blunt tone, informing the checkout clerk that the item was not properly processed.

The system has to be constantly running, awaiting input from the checkout clerk. The acknowledgement or rejection tone tells the checkout clerk to select and scan the next item. Because the checkout clerk does, in fact, wait until an accept tone or reject tone is heard before scanning the next item, the clerk is not an independent external event.

On-line system

4. A computer-controlled conveyor moves cases of empty bottles along a path to where the bottles come under a set of spouts. At the precise moment, the bottles arrive beneath the spouts, a signal passes to another computer that opens the valves for a specified amount of time to allow the bottles to fill to their capacity. When the specified amount of time has elapsed, the computerized conveyor moves the bottles along to the packaging stage.

The system has to be constantly running, awaiting the input signal from the conveyor mechanism. Upon receiving the signal, the second computer executes a command to open the valve and fill the bottles. The computer that controls the opening of the valves must respond within a guaranteed amount of time to fill the bottles before they move on to the next stage. Synchronization is required between the two independent controlling mechanisms, thus fulfilling the requirements of a real-time system.

Real-time system

5. The stock price summary system displays the stock transaction that took place on the stock market trading floor. The display contains the amount of the last sale with the symbol of the stock moments after the transaction. This information tells the buyers and sellers on the trading floor, as well as stockbrokers across the country, the most recent price paid for the stock.

> *The system has to be constantly running, awaiting input from the trading floor. The failure to display the sale does not prevent subsequent sales of the stock; thus, the system is not synchronized with any particular independent external or temporal event. The stock market reporting system is not a real-time system because the posting of the last price paid for a stock does not prevent the next sale from taking place. The results of the system are used for information purposes only and really do not have to be closely synchronized with the stock transaction taking place on the trading floor.*
>
> ***On-line system***

6. Ticketmatics is a system used to sell tickets for concerts and sporting events. The system provides a ticket agent with the seating location of the best available seats within a specified price category, and, if acceptable to the customer, the system can book the seats and print the tickets at a remote ticket office.

> *The system has to be constantly running, awaiting input from the ticket agent. The failure to display quickly the available seating or to print the tickets is necessarily not a problem. Obtaining the tickets does not have to be synchronized with any independent external or temporal event. The waiting customer will continue to wait. The action of the waiting customer is dependent on the system and does not qualify this system as a real-time system.*
>
> ***On-line system***

In considering whether a system should be an on-line system or a real-time system, I'm reminded of the story frequently told by Rear Admiral Grace Hopper of the United States Navy Reserves, Retired, when she talked about the value of information:

> I know of a chemical plant that's totally operated by computers. Information comes in from marketing and goes into a computer; a computer opens valves and pushes stuff through pipes; nice computer reports go up to the president's desk.
>
> Now suppose that two pieces of data simultaneously enter a computer at this plant. One comes from a valve and says if you don't open the valve, the plant's going to blow up. You have less than a minute to act. One hundred lives are at stake, and so is a $100 million chemical plant. At the same instant, from another part of the system, comes the information that Joe did 2 hours of overtime.[H-1]

The reaction time for the first piece of information is measured in seconds. If the process does not respond in time to the information from the pressure valve, the tank will explode. The result would be loss of product, equipment, and possibly human life.

The reaction time for the second piece of information is measured in days. Probably several days would pass before that information of time worked is needed to compute the employee's pay. If the process does not respond in time to the piece of information, the employee would experience a week or so delay in receiving the overtime pay. The delay would cause a certain amount of dissatisfaction by the employee.

If there is a need to guarantee that a piece of information is processed within a specific amount of time to synchronize the results of the processing to an independent external or temporal event, the system controlling the processing has to be a **real-time system**. Alternatively, when information is processed by a system that continuously stands ready to receive the information, but upon receiving the data, the system processes it in proper order with all of the other work active on the computer and is not obliged to respond consistently within a specific time interval, the system is simply an **on-line system.**

If the input data is permitted to accumulate and is processed at some specified period of time, the system can be a traditional **batch system.**

MODELING "FLOW OF CONTROL"
. . . when activity is not caused by the arrival of data

In considering the *flow of control* within a system, we must first look at who is doing the controlling. Control of a process can be either **external**, that is, by another process, or **internal**, that is, by its own internal logic.

When we speak about control of processes within a system, we mean that a process can take action to correct the discrepancy between the output of the system and the *criteria* for the system. **The CRITERIA is the standard of measurement used to illustrate preference to a given alternative.**[O-1] When the output does not meet the criteria, the process producing the output is changed either by an external entity or by the results of an internal monitoring activity. An external process that controls another process is referred to as a *supervisory process*, as defined earlier in Chapter 2.

CONTROL is the action that initiates, terminates, or alters process behavior in carrying out the transformation of input data into output data. A process can exercise control *internally* (controlling itself) or *externally* (controlling a subordinate process). A process operates under **INTERNAL CONTROL when a predefined plan permits the process to perform differently based on input, time, or the variation in output as compared to the criteria.** There are many techniques to plan process behavior in transforming the input data into output data. Many of these techniques apply during the design stage. However, later in this chapter, I will present finite-state processes that provide a means of specifying requirements for a process to control internally the transformation of input data into output data with respect to time or previous actions performed by the process.

EXTERNAL CONTROL is enacted by a supervisory process that creates, activates, reactivates, deactivates, or suspends the operation of another process in the system. The following section has three different ways of directing another process within a system.

Controlling a Process

Initiate/Terminate a Process. There are several ways a process can control another process. If a controlling process determines that another task has to operate simultaneously to its own execution, it can initiate a new process. The new process is a *target process* (see the definition in Chapter 4). When the supervisory process initiates the target process, the target process executes independently and may compete for the same resources with the supervisory process (unless other requirements take precedence). Upon completing its tasks, the target process will halt and relinquish all of the resources it uses to perform its tasks. These resources are then available to the operating system to allocate to other processes.

Activating/Deactivating a Process. A second way is for a supervisory process to activate a predefined target process. The target process continues to operate until the supervisory process (or some other independent process) directs it to halt. Once the target process terminates, it frees all of the resources used to accomplish its tasks. These resources are then available to the operating system to allocate to other processes.

Activating/Suspending a Process. A third type of control is for an external supervisory process to activate a target process whose activity was temporarily suspended by its own initiative. Activating a target process is commonly known as a *wake-up* of a process that has gone to *sleep*. A target process may have completed its processing, or it may have nothing to do until some external or temporal event occurs, at which time it will reactivate to continue or complete its processing. During periods of inactivity, the process simply puts itself to sleep. During the time that the process is sleeping, all of the resources that are assigned to it remain assigned to the process. The process can resume at any time and would immediately commence using the assigned resources.

Different computer vendors make the capability of temporarily suspending a process available in different ways. Some actually permit the process to execute a SLEEP command. Other vendors establish PROCESS CONTROL BLOCKS that indicate whether a process is *ready to run* or not. A process can then flag itself as *not ready* and will remain inactive until some independent process changes the status in the PROCESS CONTROL BLOCK to the *ready-to-run* state. We will designate the activation/deactivation as an "ON/OFF" control flow that passes directly from the supervisory process to the target process or passes to the target process by means of a data store.

The following data flow diagrams illustrate the various ways an independent supervisory process can control another process.

1. A supervisory process creating/activating a target process is shown in Figs. 10.1a and 10.1b.:

SUPERVISORY/TARGET PROCESSES

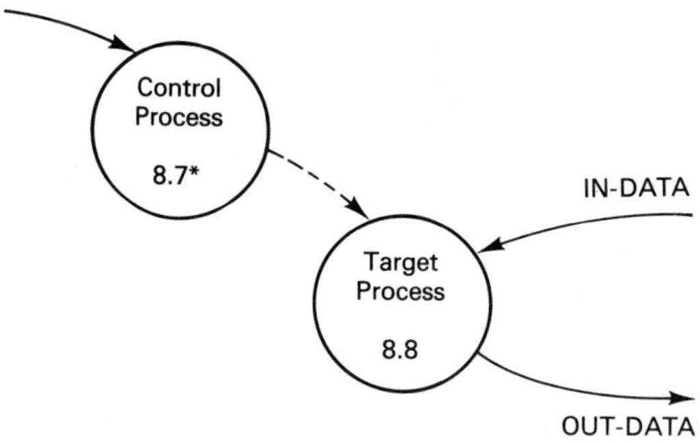

Figure 10.1a

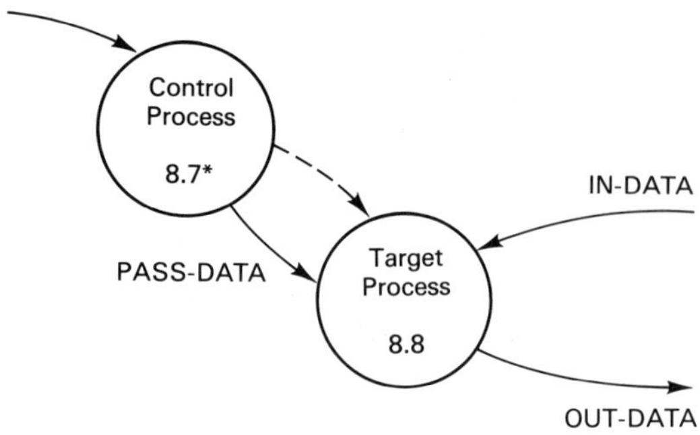

Figure 10.1b

A supervisory process creating/activating multiple target processes, each performing the exact same task, is shown in Figs. 10.1c and 10.1d.

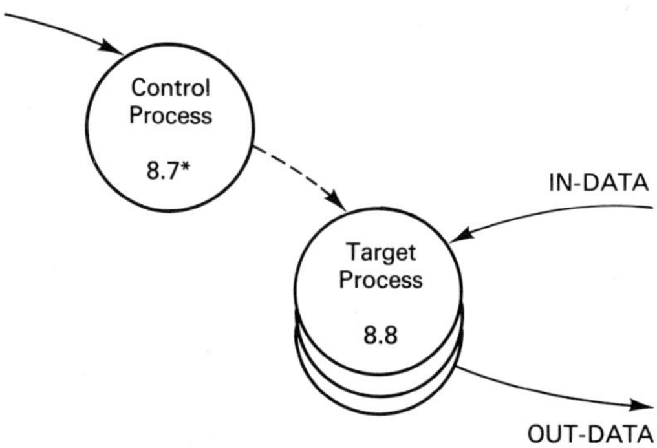

Figure 10.1c

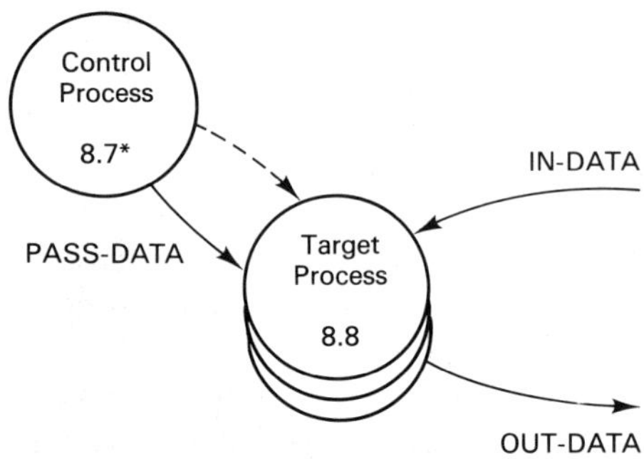

Figure 10.1d

A supervisory process activating only one of several target processes is shown in Figs. 10.1e and 10.1f.

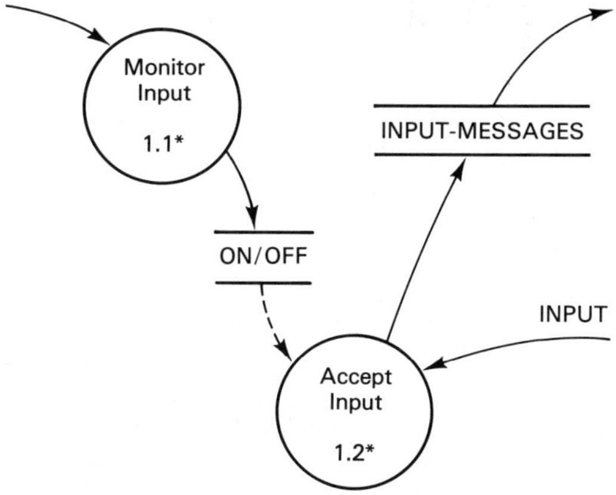

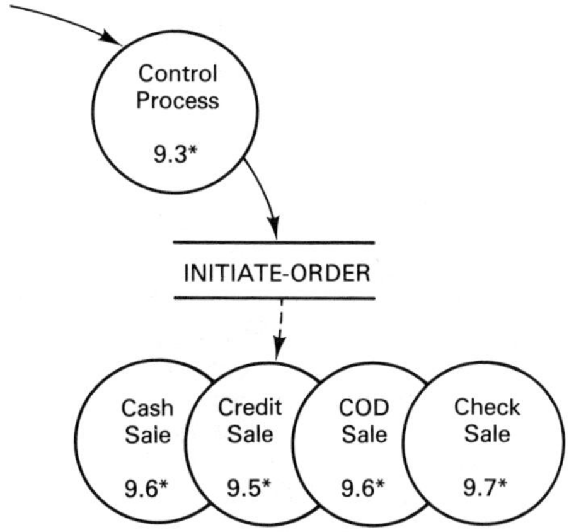

In the data dictionary:

INITIATE-ORDER = {PROCESS-NUMBER + ITEM-PURCHASED + AMT + . . . }
PROCESS-NUMBER = * #1 - Cash Sales, ON/OFF, #2 - Credit Sales, ON/OFF, #3 - COD Sales, ON/OFF, #4 - Check Sales, ON/OFF *

INITIATE-ORDER provides a means whereby additional data elements pass to a designated process activated by a control process.

Figure 10.1f

Figure 10.1e

2. A supervisory process activating a target process, which is later deactivated by the supervisory process or another process, is shown in Figs. 10.1g and 10.1h.

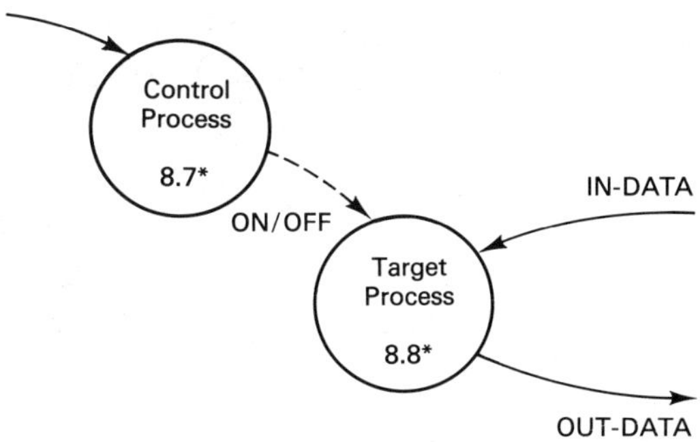

Figure 10.1g

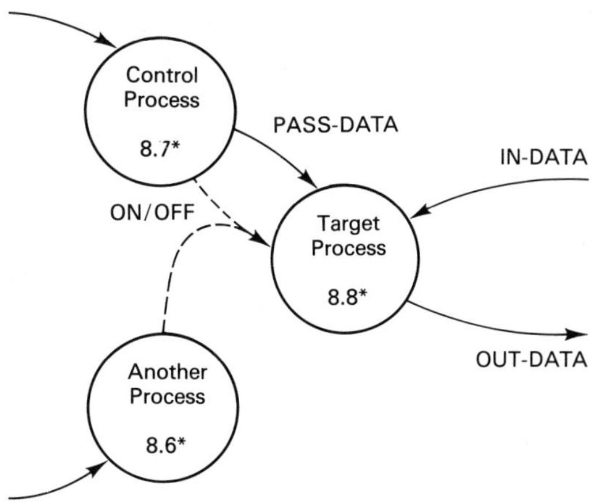

Figure 10.1h

3. A supervisory process reactivating a target process that has gone to sleep after completing its tasks is shown in Figs. 10.1i and 10.1j.

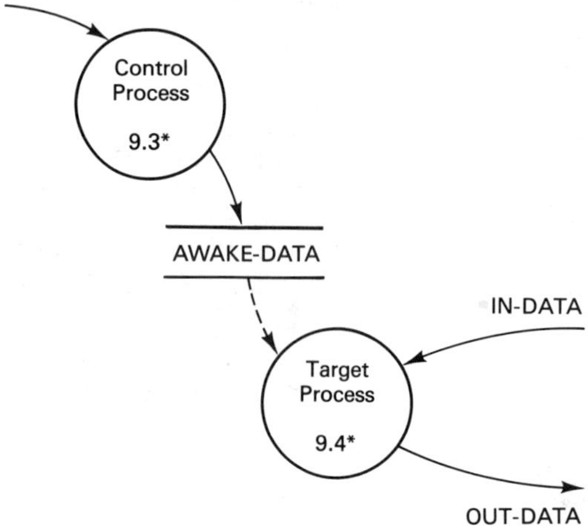

Figure 10.1i

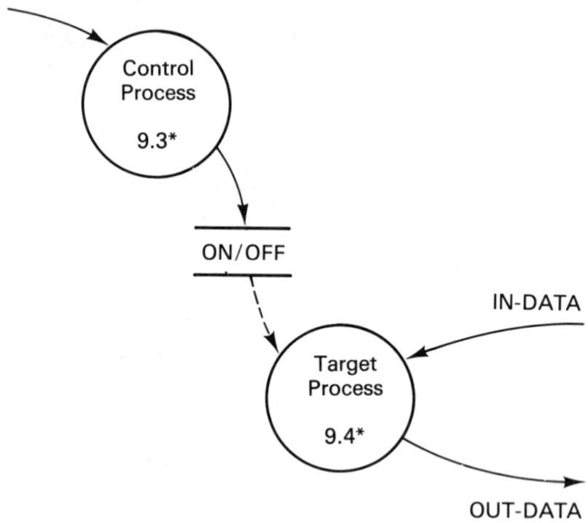

Figure 10.1j

In the data dictionary, neither the data flow ON/OFF nor the data store ON/OFF are defined. The ON/OFF is treated similarly to the data flow REJECT. Both ON/OFF and REJECT are used repeatedly throughout the structured specification and are self-defining terms that are not included in the data dictionary. Such data elements and data stores indicate the *flow of control* and, as such, are not considered input *data* elements to a process. The ON/OFF control element is never transformed into anything.

However, the data store AWAKE-DATA, used in Fig. 10.1i, consists of an elementary data element that controls the target process by using the ON/OFF signal *and* passing the target process additional input data:

AWAKE-DATA = {ON/OFF + INPUT-DEVICE-ADDR + . . . }

Another means of activating one of several processes, illustrated in Fig. 10.1f, posts an interprocessor message in a data store with a unique number that triggers the appropriate process.

INITIATE-ORDER = {PROCESS-NUMBER + CUSTOMER-NAME + . . . }

⋮

PROCESS-NUMBER = * #1 - ON/OFF for process . . . , etc,
#2 - ON/OFF for process . . . , etc,
⋮
#n - ON/OFF for process . . . , etc. *

A supervisory process can pass additional data, such as CUSTOMER-NAME + . . . , to one or more target processes. Only when there are additional (noncontrol) data elements passed between processes do we have to define the data flow or the data store in the data dictionary. ON/OFF is a control element and is excluded from the data dictionary. All other details of controlling processes are defined later in the design stage. During the analysis stage, we are concerned with specifying process-to-process relationships and not with the details of the various techniques used to control other processes.

TIME-DEPENDENT PROCESSES
. . . simultaneous processing restrictions

To further explore the process-to-process relationships, we have to consider simultaneous activation restrictions among processes. **RESTRICTIONS stipulate the internally or externally imposed guidelines that bound the problem or dictate operating constraints.**[O-1]

First of all, to begin studying how processes are permitted to overlap during their processing, we need a means to represent the duration of processing time, as well as the duration of time a process must be inactive.

A TIME-DEPENDENCY CHART (also defined in Chapter 4) **is a graphical illustration of the sequence, concurrence, and duration requirements and/or restrictions among processes in a system.** The chart shows the relationships among processes with respect to the amount of time a process is required to be active, specifies any time intervals whereby a process is required to be inactive, or specifies the overall system behavior with respect to time. Fig. 10.2 is a time-dependency chart that specifies which processes can operate simultaneously and which cannot.

TIME-DEPENDENCY CHART

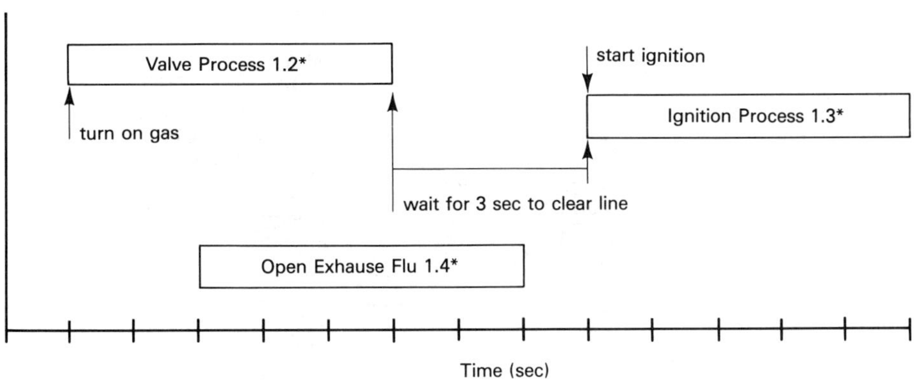

Figure 10.2

There are many techniques we can use to initiate, activate, deactivate, suspend, terminate, or otherwise control a target process that operates simultaneously or sequentially with other processes. However, most of these techniques are reserved for the design stage. The intent of the analysis stage is only to specify system processing requirements. *Flow of control* specifies a process-to-process relationship that is important primarily among real-time processes. From the requirements' standpoint, controlling target processes is a matter of specifying:

1. Processes that are initiated, activated, reactivated, deactivated, or suspended
2. The impact on the available system resources when a process is activated, reactivated, deactivated, or suspended
3. The simultaneous processing restrictions among those processes

All other details of controlling other processes are reserved for the design and programming stages of the development.

FEEDBACK AND CONTROL

. . . feedback is necessary to dynamically control processes

To know whether to change the processing behavior within a system, the supervisory process has to obtain *feedback*. **FEEDBACK is output delivered as input back**

ARCHITECTURE FOR FEEDBACK CONTROL

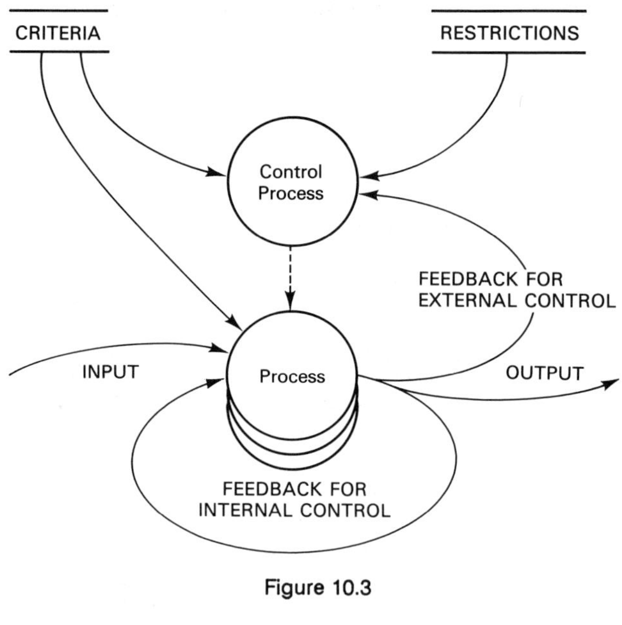

Figure 10.3

into the process from which the output was derived. The input can be a data flow or a control flow directive. A system can contain processes that use internal control, external control, or both. **FEEDBACK CONTROL is a system function that compares output with a criteria.**[O-1] Figure 10.3 illustrates the data and control flows of a system using both external and internal feedback control.

A FEEDBACK CONTROL PROCESS is one influenced by the results of either its own past behavior or by the behavior of another process.[F-2] In studying feedback control as part of system requirements, we will first define *open systems* and *closed systems*. Then, we will consider *closed systems* where internal control is used and *closed systems* where external control is used.

When real-time processes that must synchronize with independent external or temporal events are considered, the issue of controlling the rate of a data flow needs to be considered. The underlying concept used when a system controls any rate is that of a *feedback system*. **A FEEDBACK SYSTEM has a closed data flow structure that brings results from past actions of the system as inputs used to control future actions of the system,**[F-2] as shown in Fig. 10.4. A feedback system is also known as a *closed system*. **A CLOSED SYSTEM is one that is influenced by its own past behavior.**[F-2] Closed systems are *fully adaptive* to their environment, which implies the ability to deal with the full range of inputs of the environment given the existing processes that will operate upon them. Such systems meet their requirements, are stable and predictable within the limits imposed by system requirements, and exhibit invariant performance.

FEEDBACK CONTROL SYSTEM

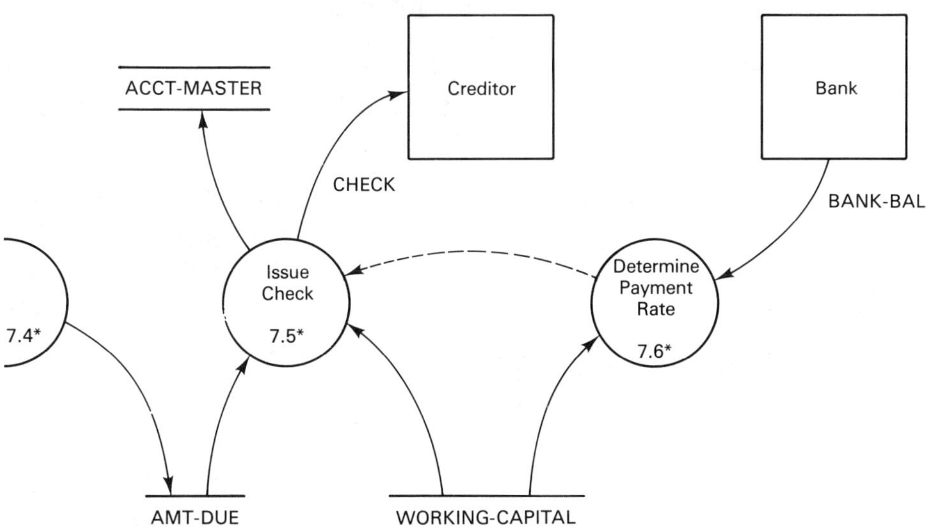

Figure 10.4

A system in which past actions do not control future actions is known as an *open system*. **An OPEN SYSTEM is one that is not aware of its own performance.**[F-2] The general business-information systems studied in the previous chapters were all open systems and, as such, did not have feedback mechanisms.

There are two classes of feedback systems. The first class is known as a *negative-feedback system*. **A NEGATIVE-FEEDBACK SYSTEM is one that seeks a goal and responds as a consequence of failing to achieve the goal.**[F-2] The important part of this definition is the established goal that the system tries to achieve. The second class is a *positive-feedback system*. **A POSITIVE-FEEDBACK SYSTEM generates activity whose action builds a result that generates still greater action.**[F-2] The important part of this definition is that the system activates processes that in turn produce results or provide service to a greater degree than the original process.

Let's consider some examples:

1. A heat thermostat is set to a particular temperature. It senses the temperature of a room and activates the heater to produce more heat to meet the goal of the preset temperature.

   Negative-feedback system

2. A checkout system in a supermarket has checkout clerks and a checkout manager. When the manager observes that a waiting line for checkout has more than three persons, he or she immediately calls in another checkout clerk to process the waiting customers.

   Positive-feedback system

Feedback systems can be contrasted as follows: A negative-feedback system increases or decreases activity as it strives to achieve a goal. A positive-feedback system exhibits growth that results in greater service or greater output. A negative-feedback system has a preset goal and controls the rate of activity of the processes to regulate the performance of the system to achieve the goal. A negative-feedback system is *goal seeking* and in its structure we find fluctuations and instabilities. A positive-feedback system manifests a structure in which we find the forces of growth.

Let's focus our attention on negative-feedback systems, where a supervisory process strives to direct the processes in a system toward achieving a goal. The control process has the capability of increasing or decreasing the processing **rate** of a target process.

Controlling the Rate of a Process

To control the rate of another process, four conditions are required:

A goal

An observed condition of the system

A way to express **the discrepancy** between the goal and the observed condition

A **statement of action** executed based on the discrepancy

In the simplest case, where a control process regulates the rate of another process, there is a single decision (rate) that controls the data flow to a specified *level*. **A LEVEL is a condition or state of the system.**[F-2] To apply the concept of levels to information systems, we need what I will call the *information level* of a system. **The INFORMATION LEVEL of a system is a measurable quantity about some aspect of the system, such as the quantity of the material being regulated or produced by a process.** The information level of a data store containing records of information is the *cardinality* of the data store (see Chapter 7). When using the information level to make decisions regulating some data flow in a system, we must remember that the discrepancy between a true system level and the information level available to the decision process always exists in principle. However, depending on the technology used to obtain the information level, the discrepancy in most systems is usually negligible.

When the level falls below the goal, the supervisory process increases the flow rate to achieve the intended level.

In the example of Fig. 10.5, the four necessary components for controlling the rate of a process are all present:

INTENDED-WATER-LEVEL is the **goal**

Monitor Water Level 4.0* **observes the condition** of the system

CONTROLLING THE RATE OF A PROCESS
(First-Order Negative-Feedback System)

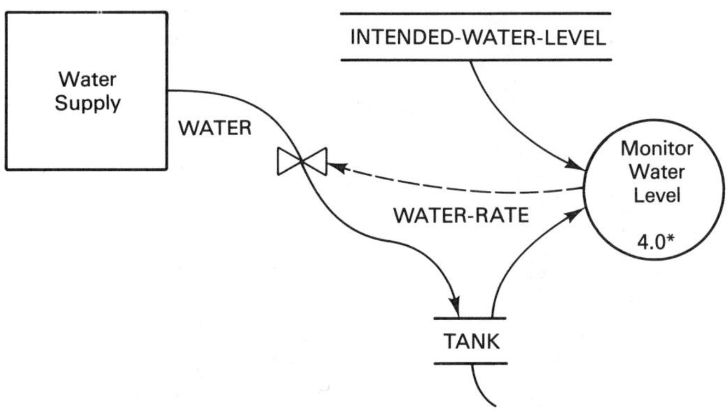

Figure 10.5

The *flow of control* between process 4.0* and WATER-RATE is the way to express the **discrepancy** between the goal and the observed condition

The mini-spec of Monitor Water Level 4.0* is the **statement describing the action** taken based on the discrepancy.

Systems that monitor both the production of the material being regulated (input) and the distribution of the material (output) usually are *second-order* negative-feedback systems. **The ORDER of a feedback system designates the number of independent levels controlled within the system.** A second-order negative-feedback structure is frequently susceptible to oscillation as the supervisory process attempts to effect the processing behavior in the system.

Figure 10.6 is an example of a second-order negative-feedback system. The system delivers water to customers upon demand from a PRIMARY-TANK. As water levels are depleted, water is pumped from the HOLDING-TANK into the PRIMARY-TANK. As water is depleted in the HOLDING-TANK, the system activates the process Well Pump 4.0*, which replenishes the supply of water from a nearby well.

The system must maintain two levels: the water level in the HOLDING-TANK and the water level in the PRIMARY-TANK. If the PUMP-RATE is not the same as the WELL-RATE or TANK-RATE, the amount of water in the PRIMARY-TANK

MUNICIPAL WATER COMPANY
(Second-Order Negative-Feedback System)

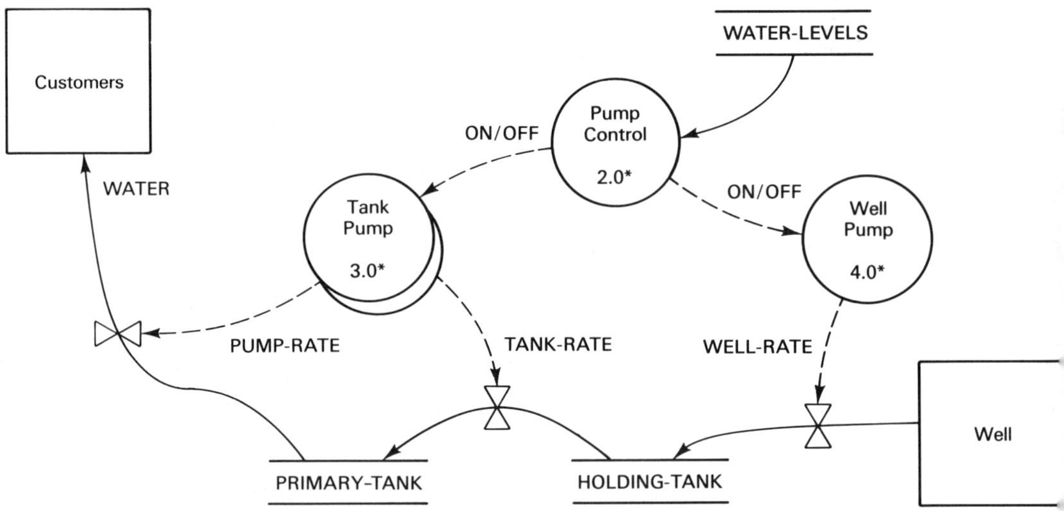

Figure 10.6

and the HOLDING-TANK will vary. From the data store WATER-LEVELS the system obtains the data element that tells the amount of water to maintain in each tank.

> WATER-LEVELS contains the **goals.**
>
> Pump Control 2.0* **observes the conditions** of the system.
>
> Input data from HOLDING-TANK and PRIMARY-TANK are compared to the data obtained from WATER-LEVELS to determine the **discrepancy** between the goal and the observed condition.
>
> The mini-specs of Well Pump 4.0* and Tank Pump 3.0* contain the **statement of action** taken to resolve the discrepancy.

A more in-depth study of feedback systems can be found in a book by Jay W. Forrester.[F-2]

The intention of including the information on feedback systems in this chapter is to provide the necessary concepts to specify the requirements for a real-time system within a structured specification. For the purposes of specifying system requirements, the analyst determines if the solution requires a feedback system. If it is determined that the solution is to be a goal-seeking structure, then the requirements will specify a negative-feedback system. The complexity and precision of the system will dictate the order of the negative-feedback system. Simpler, less costly solutions may implement a first-order negative-feedback system. A more intricate solution, and a more costly one, is specified if the system requirements call for a second-order negative-feedback system.

In construction, an architect specifies, for example, the requirements for the construction of an apartment complex. The planned services and incurred costs are a function of what the user intends the design to provide. If automatic garage-door openers and an elaborate computer-controlled security system are required, the architect's plans are more detailed and the construction will likely be more costly.

In systems development, a systems analyst specifies the requirements for the development of a system. Likewise, if the user requires services of real-time processes and wants the decision-making ability of the processes to be very sophisticated, then the structured specification is more elaborate and the cost of development and implementation will be greater.

FINITE-STATE PROCESSES
. . . when systems must respond differently to the same input data

An external control process that directs the initiation, activation, reactivation, deactivation, and suspension of another process or regulates the rate of data flow within a system is one means of implementing a control structure. Another means is to organize the internal processing logic to not only respond to the data received as

DATA FLOW DIAGRAM OF A RECURSIVE PROCESS

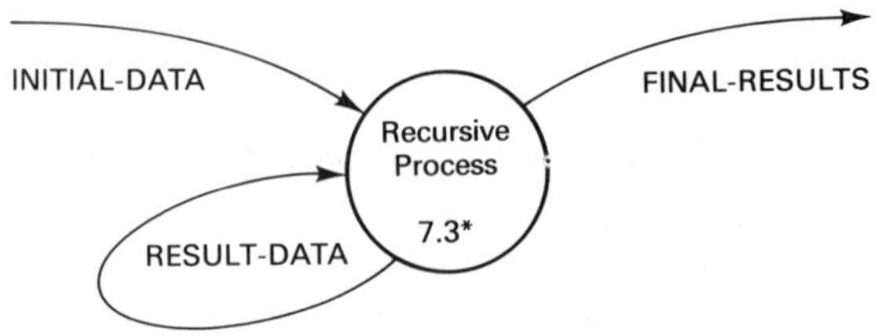

Figure 10.7

input, but also to take into account time, previous processing activity, and/or previous processing results.

When a process accepts its own output as input, it is a recursive process (discussed in Chapter 4) and is illustrated in the data flow diagram of Fig. 10.7.

The case where a process behaves differently when receiving the same input at different moments in time is one that operates in different *states*. **A STATE is an operating circumstance that has a predefined result that is triggered by an input at a particular moment in time.** Time is regarded as a set of discrete *points*. The input data flow, the resulting change in state, and the output data flow all occur at the same moment in time. A concept of a state permits us to define a different result when the same input enters a system at a different moment in time.

Arthur Gill[G-2] points out that the idea of a *state* as a basic concept in the representation of systems is not a recent discovery. A.M. Turing introduced the idea in 1936.[T-1] In 1948, the concept was employed by C.E. Shannon[S-1] in his basic work on information theory. Others such as D.A. Huffman[H-1] (1954), S.C. Kleene[K-2] (1956) and E.F. Moore[M-8] (1956), reintroduced the idea to where it was accepted as one of the most fundamental concepts in the theory of discrete systems.

All of the systems in the first nine chapters of this text respond to external events or temporal events and exhibit data-dependent behavior. A process that exhibits data-dependent behavior is readily available to transform the input data immediately upon its arrival or begin producing output data immediately upon the occurrence of the temporal or internal event. In the beginning of this chapter, I presented systems that exhibit time-dependent behavior. Time-dependent behavior is initiated at various times and performed by a process that requires a control flow to initiate or halt the execution of the process. We can use a time-dependent process to synchronize the output of the system with some independent external event or regulate an output data flow from the system. A special type of time-dependent behavior is known as state dependent behavior. In this case, a process responds to the combination of the input data flow and time.

Consider the following two cases where a person wishes to display data on a computer terminal screen. When a request to display a few lines of data from a computer terminal is issued, the computer retrieves the data and displays the entire set of data. If the user presses the "Page-Down" key (PG-DN-KEY), no action is taken and the striking of the key is basically ignored because all of the data is on the display screen. On the other hand, if a person requests to view data that takes several screens to display, the computer displays the first screen and, upon receiving the input data flow PG-DN-KEY, will display the next screen of data, etc. The input data flow to the process performing the display function is the same in both cases. The difference is the state the process is in when it receives the input PG-DN-KEY.

To define the different processing requirements for each state of a process in a structured specification, we divide the mini-spec for the process into separate states designated as a, b, c, d, etc. We write a separate mini-spec for each state. The set of all mini-specs that covers all of the states of a finite-state process forms the mini-spec for the entire finite-state process. The usual numbering format for processes and mini-specs apply, with the additional suffix designating the state a, b, c, etc. A finite-state process is illustrated in the data flow diagram of Fig. 10.8.

SIMPLE FINITE-STATE PROCESS

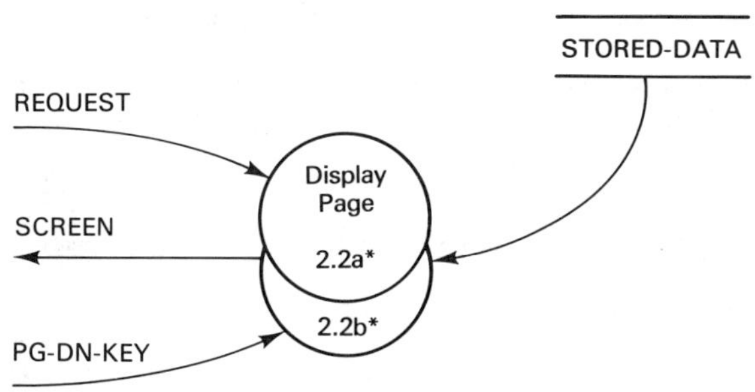

Figure 10.8

Process 2.2a* is the initial processing state and responds to the user's request to display the first page of data. If all of the data cannot be shown on the screen, the process is set to state "b" to await the PG-DN-KEY input. When the user presses the PG-DN-KEY, process 2.2b* displays the next page of output. If there is still more data to display, the process is set to state "b." When the last screen of data is displayed, the process is set to state "a" to await the user's next command.

Operating States

Most **on-line systems** operate in one state, the user state. In that state, the user is made to believe that he or she has exclusive control of the entire computer and all of its resources. If unusual conditions occur that were not anticipated, the system informs the user of the inability to process the request, gives erroneous results, locks out any further input from the user, or disconnects (*drops*) the user from the system.

In a **real-time system**, because the highest priority is continuous operation of the executing program, when an unusual condition is encountered, the process changes state and it may respond differently to the exact same input than it did previously. The response may be undesirable. However, the system ultimately resets itself to the initial state and the user can attempt the request again.

An indication that a system needs a finite-state process is if the system requirements demand that it respond differently to the exact same input when the input is received at different moments in time. In the previous case of the display screen, the system does not respond to the PG-DN-KEY unless the system has displayed a screen of data and there is additional data to display. The two basic states that the display process can exist in are "initial display of data" and "more data remaining to be displayed."

For another example of a finite-state process consider an automated toll booth that is a simple real-time system that operates in many different states. If the toll is \$.50, consider the circumstances described in Fig. 10.9.

AUTOMATIC TOLL BOOTH

(FINITE-STATE PROCESS)

For brevity, assume the system does not accept pennies or half dollars. The application software processes one coin at a time and does not provide change.

INPUT	RESULT (Output)	NEXT STATE	INPUT	RESULT (Output)	NEXT STATE
STATE 0 (INITIAL STATE - Cumulated total of \$.00):			**STATE 5** (Cumulated total of \$.25):		
CLOSE GATE, RED LIGHT					
Nickel	None	STATE 1	Nickel	None	STATE 6
Dime	None	STATE 2	Dime	None	STATE 7
Quarter	None	STATE 5	Quarter	OPEN GATE, GREEN LIGHT	STATE 0
STATE 1 (Cumulated total of \$.05):			**STATE 6** (Cumulated total of \$.30):		
Nickel	None	STATE 2	Nickel	None	STATE 7
Dime	None	STATE 3	Dime	None	STATE 8
Quarter	None	STATE 6	Quarter	OPEN GATE, GREEN LIGHT	STATE 0
STATE 2 (Cumulated total of \$.10):			**STATE 7** (Cumulated total of \$.35):		
Nickel	None	STATE 3	Nickel	None	STATE 8
Dime	None	STATE 4	Dime	None	STATE 9
Quarter	None	STATE 7	Quarter	OPEN GATE, GREEN LIGHT	STATE 0
STATE 3 (Cumulated total of \$.15):			**STATE 8** (Cumulated total of \$.40):		
Nickel	None	STATE 4	Nickel	None	STATE 9
Dime	None	STATE 5	Dime	OPEN GATE, GREEN LIGHT	STATE 0
Quarter	None	STATE 8	Quarter	OPEN GATE, GREEN LIGHT	STATE 0
STATE 4 (Cumulated total of \$.20):			**STATE 9** (Cumulated total of \$.45):		
Nickel	None	STATE 5	Nickel	OPEN GATE, GREEN LIGHT	STATE 0
Dime	None	STATE 6	Dime	OPEN GATE, GREEN LIGHT	STATE 0
Quarter	None	STATE 9	Quarter	OPEN GATE, GREEN LIGHT	STATE 0

Figure 10.9

In Fig. 10.9, we can see that the inputs (nickel, dime, quarter) can be the same in many instances, however, the results may not be the same, depending on the state of the process when it receives the input. If you consider this process as a black-box that responds to events, you can call it a *finite-state process*. **A FINITE-STATE PROCESS (also known as a finite-state machine) is one that has**

A finite set of states

A finite set of input data flows

A finite set of output data flows

A mapping of the input and present state into next state

A mapping of the input and present state into the output data flow(s)[M-7]

The data flow diagram notation for a finite-state process is shown in Fig. 10.10. Notice that although the process has many distinct states, it actually has only one mini-spec (MINI-SPEC 2.1) made up of many sections (a, b, c, d, and e). The several states in the process are each defined with a unique processing policy. The states are denoted as 2.1a*, 2.1b*, 2.1c*, 2.1d*, etc. In this manner, a separate mini-spec section can be created for each state, and the collection of mini-specs of all states defines a single process. Other useful tools to organize the logic of a finite-state process are state-transition tables and/or state-transition diagrams, which are discussed later in this chapter.

In the case where there is a unique set of input and output data flows for each state, the data dictionary defines the set that applies to each separate state.

When a process is in a particular state, it only responds to the input data flow as defined for that particular state. Thus, if the process is in state "b" and the input data

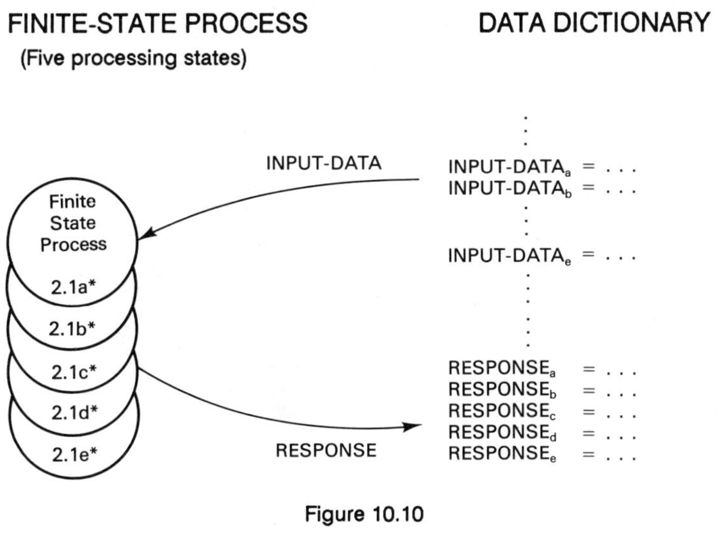

Figure 10.10

flow for state "a" arrives to be processed, it is ignored.

In the following example of the Donut Machine, a customer initially enters the COINS and his or her CHOICE of product. If the product is available, the system (or process) provides the PRODUCT as output, calculates the CHANGE, and provides the CHANGE if required. However, if the PRODUCT is not available, the process sends out an indication that the PRODUCT is "SOLD OUT." The system awaits the customer's second input, which may be an ALTERNATE-CHOICE or COIN-RETURN-REQUEST.

In this example, the process or machine operates in one of two states: *initial state*, where it awaits the customer's PAYMENT and CHOICE, and the *alternate selection state*, where it just awaits the customer's ALTERNATE-CHOICE or request for refund. A partial data dictionary follows. The partial data flow diagram is shown in Fig. 10.11 and the partial mini-specs in Fig. 10.12.

PARTIAL DATA DICTIONARY (DONUT MACHINE)

CHANGE	= *The difference between accumulated values of all COINS and the cost of the chosen product. *
CHOICE	= ["PLAIN" | "CHOCOLATE" | "PEANUT"]
COINS	= * Accumulated value of all money entered into the Donut Machine. *
PRODUCT	= [DONUT-PLAIN | DONUT-CHOC | DONUT-PEANUT]
PRODUCT-CASE	= {PRODUCT-TYPE + PRODUCT}
	= * Storage compartment for donuts organized by PRODUCT-TYPE. *
PRODUCT-TYPE	= ["PLAIN" | "CHOCOLATE" | "PEANUT"]
PURCHASE-REQUEST$_a$	= COINS + CHOICE
PURCHASE-REQUEST$_b$	= [CHOICE | REFUND]
REFUND	= * Request signal to return the customer's COINS. *
SALE$_a$	= PRODUCT + (CHANGE)
SALE$_b$	= SOLD-OUT-DISPLAY
SALE$_c$	= [PRODUCT + (CHANGE) | COINS]
SOLD-OUT-DISPLAY	= * Indicator informing the customer that his or her choice is not available. *

PARTIAL DATA FLOW DIAGRAM (DONUT MACHINE)

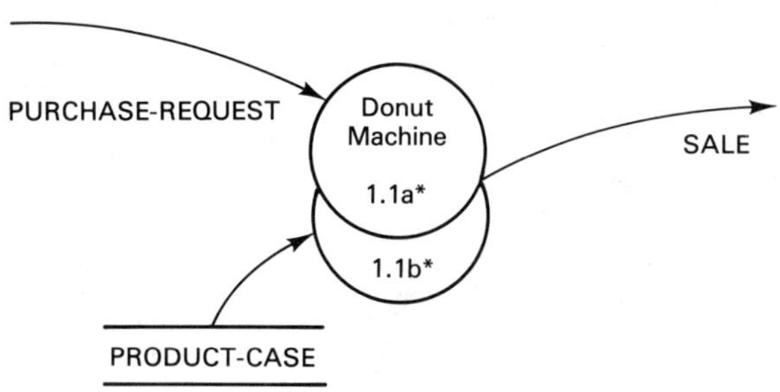

Figure 10.11

PARTIAL MINI-SPECIFICATIONS (DONUT MACHINE)

Donut Machine MINI-SPECIFICATION 1.1a*

Initial state a.

```
Input PURCHASE-REQUEST from Customer.
While COINS < $.40,
    Input COINS from Customer.
Endwhile.
Determine if CHOICE is available from PRODUCT-CASE.
If CHOICE is available,
    Select PRODUCT-TYPE = CHOICE from PRODUCT-CASE.
    If COINS > $.40,
        Compute CHANGE = COINS - $.40
        Set SALE = PRODUCT + CHANGE.
    Else Set SALE = PRODUCT.
        Output SALE to Customer.
        Set STATE-1.1 to a.
    Endif.
Else Set SALE = SOLD-OUT-DISPLAY
    Output SALE to Customer.
    Set STATE-1.1 to b.
Endif.
Stop.
```

Donut Machine MINI-SPECIFICATION 1.1b*

State b.

```
Input PURCHASE-REQUEST from Customer.
Case 1 (Alternate CHOICE is entered):
    Determine if CHOICE is available from PRODUCT-CASE.
    If CHOICE is available,
        Select PRODUCT-TYPE = CHOICE from PRODUCT-CASE.
        If COINS > $.40,
            Compute CHANGE = COINS - $.40
            Set SALE = PRODUCT + CHANGE.
        Else Set SALE = PRODUCT.
            Output SALE to Customer.
            Set STATE-1.1 to a.
        Endif.
    Else Set SALE = SOLD-OUT-DISPLAY
        Output SALE to Customer.
        Set STATE-1.1 to b.
    Endif.
    Stop.
Endcase 1.
Case 2 (REFUND is requested):
    Set SALE = COINS.
    Output SALE to Customer.
    Set STATE-1.1 to a.
    Stop.
Endcases.
```

Figure 10.12

State-Transition Tables

A way of tabulating the pattern of system behaviors is to specify the input, the change in state, and the resulting action in a *state-transition table.* **A STATE-TRANSITION TABLE has all possible input[1], process states, and the respective response (action) made by the process when a particular input is recognized in each state.** The process response is characterized by both the output provided and the change (if any) in the state of the process.

Suppose that you wanted to represent the following requirement for a finite-state process:

> There are four possible inputs or conditions: $\mathbf{I_1}$, $\mathbf{I_2}$, $\mathbf{I_3}$, and $\mathbf{I_4}$.
>
> There are five possible states: **a, b, c, d,** and **e.**
>
> There are five possible outputs or actions: $\mathbf{O_1}$, $\mathbf{O_2}$, $\mathbf{O_3}$, $\mathbf{O_4}$, and $\mathbf{O_5}$.

The rules governing the operation of a finite-state process are as follows:

> When the process receives input $\mathbf{I_1}$ while in state **a,** it outputs $\mathbf{O_1}$ and changes the state to **b.**
>
> When the process receives input $\mathbf{I_4}$ while in state **a,** it outputs $\mathbf{O_5}$ and changes the state to **d.**
>
> When the process receives input $\mathbf{I_1}$ while in state **b,** it outputs $\mathbf{O_2}$ and remains in state **b.**
>
> When the process receives input $\mathbf{I_2}$ while in state **b,** it outputs $\mathbf{O_3}$ and $\mathbf{O_4}$ and changes the state to **c.**
>
> When the process receives input $\mathbf{I_2}$ while in state **c,** it outputs $\mathbf{O_2}$ and changes the state to **e.**
>
> When the process receives input $\mathbf{I_3}$ while in state **c,** it outputs $\mathbf{O_4}$ and changes the state to **d.**
>
> When the process receives input $\mathbf{I_4}$ while in state **c,** it provides no output and changes the state to **d.**
>
> When the process receives input $\mathbf{I_1}$ while in state **d,** it outputs $\mathbf{O_1}$ and changes the state to **a.**
>
> In state **e,** if the process receives input $\mathbf{I_1}$ within 30 seconds, the process provides no output and changes the state to **d.** If no input is received during the 30 seconds, it outputs $\mathbf{O_4}$ and changes the state to **c.**

---

[1] Input can be one or more input data flows or some condition that is recognized by the system.

STATE-TRANSITION TABLE

INPUT or CONDITION:	a	b	c	d	e
I_1	O_1 --->b	O_2 --->b		O_1 --->a	O_2 --->d
I_2		O_3,O_4 --->c	O_2 --->e		
I_3			O_4 --->d		
I_4	O_5 --->d		None --->d		
timeout 30 secs					O_4 --->c

(Columns a–e: ← STATES →)

The data dictionary contains the detailed definition for the input or condition (I_n) and the output or action (O_m).

Figure 10.13

Compare the narrative text explanation to the state-transition table shown in Fig. 10.13.

Graphical Aids for State Processes

Another way to illustrate the behavior of a finite-state process is with a state-transition diagram. There are two different graphical formats used to illustrate state transitions: the *bubble form* and the *rectangular form*.

In the bubble form, also known as a Moore diagram, a circle or bubble represents each state. If it is possible to transition from state a to state b, then we draw an arc from the state a bubble to the state b bubble. Along the arc, we list the condition or input data flow that caused the change in state along with any action or output response. See Fig. 10.14.

There can be several actions or outputs generated as a result of recognizing a condition or upon receiving an input, followed by a change in state. The state-transition diagram is *not* intended to show *all* necessary data flows and data stores, but only the input stimulus (if any) or condition, the resulting output response (if any) or action, and the change of state (if any). Not every input or condition is valid for every state. If a condition or input occurs during a state for which it is invalid, the

BASIC COMPONENTS OF STATE-TRANSITION DIAGRAM
(BUBBLE FORM)

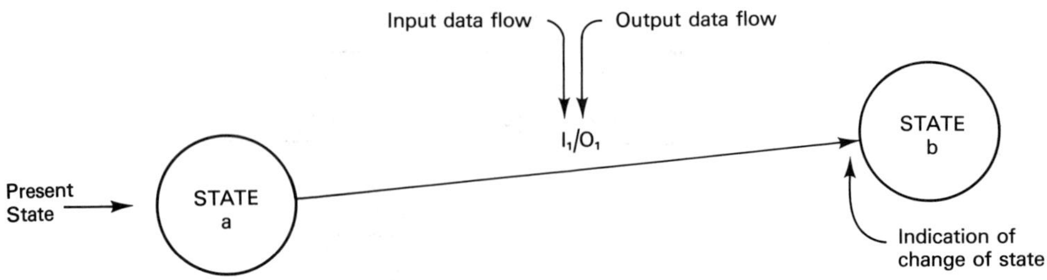

Figure 10.14

process ignores the condition or input.

Notice that an input stimulus presented to the system by a source when an external event occurs is not the only way to activate a finite-state process. A condition such as a temporal event can occur during one state that causes the process to change states. Such a phenomenon is the typical *timeout* condition. **A TIMEOUT is a**

STATE-TRANSITION DIAGRAM
(Bubble Form)

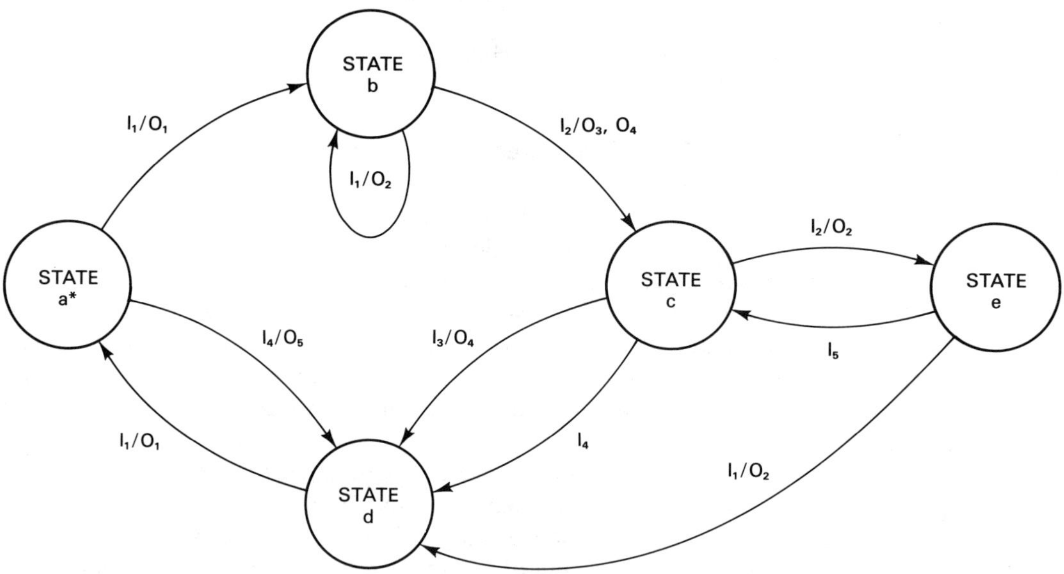

*Initial state of the process.

Figure 10.15

temporal event that initiates action if no anticipated action or response from a process or a source is received within a specific time interval. A timeout occurs when a process waits for input data for a designated interval of time, and after that interval has elapsed and the action has not occurred or the input data flow is not received, the process can produce output and/or can change states. If the process changes states, it is likely to respond differently to the input data flow than it would have if it were received prior to the change in state. Figure 10.15 is an illustration of the set of inputs, outputs, and corresponding changes in state, specified in the table of Fig. 10.13.

Another form of illustrating the behavior of a finite-state process is by using horizontal rectangles in a vertical arrangement with vertical lines drawn to represent the changes in the state, together with the input data flows or conditions that stimulate the process, the resulting output data flow(s) or actions and the corresponding changes in state (if any), as shown in Fig. 10.16. A long narrow rectangle represents a state. A vertical arrow represents the transition from one state to another. Along the vertical arrow, two items are listed. The first ($\mathbf{I}_n$) is used to represent the arrival of an input data flow or some other condition that initiates the processing in state **x**. The second ($\mathbf{O}_m$) represents the output data flow or resulting action produced by the process while in state x. The system can then set the process to a new state (**y**) or remain in the current state.

BASIC COMPONENTS OF STATE-TRANSITION DIAGRAM

(Rectangular Form)

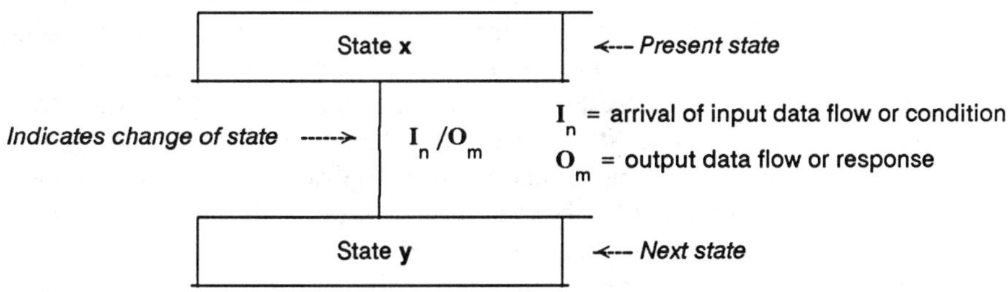

Figure 10.16

The example of a rectangular form state-transition diagram shown in Fig. 10.17 illustrates the same system behavior that was shown in the bubble form of Fig. 10.15.

The movement from one state to another takes place when a condition occurs or an input $\mathbf{I}_n$ is received. The process also produces an action or output $\mathbf{O}_m$. Notice in the example that in state b the process accepts input $\mathbf{I}_1$ and produces the output $\mathbf{O}_2$ with no change in state. Meanwhile, in other states, one condition or input stimulus causes several output responses or none, and may or may not cause a change in state. A timeout condition exists during state **e**, whereby if input $\mathbf{I}_1$ is received during the

STATE-TRANSITION DIAGRAM
(Rectangular Form)

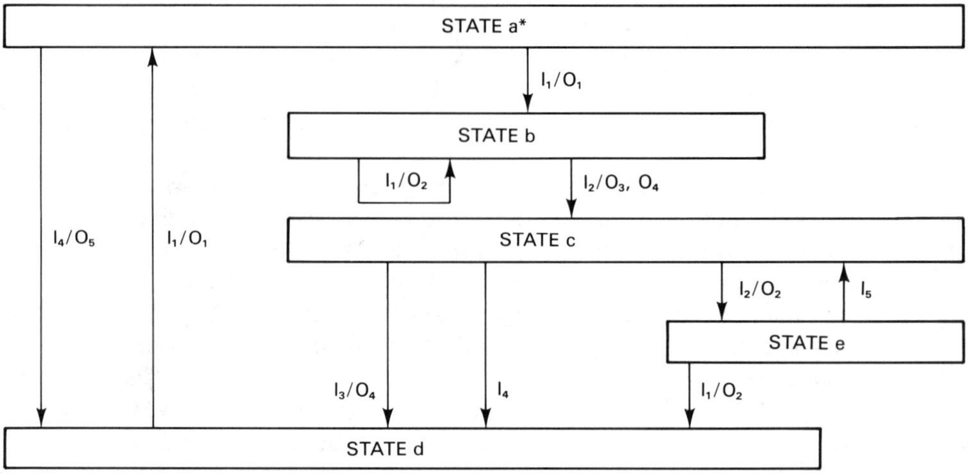

Figure 10.17

waiting interval (30 sec), the process outputs O_2 and changes states to state **d.** If the process does not receive the input within the specified interval of time while in state **e**, the process outputs O_4 and changes its state to **c.**

All three examples, the state-transition table, the bubble form of the state-transition diagram, and the rectangular form of the state-transition diagram provide a better means of communicating the requirements for a finite-state process than the narrative text initially presented. One advantage in using the state-transition diagram (bubble form) is that it is easier to update, because the diagram is capable of withstanding more changes, such as additional states, etc., before it has to be redrawn. Some analysts feel that the bubble form may be confusing because data flow diagrams also are drawn using bubbles, and they thus prefer the rectangular form. None of the methods provides a convenient way to list the data flow elements used as input stimulus and the output response. The convention of using the I_n and the O_m notation is suggested together with a *legend table* appended to the diagram. **A LEGEND TABLE is a listing of the abbreviations used on the diagram.** Abbreviations are used for the input data flows or conditions (I_n) and output data flows or actions (O_m) used on data flow diagrams and state-transition diagrams.

States and Subprocesses

Just as "parent" processes can be decomposed into "child" processes, which in turn can be decomposed into lower-level "child" processes, until we reach all functional primitives, a single finite-state process can contain a finite-state process. Granted that the system requirements can become rather complex, nevertheless, the structured specifications still have to be written so that the users can verify the processing performed by the system. The structured specification can also serve as the basis for preparing an exhaustive test plan.

In the case of real-time systems, many system requirements are written that are *not reviewed by the users,* such as electronic switching systems, monitor and control systems, routing and traffic control, etc. In these cases, the systems analyst or software engineer must subject the structured specification to a *peer review.* During the reviews other systems analysts, designers, engineers, and specialists review the documents. Often professionals from an independent systems test group participate in the reviews, representing the users of the system.

The systems designers design and build the system according to the system requirements as documented in the structured specifications. The theoretical concepts of finite-state machines, as published by Arthur GillG-2 can be applied during the analysis stage to prepare system requirements for real-time systems. The concepts of transient, persistent, and isolated states applied to successive levels of data flow diagrams provide a structure from which exhaustive test plans can be designed. Before I give an example, let's look at the definitions of the three types of states that can exist:

> **A TRANSIENT STATE is a "parent" state that contains a finite-state process that can lead into at least one other "parent" state, but cannot be reached once it is left.**
>
> **A PERSISTENT STATE is a "parent" state that contains a finite-state process that can be reached from at least one other "parent" state, but cannot be left once it is reached.**
>
> **AN ISOLATED STATE is a "parent" state that cannot lead into any other "parent" state and cannot be reached by any other "parent" state.**

From a state-transition diagram, not only can users or designers determine the accessibility of particular processes, but the system test group, being independent from the developers, uses the structured specification to prepare test plans and procedures to exhaustively exercise the system to verify that it performs as requirements specify. Without going into material that is beyond the scope of this text, be advised that there are theorems that use the above-mentioned state-transition structures to specify the scope of input sequences required to test systems that implement levels of finite-state processes.[G-2]

I offer one example theorem, without proof:

> *Let j and k be two states in an n-state process. If state k is at all reachable from state j, it is reachable by applying an input sequence whose length is at most n - 1.*[G-2]

Other theories exist that specify the minimal input sequence to test the logical continuity between two states. The intent of this text is only to provide the framework and structure to specify system requirements. One group that can take advantage of systems specified according to the theory of finite-state machines is the independent system test group.

During the analysis stage of a real-time system, there can be more than one set of tasks that function as a finite-state process. In other words, we may find a particular state within a finite-state process composed of at least one other task that is also a finite-state process. A notation for specifying finite-state processes within a higher-level finite-state process is useful to the two intended recipients of the document:

1. The structured specifications provide **systems designers** with the requirements in various levels of detail. The systems designers then provide the designs to carry out the requirements.
2. The structured specifications provide the **independent system test group** with the requirements of the system in various levels of detail. The test group provides the written test plans that exhaustively test the system, verifying the quality of the system before it is delivered to the users.

To organize the processing requirements for a finite-state process to operate within a finite-state process, the data flow diagram of Fig. 10.18 simply shows the higher-level process as a "parent" finite-state process.

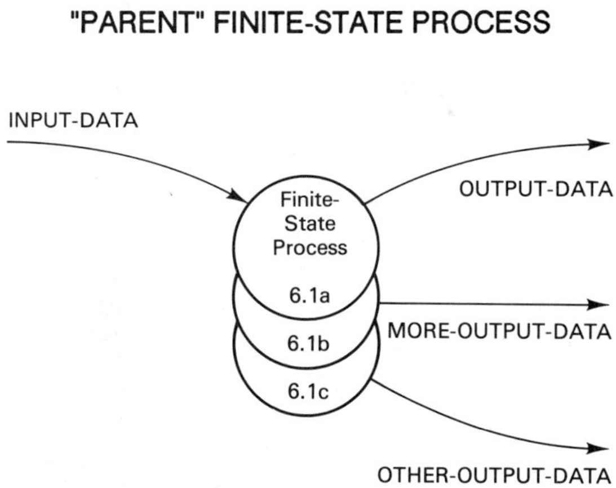

Figure 10.18

Note that process 6.1 is *not* a functional primitive, and hence there exist diagrams 6.1a, 6.1b, and 6.1c as shown in Fig. 10.19.

DATA FLOW DIAGRAM 6.1

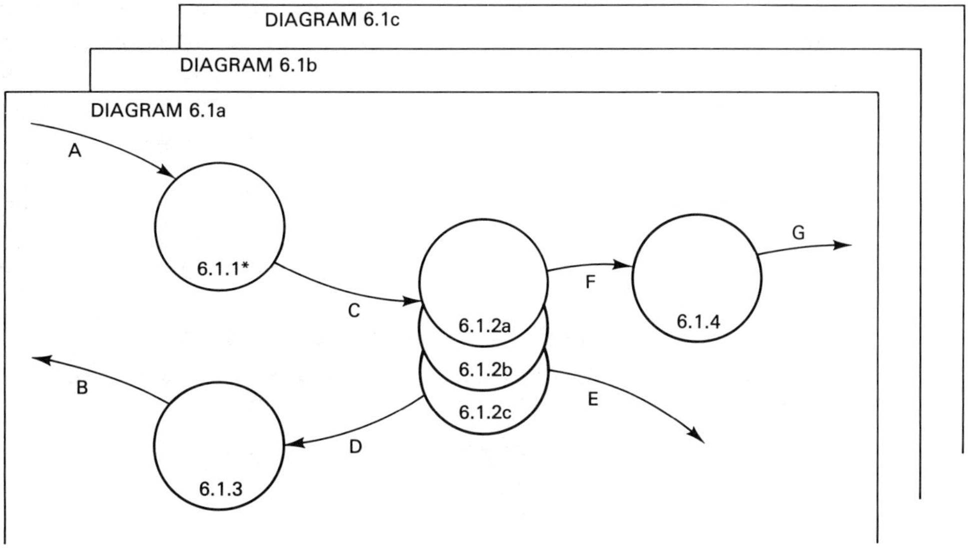

Figure 10.19

STATE-TRANSITION TABLES WITH SUBTABLES

INPUTS or CONDITIONS:	subtable a: a			subtable b: b				subtable c: c	
	a	b	c	a	b	c	d	a	
I_1	O_1 →(a,a)				O_4 →(b,c)				
I_2	O_2 →(a,b)	O_3 →(a,c)						O_7 →(a,a)	
I_3				O_5 →(b,b)	→(a,a)				
I_n	⋮ O_8 →(a,b)	⋮	⋮	⋮	⋮	⋮	⋮	⋮	⋮

O_8 = Output data flows

--> (a,b) = Indicates a change of state to subtable a, state b.

Figure 10.20

There can be some input data flows (I_n) that only apply to a specific "parent" process (subtable) and others that apply to any process within any subtable. Likewise, the output data flows (O_m) can also apply to just one "parent" process (subtable) and even a particular process (state) within that process. Other output data flows can apply to several "parent" processes (subtables) and/or states.

State transitions can apply to states within a subtable, and sometimes there can be a transition made to a state in another "parent" process (subtable). See Fig. 10.20.

The bubble form illustration of a finite-state process has three distinct branches: *reflecting branches, converging branches, and diverging branches*, as shown in Fig. 10.21.

A REFLECTING BRANCH is one that originates and terminates in the same state.

A CONVERGING BRANCH is one that points toward the state.

A DIVERGING BRANCH is one that points toward another state.

FINITE-STATE BRANCHES

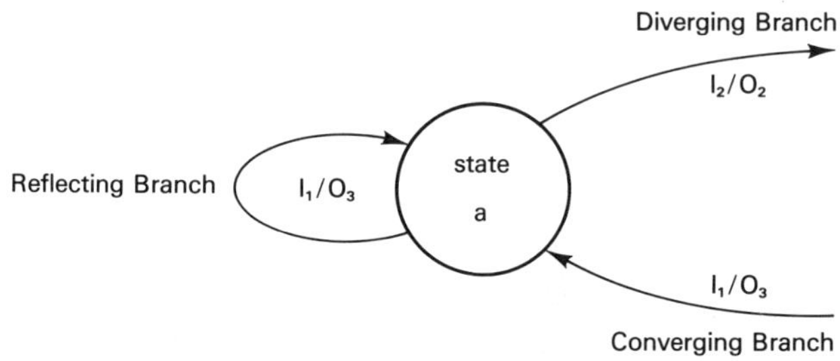

Figure 10.21

AUTOMATIC VENDING SYSTEM
. . . example of structured specification with finite-state requirements

Given the preceding definitions and tools for specifying requirements, let's consider the following example of an Automatic Vending System (AVS).

The AVS system receives the customer's MONEY and a PURCHASE-REQUEST for a can of soda (PRODUCT). The system provides the requested type of soda and computes the appropriate amount of change (if any). If the system determines that the type of soda requested is not available, it illuminates a display informing the customer that the selection is "sold out" and asks the customer to make an alternative selection. The customer can choose another product or press the coin refund selector. If the customer makes an alternative selection, the system provides the requested product (if it is available). If the customer requests a refund, the system returns the customer's money.

The following structured specification models the Automatic Vending System. Assume that this Automatic Vending System only accepts nickels, dimes, and quarters, and returns the customer's coins upon request. Change is composed of nickels and dimes only.

The context diagram is shown in Fig. 10.22. The data flow diagram is shown in Fig. 10.23. The data dictionary is shown in Fig. 10.24 and the mini-specs in Fig. 10.25.

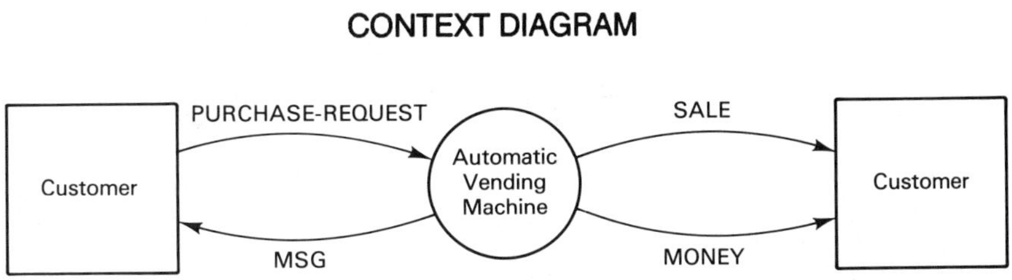

Figure 10.22

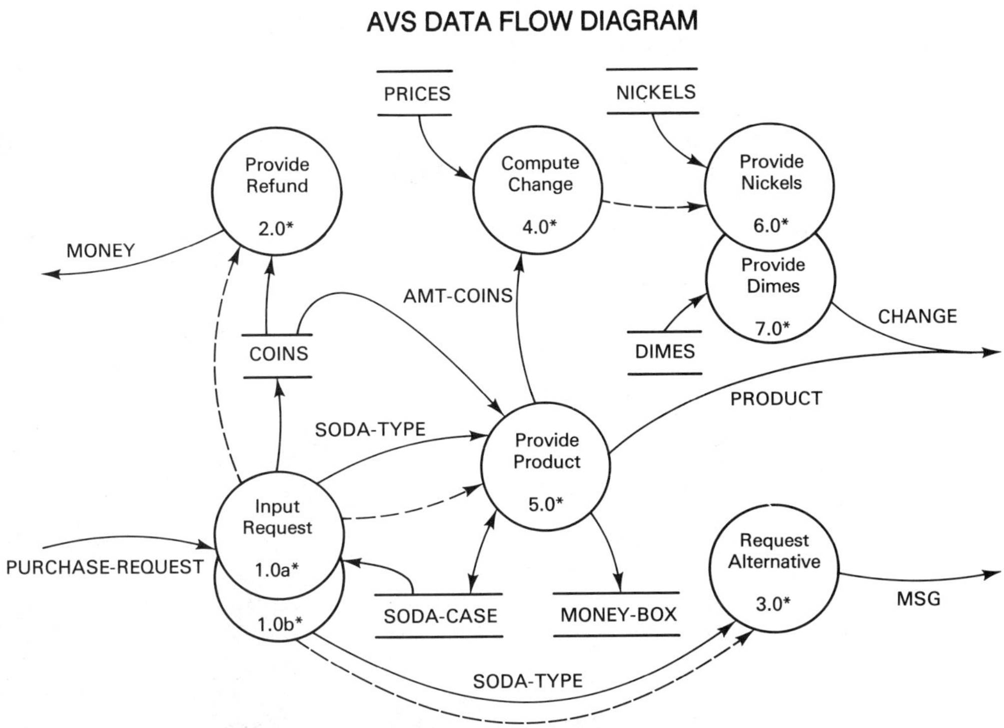

Figure 10.23

DATA DICTIONARY

AMT-COINS	= $_0\{\text{TEN-CENTS}\}^2 + {}_0\{\text{FIVE-CENTS}\}^1$
	= * Total amount of PAYMENT received from the Customer. *
CHANGE	= * Amount of US Currency = PAYMENT - PRICE *
COINS	= {PAYMENT}
	= * Temporary holding area for PAYMENT. *
DIMES	= {TEN-CENTS}
FIVE-CENTS	= * US Coin = $.05 *
MONEY	= * PAYMENT returned to the Customer. *
MONEY-BOX	= * Coin box containing the PAYMENT from all purchases. *
MSG	= SODA-TYPE + "SOLD OUT - PLEASE MAKE ANOTHER SELECTION"
NICKELS	= {FIVE-CENTS}
PAYMENT	= Amount of currency supplied by the Customer. *
PRICE	= * Purchase price of one product. *
PRICES	= {PRICE}
PRODUCT	= * 12-oz can of SODA-TYPE. *
PURCHASE-REQUEST$_a$	= PAYMENT + SODA-TYPE
PURCHASE-REQUEST$_b$	= [SODA-TYPE | "REFUND"]
SALE	= PRODUCT + $_0\{\text{CHANGE}\}^3$
SODA-CASE	= {SODA-TYPE + SODA-TYPE-AVAIL + {PRODUCT}}
SODA-TYPE-AVAIL	= ["YES" | "NO"]
SODA-TYPE	= ["Cola" | "Decaffinated Cola" | "Cherry Cola" | "Diet Cola"]
TEN-CENTS	= * US Coin = $.10 *

Figure 10.24

MINI-SPECIFICATIONS

Input Request MINI-SPECIFICATION 1.0a

STATE a - INITIAL STATE of a Finite-State process 1.0

Input PURCHASE-REQUEST$_a$.
Output PAYMENT to COINS.
Locate SODA-TYPE entry in SODA-CASE.
Input SODA-TYPE-AVAIL from SODA-CASE.
If SODA-TYPE-AVAIL = "YES",
    Initiate Provide Product.
    Output SODA-TYPE to Provide Product.
    Change STATE-1.0 to a.
Else Initiate Request Alternative.
    Output SODA-TYPE to Request Alternative.
    Change STATE-1.0 to b.
Endif.
Wait.
Processor Requirement: . . .
⋮
Other Requirements: . . .

Note the STATE-1.0 variable is not in the data dictionary because it is a local variable for the process 1.0.

Figure 10.25

MINI-SPECIFICATIONS (Cont.)

Input Request MINI-SPECIFICATION 1.0b

STATE b - Out-of-Stock condition exists for requested SODA-TYPE.

Input PURCHASE-REQUESTb.
If PURCHASE-REQUESTb = REFUND",
    Initiate Provide Refund.
    Change STATE-1.0 to a.
    Wait.
Endif.
Locate SODA-TYPE entry in SODA-CASE.
Input SODA-TYPE-AVAIL from SODA-CASE.
If SODA-TYPE-AVAIL = "YES",
    Initiate Provide Product.
    Output SODA-TYPE to Provide Product.
    Change STATE-1.0 to a.
    Wait.
Else Initiate Request Alternative.
    Output SODA-TYPE to Request Alternative.
    Change STATE-1.0 to b.
    Wait.
Endif.
Wait.

Processor Requirement: . . .
⋮
Other Requirements: . . .

Provide Refund MINI-SPECIFICATION 2.0

Input PAYMENT from COINS.
MONEY = PAYMENT.
Output MONEY to Customer.
Stop.

Processor Requirement: . . .
⋮
Other Requirements: . . .

Request Alternative MINI-SPECIFICATION 3.0

Input SODA-TYPE from Input Request.
Output MSG to Customer.
Stop.

Processor Requirement: . . .
⋮
Other Requirements: . . .

Figure 10.25 (cont.)

MINI-SPECIFICATIONS (Cont.)

Compute Change MINI-SPECIFICATION 4.0

Input AMT-COINS from Provide Product.
Input PRICE from PRICES.
AMT-RETURNED = PAYMENT - PRICE
Compute D = number of TEN-CENTS in AMT-RETURNED.
Compute N = number of FIVE-CENTS in (AMT-RETURNED - (D * $.10)).
Initiate Provide Nickels N times.
Initiate Provide Dimes D times.
Stop.

Processor Requirement: . . .
⋮
Other Requirements: . . .

Provide Product MINI-SPECIFICATION 5.0

Input SODA-TYPE from Input Request.
Locate SODA-TYPE in SODA-CASE.
Input PRODUCT from SODA-CASE.
If PRODUCT was the last one,
    SODA-TYPE-AVAIL = "NO",
    Output SODA-TYPE-AVAIL to SODA-CASE.
Endif.
Output PRODUCT to Customer.
Input PAYMENT from COINS.
AMT-COINS = PAYMENT.
Output AMT-COINS to Compute Change.
Output PAYMENT to MONEY-BOX.
Stop.

Processor Requirement: . . .
⋮
Other Requirements: . . .

Provide Nickels MINI-SPECIFICATION 6.0

Input FIVE-CENTS from NICKELS.
CHANGE = FIVE-CENTS
Output CHANGE to Customer.
Stop.

Processor Requirement: . . .
⋮
Other Requirements: . . .

Figure 10.25 (cont.)

MINI-SPECIFICATIONS (Cont.)

Provide Dimes MINI-SPECIFICATION 7.0

Input TEN-CENTS from DIMES.
CHANGE = TEN-CENTS
Output CHANGE to Customer.
Stop.

Processor Requirement: . . .

⋮

Other Requirements: . . .

Figure 10.25 (cont.)

A state-transition diagram can also apply in many areas of design such as specifying accessibility to subtending menus in interactive user/system dialogues. Menu structures are often viewed as a design consideration and are usually not modeled as part of the structured specification. The structured specification includes the requirements of the system and not the design strategies. In keeping with that objective, a data flow diagram indicating that one of several target processes are activated by a supervisory process is sufficient during the requirements stage. See Fig. 10.26. The screen displays that guide a user in choosing a process from a set of menus is a design issue. The design strategy can be shown using a state-transition diagram.

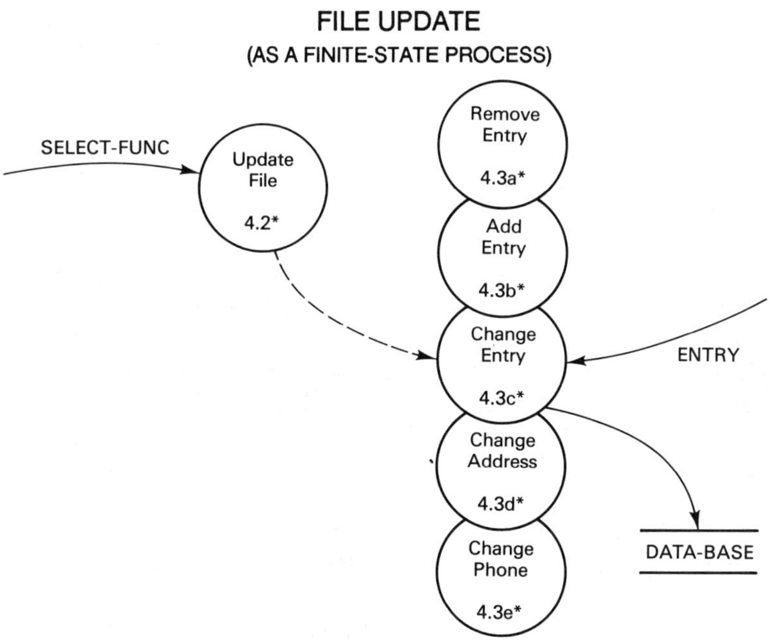

Figure 10.26

Analysis of the Automatic Vending System

Input Request 1.0* is a finite-state process that has two states, as documented in the mini-specs. When the process receives the input PURCHASE-REQUEST, it initiates one of three subsequent processes:

Provide Refund 2.0*
Request Alternative 3.0*
Provide Product 5.0*

In the first case, the supervisory process Input Request 1.0* and the target process Provide Refund 2.0* both use the data store COINS. The supervisory process receives PAYMENT as part of PURCHASE-REQUEST and places it into the COINS data store. The target process retrieves PAYMENT from the COINS data store .

In the case of Request Alternative 3.0*, the supervisory process Input Request 1.0* initiates the target process Request Alternative 3.0*, and passes the data flow element SODA-TYPE to it. The Request Alternative 3.0* process may appear to be a simple process from the mini-spec 3.0*, however, it is responsible for interfacing with the customer by using some available technology. With this type of partitioning, when there is a modification to the vending system (such as a new piece of equipment that permits the system to flash an alphabetic display stream to the customer instead of just lighting a light under the name of the PRODUCT), only one process has to be changed.

In the case of Provide Product 5.0*, the supervisory process Input Request 1.0* initiates the target process Provide Product 5.0*, passes the data flow element SODA-TYPE to it, and shares the data store SODA-CASE with it.

Consider processes Compute Change 4.0*, Provide Nickels 6.0*, and Provide Dimes 7.0*. Process Compute Change 4.0* dispatches either or both processes, Provide Nickels 6.0* and Provide Dimes 7.0*. The data flow diagram in Fig. 10.23 shows that the processes Provide Nickels 6.0* and Provide Dimes 7.0* are under the control of Compute Change 4.0*, and are dispatched simultaneously and independently from each other. The two processes Provide Nickels 6.0* and Provide Dimes 7.0* are similar processes whose individual outputs combine to produce the single output data flow CHANGE.

SUMMARY
. . . study the relationship between the system and its' environment

In this chapter, we covered the special conditions and tools used when dealing with real-time systems. The distinctive characteristic in real-time systems is the need to include the *flow of control* in the system specifications. I presented three aspects of modeling control within on-line and real-time systems:

> Supervisory processes are used to synchronize the outputs from target processes with independent external or temporal events outside the system.
>
> Finite-state processes are means to indicate the time-dependency behavior of a single process or a structure of processes within a system whose behavior is a function of time as well as the input data.
>
> Feedback systems are a means of specifying the requirements for a process or set of processes to control the system from within.

All of these techniques can be used to specify the special behavior that a system has to exhibit to synchronize its results with actual *real-world* events. These techniques allow an analyst not only to specify process-to-process relationship requirements of a system, but also specify system behavior with respect to time. Further details of how processes perform the necessary activities to synchronize its outputs to independent external, temporal, or internal events fall to the systems designer. It is in system design that state-transition diagrams and other design tools specify greater level of detail undertaken by a system.

In the analysis stage, we must avoid doing the system design; specify only the system requirements. The tools and techniques presented in this chapter, as well as throughout the text, have viewed the system as an *event-driven system*. **An EVENT-DRIVEN SYSTEM is one that produces results upon the arrival of a stimulus, an occurrence of time, or the recognition of an abnormal condition within the system.** Most commercial information systems and real-time systems are driven or directed by external, temporal, or internal events; thus, they are event-driven systems. Other types of systems exist, but they are beyond the scope of this text.

QUESTIONS

1. Explain the differences between batch systems, on-line systems, and real-time systems.
2. What are criteria?
3. Explain the difference between external control, and internal control and give an example technique that illustrates each type.
4. What is required for a control process to control a target process?
5. What is a feedback system?
6. What are the different types of feedback systems?

7. What is a restriction and what tools are available to specify a restriction in system requirements?
8. What is a finite-state process?
9. What are the methods to document requirements for finite-state processes?
10. What type of behavior do systems exhibit?
11. From the structured specification example of the Automatic Vending System, determine the following:

    a. When the customer inserts PAYMENT into the vending system is the PAYMENT put into the MONEY-BOX *before* or *after* the system provides the can of soda (PRODUCT) to the customer?

    b. What does the customer receive first, the CHANGE or the can of soda (PRODUCT)?

    c. If the customer's first choice is not available and the SODA-CASE is in fact empty, does the process Request Alternative 1.0b* only tell the customer that the selection is not available or does it tell the customer that all SODA-TYPEs are not available?

12. Provide the data flow diagram for a toll booth process that collects tolls and automatically returns the correct amount of change by dispatching nickels from a NICKEL-CONTAINER and dimes from a DIME-CONTAINER to the patron.
13. Determine which of the following systems are **real-time systems, on-line systems**, or **batch systems**:

    a. An auto rental agent faced with an unexpectedly high number of rental sales enters a request into an Auto Inventory System that indicates the number of cars at each rental location in the city, accepts the agent's request for additional cars, sends the request to the headquarters' location, and sends a delivery notice to a staff of drivers who bring the needed cars to the requesting location.

    b. A hospital admissions clerk with a seriously ill patient enters a transaction into the Hospital Room Inventory System to find the location of an available room.

    c. An Air Traffic Control System receives the coordinates of aircraft in the area and computes their trajectories to notify an air traffic controller of an impending collision.

    d. A Stock Market Summary System prints the results of the trading that has taken place during the day.

e. A Current Odds Posting System summarizes all of the bets that are made on a particular race and dynamically changes the odds as more and more customers place their bets.

f. A Page Printing System obtains text from a word processing file, formats it according to the embedded format commands, and drives a laser scanner that charges a plate. As an independently controlled feeder moves paper past the laser-charged plate at the constant rate of 20 sheets per minute, the printed image appears on the sheet.

14. What is the distinguishing characteristic of structured specification for a real-time system?

15. a. What is the predominant objective of a data-processing system?

b. What is the predominant objective of a real-time processing system?

16. a. Provide a data flow diagram that illustrates a person turning a Lamp process "ON" or "OFF" by means of an ON/OFF switch. The Lamp process has ELECTRICITY as input and LIGHT and HEAT as output.

b. Provide a data flow diagram that illustrates a person not only turning a Lamp process "ON" or "OFF," but also varying the amount of ELECTRICITY that flows to the Lamp by a dimmer process. The lamp process produces LIGHT and HEAT as output.

17. a. Provide an example of a simple system that exhibits data-dependent behavior.

b. Provide an example of a simple system that exhibits time-dependent behavior.

18. Given an on-line display system that uses function keys in the following manner:

F1 = Display a "help" explanation to the user
F2 = Display the user's options
F3 = Return to the main menu of the system
F4 = End process and return the user to the operating system

The system has the following five processes: Input Key 1.0*, Display "HELP" 2.0*, Display Options 3.0*, Return to Main Menu 4.0*, and Return to Operating System 5.0*.

a. Provide a simplified data flow diagram that receives the input data flow:

FUNC-KEY = [F1 | F2 | F3 | F4]

and the output data flow:

NEW-SCREEN = [HELP-SCREEN | OPTIONS-SCREEN | MAIN-MENU | "A>"].

b. What type of behavior does the system exhibit?

19. What type of behavior does a system exhibit that responds to an anomalous event?

20. What is the intended audience of a structured specification and what role does the structured specification fulfill?

PROBLEMS

1. Consider a shipping system that has a supervisory task that receives a MAIL-REQUEST from a customer and based on the URGENCY determines what process to use to mail the PACKAGE. The supervisory process sends the data flow SHIPMENT (containing the PACKAGE and necessary SHIP-TO-INFO) to either Bulk Mail (slowest delivery, 4-7 days), First Class (3-4 days), Rapid Mail (2 days), or Overnight Mail (guaranteed overnight delivery).

   Each target process properly addresses the PACKAGE, seals the PACKAGE, looks up the cost in PRICE-TABLES, posts the costs into the CHARGES file, places the receipt (for overnight mail only) in a RECEIPTS file, and outputs the item as MAIL to one of four delivery bins (sinks). The bulk-mail trunk picks up mail only once a day. First class mail is picked up four times a day (6:00 A.M., noon, 4:00 P.M., and 9:00 P.M.). Rapid mail leaves every three hours. Overnight mail leaves twice a day (noon and 6:00 P.M.). In the case of the overnight mail, if the process that prepares the MAIL does not work fast enough, the MAIL may miss the pickup. If the item is not delivered on time, the customer is eligible for a refund. All pickups leave promptly at the scheduled times and do not wait for any tasks to complete.

   a. Prepare a level-1 data flow diagram to represent the shipping system.
   b. Is this system considered a real-time system? Explain your answer.

2. Construct a time-dependency chart to show the following tasks which are necessary to build and start a fire in a fireplace using a gas starter and a match in one minute.

   Wood must first be placed into the fireplace by opening the glass doors and chain screen, OPEN-DOORS (4 secs) and OPEN-SCREEN (4 secs), and placing the wood on the grate, STACK-WOOD (30 secs). Sometime after the glass doors are opened and the gas starter is turned "on", the chimney flu must be opened. Opening the flu, OPEN-FLU (6 secs), can take place simultaneously with the STACK-WOOD task.

   Only after the wood is stacked and the flu is opened do we begin the careful job of lighting the gas starter. A match is lit, LIGHT-MATCH (3 secs), and the gas starter is turned "on", OPEN-GAS, for a maximum of 15 seconds unless the gas is ignited. We wait the first 3 seconds to allow the gas to begin flowing and then apply the lit match (APPLY-MATCH) to the starter for the remaining 12 seconds or until the gas is ignited. If the gas starter is not ignited within 12 seconds, the gas starter is turned "off," and we must wait at least 30 seconds to clear the collected gas fumes from the fireplace before repeating the tasks LIGHT-MATCH, OPEN-GAS, and IGNITE-GAS. Once the gas is ignited, we close the screen, CLOSE-SCREEN (4 secs), and the process is completed.

Prepare the time-dependency chart for one complete process, and show the tasks necessary if the gas starter fails to start the first time.

3. Construct the portion of a data flow diagram that uses a finite-state process to accept the dialed digits from a customer's telephone when a call is placed: area code (AC), dialed number (DN), and account number to bill the charges (ACCT#) to a credit card. The process inputs the CALLING# and the dialed digits and outputs the area code, the seven-digit dialed number, and the billing information, BILL-INFO = AC-BILLED + DN-BILLED + (ID#-BILLED) + TYPE, where TYPE = ["local"|"Time and Distance"|"$.50"|"IN-WATS"].

ACCT# is either the originating telephone number (CALLING#), the telephone number being called (CALLED#), or the credit card number (CARD#). All three numbers include the area code (aaa) and seven-digit telephone number (xxx-xxxx), and in the case of the CARD#, an additional four digit ID# (iiii) is included.

In placing a call, the first digit dialed is either a "1" for long distance calls, "0" for credit card calls, or any digit except "1" or "0" for local calls. *For simplicity, no operator-assisted or payphone calls are allowed in this problem.*

The finite-state process determines billing information by the following:

DIGITS DIALED	TYPE OF BILLING (BILL-INFO)
xxx-xxxx	Set: AC-BILLED to local area code, DN-BILLED to CALLING#, "local"
1-aaa-xxx-xxxx	Set: AC-BILLED to aaa, DN-BILLED to CALLING#, "Time and Distance"
1-800-xxx-xxxx	Set: aaa = 800, then AC-BILLED = 800, DN-BILLED = xxx-xxxx, "IN-WATS"
1-900-xxx-xxxx	Set: AC-BILLED to local area code, DN-BILLED = xxx-xxxx, "$.50"
0-aaa-xxx-xxxx -ccc-yyy-yyyy-iiii	Set: AC-BILLED to ccc, DN-BILLED to yyy-yyyy, ID# to iiii, if aaa is the local area code then "local", if aaa is 900 then "$.50", else "time & distance"

Figure 10.27

a. Prepare mini-specs for the finite-state process described above.
b. Construct a state-transition table for the process.
c. Construct a state-transition diagram (bubble form) for the process.
d. Construct a state-transition diagram (rectangular form) for the process.

4. Provide a narrative description of a practical example of a feedback control system. Provide a data flow diagram of the feedback system that accepts input data from a source, regulates the rate of flow of an item to a sink, and monitors the rate of flow of the output by processing additional input as feedback. Also, provide the mini-specs for the processes and data dictionary entries for all data.

11

CONVENIENT AUTO RENTAL SYSTEM

The Convenient Auto Rental (CAR) Company has a system called the Convenient Auto Rental System (CARS) that offers great convenience as well as new services to the auto rental customer. All of CAR's autos come equipped with a mobile phone. All calls made from the mobile phone are directly billed to a customer's credit card or a special AUTO-CARD that the customer receives when he or she rents the auto. The AUTO-CARD is similar to a credit card, and the customer uses it to enter and exit the parking lot as well as to authorize the use of the mobile phone when a credit card is not used. The AUTO-CARD is also used when autos are refueled upon their return. Every auto has its own AUTO-CARD with a magnetic strip that is encoded with a unique AUTO-CARD# and the license plate number. All charges made with the AUTO-CARD are applied to the customer's bill upon check-out.

CARS takes reservation for autos, issues autos and AUTO-CARDs, orders autos from the Central Auto Pool based on an anticipated level of business, accumulates phone charges from the Telephone Company, computes and issues bills to customers, controls access to the parking lot by means of automatic gates, and schedules periodical maintenance for each auto. The CONTEXT DIAGRAM of Fig. 11.1 illustrates the interfaces between CARS and all external sources and sinks.

The interfaces between CARS and the Fuel Pump and the Telephone Company are real-time interfaces. The interface between CARS and the Parking Lot Automatic Gate is an on-line interface that requires a high-priority response.

The first real-time function of CARS is to control the fuel pumps used to refuel an auto when the customer returns it. The attendant or customer inputs the AUTO-CARD into the Fuel Pump, which initiates the flow of FUEL. When the tank is near full, the automatic pump sends a "Stop Pump" signal to CARS as a PUMP-REQuest. The first time this happens, CARS restarts the flow of FUEL but at a slower rate to make sure the tank is full. The second time the automatic pump issues the "Stop Pump" signal, CARS stops the flow of FUEL, computes AMT-FUEL, and stores the

CARS CONTEXT DIAGRAM

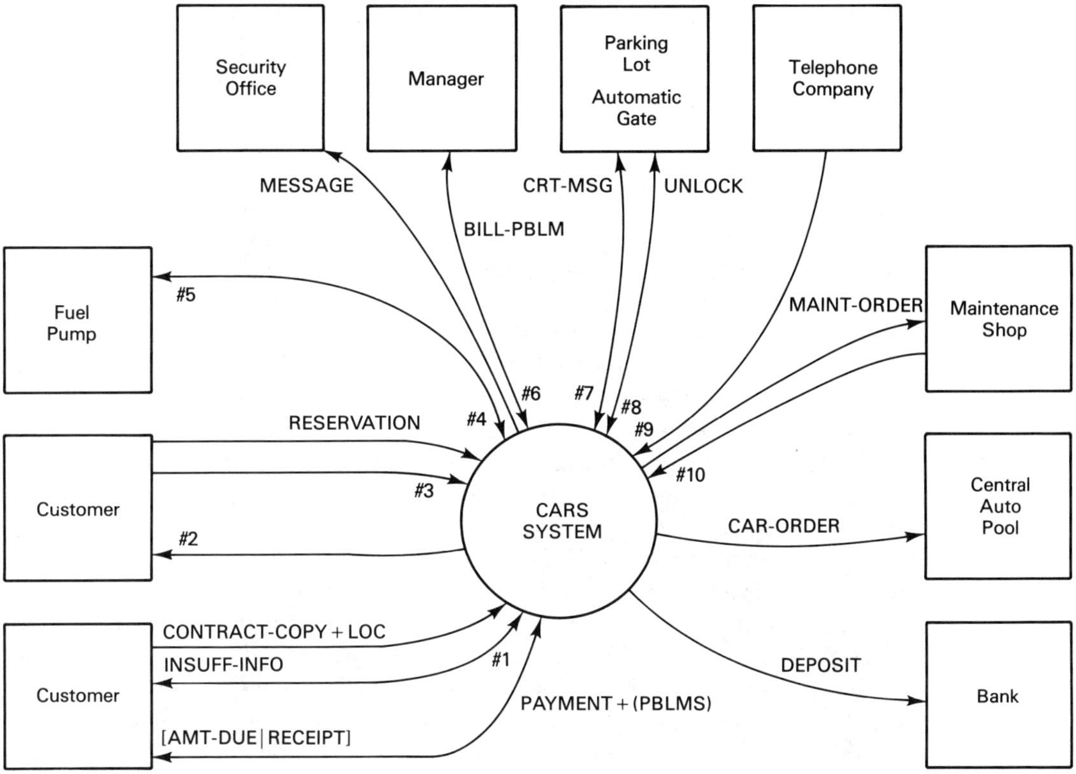

LEGEND TABLE

#1	(FUEL-LEV) + (CURR-MILEAGE) + CUSTM-NAME
#2	[AUTO-CARD + CONTRACT-COPY + LOC + KEYS | APOLOGY]
#3	[CUSTM-NAME + CUSTM-ADDR | INSURANCE + DRIVER-LIC#]
#4	[AUTO-CARD# + PUMP# | PUMP-REQ | PUMP-ON]
#5	[FUEL | UNLOCK-CARD]
#6	APPROVAL
#7	CURR-MILEAGE + FUEL-LEV
#8	AUTO-CARD + [ENTER-REQ | EXIT-REQ] + GATE
#9	DETAIL-CALL-RECORD
#10	AUTO-CARD + LOC + KEYS + MILEAGE

Figure 11.1

information in GAS-CHARGES for inclusion in the customer's bill. The system then unlocks the customer's AUTO-CARD by sending an UNLOCK-CARD message to the fuel pump. Because gasoline is a dangerous substance and its flow is regulated by the system, CARS processes the signals from the automatic pump with the highest priority. The system must synchronize the filling of the tank and the pumping of the fuel, and therefore performs a real-time function.

The second high-priority function of CARS is an on-line function of controlling access to the parking lot. When a customer inserts his or her AUTO-CARD in the automatic gate, the system is interrupted and a request is made to immediately verify the customer and decide to open the gate. If the system cannot determine if the auto has been properly checked out, it does not open the gate and notifies security by sending a MESSAGE to the Security Office. The access function has to respond rapidly or else the system will impede the traffic flow of autos returning. Because of safety, the priority of this function is the second highest priority of the system's processes. Because the operation of the gate regulates traffic and the traffic is under the control of drivers that are dependent on the gate controlled by the CARS, the function is a high-priority on-line activity, not a real-time activity.

A third high-priority task is a real-time function that is used when a customer completes a mobile phone call somewhere in the city. Upon completion of the phone call, the Telephone Company immediately issues a DETAIL-CALL-RECORD, which must be retrieved immediately by CARS and filed into the data store (PHONE-CALLS) used in computing the customer's bill. Failure to immediately process the DETAIL-CALL-RECORD can cause the record to be overwritten by another incoming record. If the record is overwritten, the CAR company loses the billing data and the revenue it represents. The Telephone Company is capable of sending seven DETAIL-CALL-RECORDs per second and the arrival of the records is independent of the system's ability to process the records. Therefore, processing DETAIL-CALL-RECORDs is a high-priority real-time task in the system.

The system also is responsible for many business functions. It takes advanced RESERVATIONs from customers. When a customer arrives for an auto, he or she simply gives the counter clerk his or her name (CUSTM-NAME), address (CUSTM-ADDR), driver's license number (DRIVER-LIC#), and tell whether or not they want insurance (INSURANCE). The clerk then issues a vehicle by preparing the CONTRACT, marking the location (LOC) of the auto on the envelope of the CONTRACT, and giving the customer the AUTO-CARD, CONTRACT, and KEYS. The customer locates the auto in the parking lot, drives the car to the departure gate, and inserts the AUTO-CARD into the slot at the automatic gate. The automatic gate reads the magnetic strip on the AUTO-CARD and sends an EXIT-REQuest to the system, which verifies the AUTO-CARD and sends a signal (UNLOCK) to open the gate.

To return a rental car, the customer pulls up to the automatic gate and inserts the AUTO-CARD into a slot. CARS immediately verifies the AUTO-CARD and opens the gate by sending it an UNLOCK message. The driver pulls up to the next station and is greeted by a message on a cathode-ray tube terminal (CRT-MSG). The driver has to only enter the current mileage (CURR-MILEAGE) on the auto's odometer and the reading from the fuel gauge (FUEL-LEV). He or she proceeds to park the car, leave the KEYS and AUTO-CARD in the auto, bring the CONTRACT to the service desk and tell the clerk the location (LOC) of the auto. Because delays at the gate can cause traffic problems, gate control is required to be a high-priority on-line function. Gate control is the second highest-priority function compared to the other system functions. By the time the customer arrives at the service desk, the system computes the customer's bill and is ready to request the AMT-DUE. The customer provides PAYMENT + (PBLMS) and the system issues a RECEIPT.

During the computation of the customer's bill, the system checks for several routine maintenance items, such as engine oil change, transmission oil change, brake linings, tune-up, etc. If the system determines that routine service is necessary, it generates a MAINT-ORDER and sends it to the Maintenance Shop. After maintenance is performed, the vehicle is washed, refueled, and placed back into service. The service person brings the KEYS, AUTO-CARD, and the car's location (LOC) to the service counter.

At 6:00 P.M. every evening, the system executes a process that determines how many autos are available for rent and the number of reservations for the following day. If there is an insufficient number of autos for the following day, the system generates an "Immediate Shortage" CAR-ORDER and sends it to the Central Auto Pool. Drivers bring in additional autos by 8:00 A.M. the next morning. A minimum inventory of six "Compact," four "Intermediate," two "Full-Size," and one "Luxury" auto is used to determine shortages when placing a CAR-ORDER.

In this example, the following assumptions apply:

1. The customer makes all payments in cash.
2. All autos use unleaded gas only.
3. All maintenance of autos is done off premises and outside the system.
4. All auto rentals are on a "Reservation-only" basis.
5. All sources provide "truthful" inputs to the system.
6. Customer names are not unique; Customer name and address are unique.
7. Every auto has its own unique AUTO-CARD and AUTO-LIC#.

Figure 11.2 contains the level-1 data flow diagram. On that diagram three processes are broken down further to level-2 diagrams, namely DISPATCH AUTOS process 1.0, PARKING LOT ACCESS process 7.0, and BILL CUSTOMER process 8.0 shown in Figs. 11.3, 11.4, and 11.5 respectively.

This Chapter contains a complete analysis of the system and has a data dictionary and mini-specifications (properly organized with one mini-spec per page). A logical analysis, object analysis, and event analysis are all performed with examples of an E-R diagram, input/output processing chart, and an event diagram. The system contains real-time characteristics modeled with real-time notation and tools.

CARS DATA FLOW DIAGRAM 0
(Level 1)

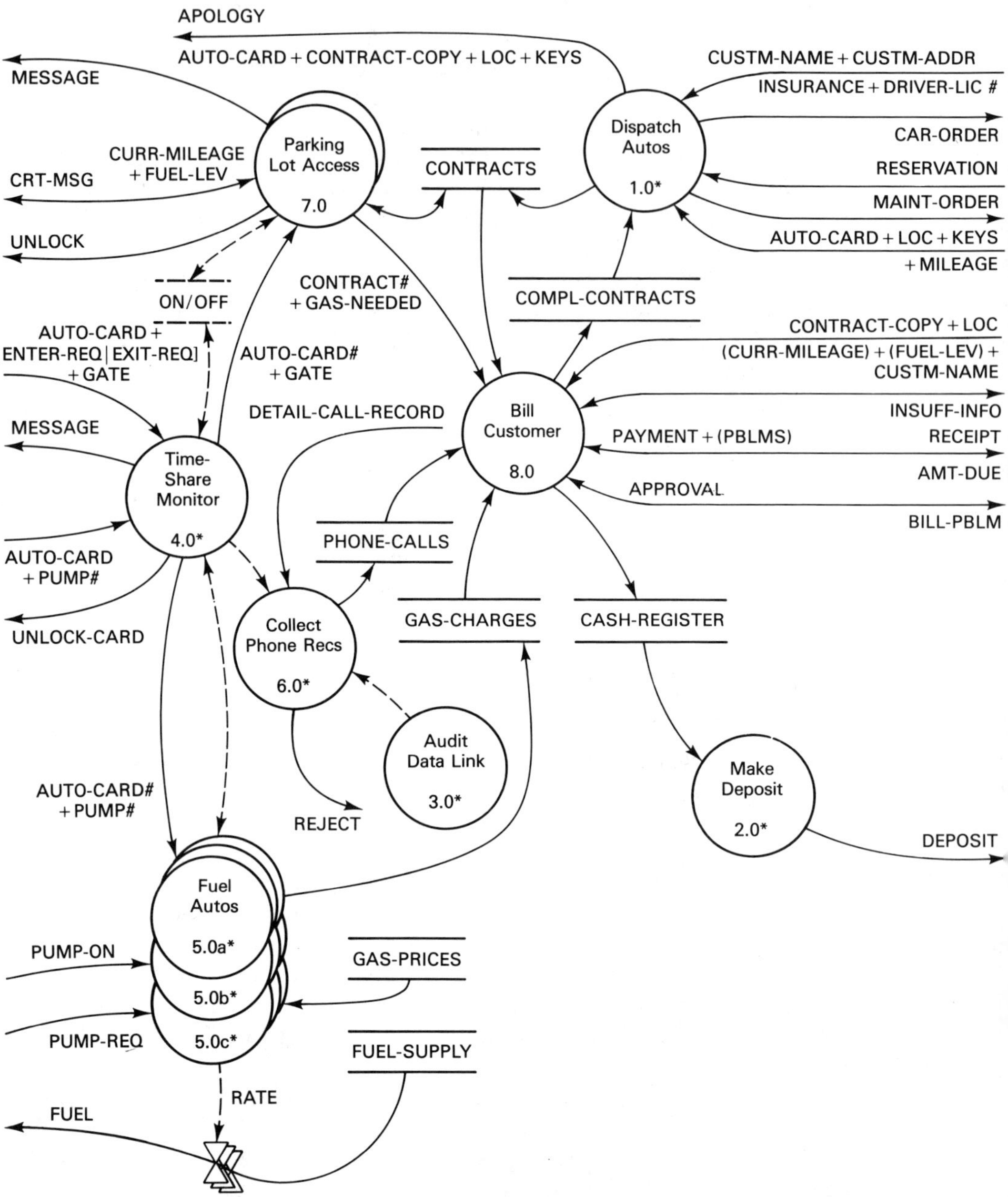

Figure 11.2

DISPATCH AUTOS - DATA FLOW DIAGRAM 1.0
(Level 2)

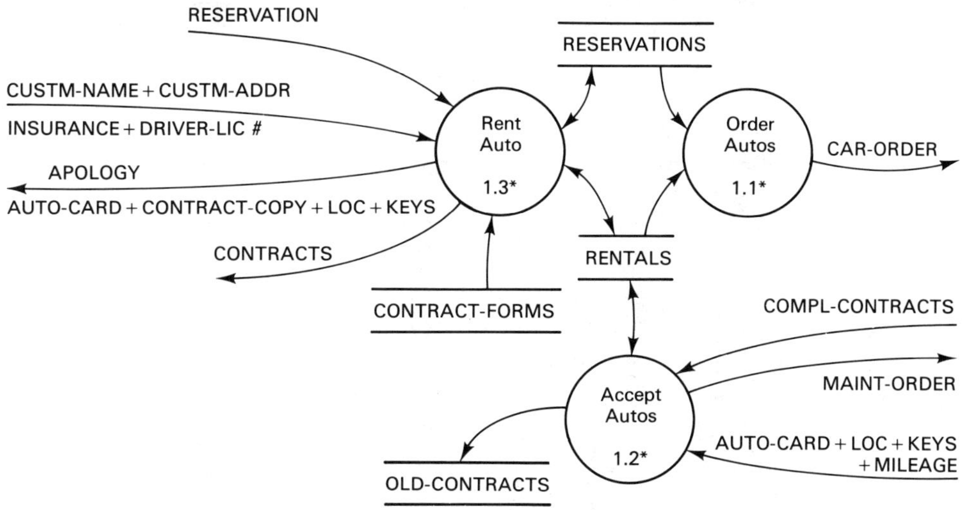

Figure 11.3

PARKING LOT ACCESS - DATA FLOW DIAGRAM 7.0
(Level 2)

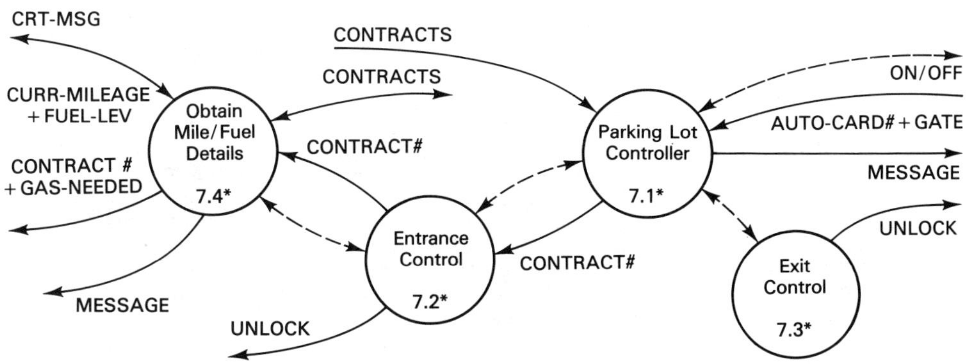

Figure 11.4

BILL CUSTOMER - DATA FLOW DIAGRAM 8.0
(Level 2)

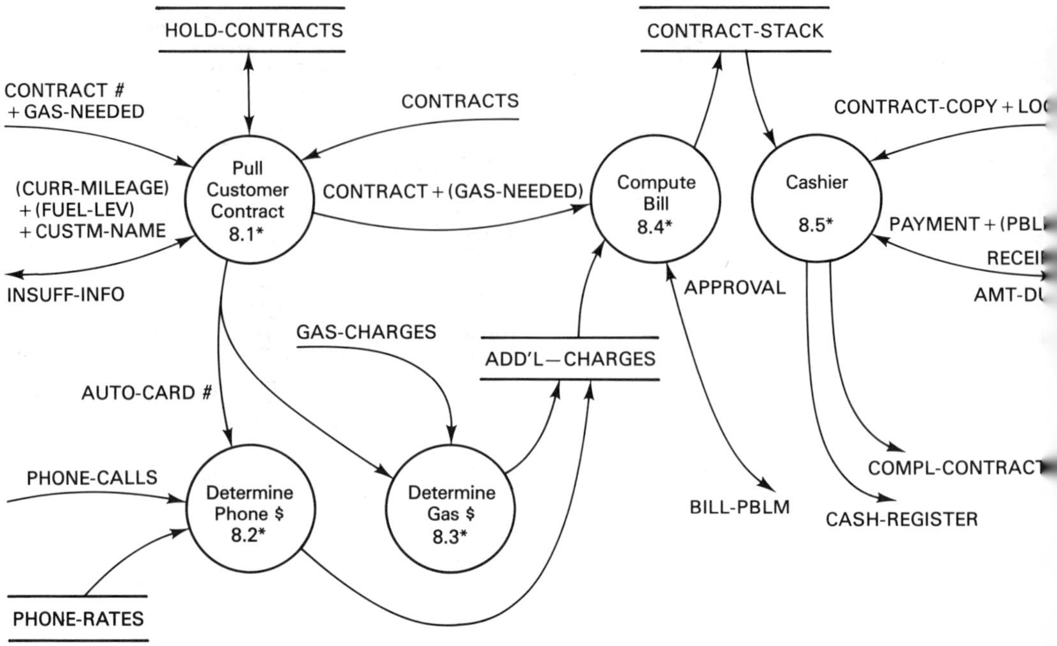

Figure 11.5

Figure 11.6 contains the data dictionary for the system followed by all of the mini-specs in Fig. 11.7.

DATA DICTIONARY

ADD'L-CHARGES	= {AUTO-CARD# + REASON + AMT$}
AMT$	= * Dollar amount *
AMT-CALL	= * Amount due for the phone call. *
AMT-FUEL	= * Amount of gasoline used to refuel a returned auto. *
AMT$-GAS	= * Amount in dollars of gas to refuel returned auto. *
APOLOGY	= ["Sorry, no autos are available." | "Sorry, you do not have a reservation."]
APPROVAL	= [zero | [GAS$ | MILES]]
AUTO-CARD	= AUTO-CARD# + AUTO-LIC#
AUTO-CARD#	= * Identification number associated to the rented auto.*
AUTO-LIC#	= * License number of the auto. *
AUTO-TYPE	= ["COMPACT" | "INTERMEDIATE" | "FULL SIZE" | "LUXURY"]
AVAILABILITY	= ["AVAILABLE" | "RENTED" | "BEING SERVICED"]
BILL-PBLM	= ["GAS NOT FOUND" | "MILEAGE ERROR"]
BRAKE-DATE	= * Date of the last brake linings change. *

Figure 11.6

DATA DICTIONARY (cont.)

CALLED-PHONE#	= * The phone number called from the mobile phone in the rented auto. *
CAR-KEYS	= * Ignition and door keys for an auto. *
CAR-ORDER	= RENTAL-AGENCY + ORDER-DATE + {QTY + TYPE-NEEDED + "IMMEDIATE"}
CASH-REGISTER	= {CASH}
COMPL-CONTRACT	= CONTRACT + DATE-RETURNED + (PBLMS) + TOTALS + "PAID" + LOC
COMPL-CONTRACTS	= * Filed copies of COMPL-CONTRACTs organized by DATE-RETURNED. *
CONTRACT	= CONTRACT# + CUSTM-NAME + CUSTM-ADDR + DRIVER-LIC#
	+ DATE-RENTAL + INSURANCE + RENTAL-RATE + MILEAGE + FUEL-LEV
	+ AUTO-TYPE + AUTO-LIC# + AUTO-CARD# + CURR-MILEAGE
CONTRACT-COPY	= CONTRACT
	= * Copy of of the customer's CONTRACT. *
CONTRACT-FORM	= CONTRACT# + RENTAL-RATES
CONTRACT-FORMS	= {CONTRACT# + RENTAL-RATES}
CONTRACT-STACK	= {CONTRACT + TOTALS}
	= * Organized by CONTRACT#)
CONTRACTS	= {CONTRACT# + CUSTM-NAME + CUSTM-ADDR + DRIVER-LIC#
	+ DATE-RENTAL + INSURANCE + RENTAL-RATE + MILEAGE + FUEL-LEV
	+ AUTO-TYPE + AUTO-LIC# + AUTO-CARD# + CURR-MILEAGE}
	= * Alternate key is AUTO-CARD#. *
CRT-MSG	= "Enter in the current mileage from the odometer and the amount of fuel in
	the gas tank: Full, 7/8, 3/4, 5/8, 1/2, 3/8, 1/4, 1/8, Empty"
CURR-MILEAGE	= * Mileage on the auto at the end of the rental period or "UNKNOWN." *
CUSTM-ADDR	= * Customer's address *
CUSTM-NAME	= * Customer's name *
DATE-CALL	= * The date of the phone call. *
DATE-RENTAL	= * Date the rental period began. *
DEPOSIT	= RENTAL-AGENCY + CASH
DETAIL-CALL-RECORD	= [AUTO-CARD# | "Credit card call"] + DATE-CALL + START-TIME
	+ FINISH-TIME + MOBILE-PHONE# + CALLED-PHONE#
DRIVER-LIC#	= DRIVER-STATE + LIC-ID# + EXPIRATION-DATE
DRIVER-STATE	= * State that issued driver license. *
DURATION	= * Time rounded to nearest minute (FINISH-TIME - START-TIME). *
ENTER-REQ	= AUTO-CARD# + AUTO-LIC#
	= * Request to enter parking lot for returned autos, activated by passing the
	AUTO-CARD through the automatic gate card reader. *
EXIT-REQ	= AUTO-CARD# + AUTO-LIC#
	= * Request to exit parking lot, activated by passing the AUTO-CARD through
	the automatic gate card reader. *
FINISH-TIME	= * The time the phone call was ended. *
FUEL	= * Gasoline *
FUEL-LEVEL	= ["FULL" | "7/8" | "3/4" | "5/8" | "1/2" | "3/8" | "1/4" | "1/8" | "EMPTY" | "UNKNOWN"]
FUEL-SUPPLY	= {FUEL}
GAS-CHARGES	= {AUTO-CARD# + AMT$-GAS + AMT-FUEL}
GAS-NEEDED	= ["YES" | "NO" | "UNKNOWN"]
GAS-PRICE	= * Price per gallon of fuel. *
GAS-PRICES	= {PUMP# + GAS-PRICE}
GATE	= ["1" | "2"]
	= * GATE 1 = EXIT and GATE 2 = ENTER. *
HOLD-CONTRACTS	= {CONTRACT + GAS-NEEDED}
	= * Temporary holding area for CONTRACTs, organized by CUSTM-NAME.*
INSUFF-INFO	= {["Need current mileage" | "Need auto refueled using AUTO-CARD"]}

Figure 11.6 (cont.)

DATA DICTIONARY (cont.)

INSURANCE	= ["YES" + SIGNATURE | "NO" + SIGNATURE]
	= * Whether the customer buys temporary collision insurance. *
KEYS	= AUTO-TYPE + AUTO-LIC# + CAR-KEYS + RENTAL-AGENCY
LOC	= * Location (parking space number) where the rental vehicle is parked. *
MAINT-ORDER	= AUTO-LIC# + ODOMETER-READING + LOC + {["Oil change" | "Transmission oil change" | "Brakes change" | "Tune-up" | PBLMS]}
MESSAGE	= * Message sent to the security office by the system. *
MILEAGE	= * Mileage reading from the auto's odometer when the rental began. *
MOBILE-PHONE#	= * The seven-digit phone number of the mobile phone in the rented auto. *
ODOMETER-READING	= * The mileage indicated on the auto's odometer. *
OIL-DATE	= * Date of the last oil change. *
OLD-CONTRACTS	= * File of COMPL-CONTRACTs organized by DATE-RETURNED. *
PBLMS	= * Any complaints regarding the rented vehicle from the customer. *
PHONE-CALLS	= {AUTO-CARD# + DATE-CALL + START-TIME + FINISH-TIME + MOBILE-PHONE# + CALLED-PHONE#}
PHONE-RATES	= {DURATION + AMT-CALL}
PUMP#	= * Identification number of the gas pump used to refuel returned auto. *
PUMP-ON	= * Verification signal that the fuel pump is running. *
PUMP-REQ$_a$	= "START"
PUMP-REQ$_b$	= ["STOP" | "EMERGENCY STOP"]
PUMP-REQ$_c$	= ["STOP" | "EMERGENCY STOP"]
QTY	= * Quantity *
RATE	= ["FULL" | "QUARTER"]
	= * Rate of pumping fuel: FULL = 4 gal/min, QUARTER = 1 gal/min. *
REASON	= ["PHONE CALL" | "GAS" | AMT-FUEL]
RECEIPT	= CONTRACT-COPY + TOTALS + "PAID"
RENTAL	= AUTO-TYPE + AUTO-LIC# + MILEAGE + OIL-DATE + OIL-MILES + TRANS-DATE + TRANS-MILES + BRAKE-DATE + BRAKE-MILES + TUNE-UP-DATE + TUNE-UP-MILES + LOC + KEYS + AUTO-CARD# + AVAILABILITY
RENTALS	= {AUTO-TYPE + AUTO-LIC# + MILEAGE + OIL-DATE + OIL-MILES + TRANS-DATE + TRANS-MILES + BRAKE-DATE + BRAKE-MILES + TUNE-UP-DATE + TUNE-UP-MILES + LOC + KEYS + AUTO-CARD# + AVAILABILITY}
RENTAL-AGENCY	= * Contains the name, address, and phone number of the rental agency. *
RENTAL-DATE	= * Date customer needs an auto. *
RENTAL-DAYS	= * Number of days customer wishes to rent auto. *
RENTAL-RATE	= DAILY$ + MILE$ + INSURANCE$
RENTAL-RATES	= {AUTO-TYPE + RENTAL-RATE}
RESERVATION	= CUSTM-NAME + CUSTM-ADDR + RENTAL-DATE + [AUTO-TYPE + RENTAL-DAYS | "CANCEL"]
RESERVATIONS	= {CUSTM-NAME + CUSTM-ADDR + AUTO-TYPE + RENTAL-DATE + RENTAL-DAYS}
SIGNATURE	= * Customer's signature indicating approval or acceptance. *
START-TIME	= * The time the called party answered the phone called. *
TOTALS	= * Total amount due from the customer. *
TRANS-DATE	= * Date of the last transmission oil change. *
TUNE-UP-DATE	= * Date of the last tune-up. *
UNLOCK-CARD	= * Unlock the release mechanism of the fuel pump to return AUTO-CARD. *

Figure 11.6 (cont.)

MINI-SPECIFICATIONS

Order Autos MINI-SPECIFICATION 1.1

```
* Determine how many of each type of auto are still leaving tonight.
Repeat
    Search RESERVATIONS for RENTAL-DATE = CURRENT-DATE.
    If AUTO-TYPE = "COMPACT", then Add 1 to C-GOING.
    If AUTO-TYPE = "INTERMEDIATE", then Add 1 to I-GOING.
    If AUTO-TYPE = "FULL SIZE", then Add 1 to F-GOING.
    If AUTO-TYPE = "LUXURY", then Add 1 to L-GOING.
Until all entries in RESERVATIONS.

* Determine how many of each type of auto are reserved for tomorrow.
Repeat
    Search RESERVATIONS for RENTAL-DATE = CURRENT-DATE + 1.
    If AUTO-TYPE = "COMPACT", then Add 1 to C-NEEDED.
    If AUTO-TYPE = "INTERMEDIATE", then Add 1 to I-NEEDED.
    If AUTO-TYPE = "FULL SIZE", then Add 1 to F-NEEDED.
    If AUTO-TYPE = "LUXURY", then Add 1 to L-NEEDED.
Until all entries in RESERVATIONS.

* Determine how many of each type of auto we have for rent.
Repeat
    Search RENTALS for AVAILABILITY = "AVAILABLE".
    If AUTO-TYPE = "COMPACT", then Add 1 to C-HAVE.
    If AUTO-TYPE = "INTERMEDIATE", then Add 1 to I-HAVE.
    If AUTO-TYPE = "FULL SIZE", then Add 1 to F-HAVE.
    If AUTO-TYPE = "LUXURY", then Add 1 to L-HAVE.
Until all entries in RENTALS.

* Determine how many of each type of auto to order.
Set C-ORDER = (C-NEEDED + 6) - (C-HAVE - C-GOING)
Set I-ORDER = (I-NEEDED + 4) - (I-HAVE - I-GOING)
Set F-ORDER = (F-NEEDED + 2) - (F-HAVE - F-GOING)
Set L-ORDER = (L-NEEDED + 1) - (L-HAVE - L-GOING)

* Place order.
If C-ORDER > 0, then QTY = C-ORDER, TYPE-NEEDED = "COMPACT".
If I-ORDER > 0, then QTY = I-ORDER, TYPE-NEEDED = "INTERMEDIATE".
If F-ORDER > 0, then QTY = F-ORDER, TYPE-NEEDED = "FULL SIZE".
If L-ORDER > 0, then QTY = L-ORDER, TYPE-NEEDED = "LUXURY"
If C-ORDER > 0, or I-ORDER > 0, or F-ORDER > 0, or L-ORDER > 0, then
    Set ORDER-DATE = CURRENT-DATE,
    Output CAR-ORDER to Central Auto Pool.
Stop.
```

<u>Processor Requirements</u>: Manual process performed by the agency's manager at approximately 6:00 P.M.
<u>Security Requirements</u>: N/A.
<u>Time Requirements</u>: N/A.
<u>Processing Priority</u>: Low.
<u>Performance Requirements</u>: N/A.
<u>Safety Requirements</u>: N/A.
<u>Other Requirements</u>: None.

Figure 11.7

MINI-SPECIFICATIONS (Cont.)

Accept Autos MINI-SPECIFICATION 1.2

Case 1 (*Put an auto back into service after maintenance*):
Input AUTO-CARD + LOC + KEYS + MILEAGE from Maintenance Shop.
Locate AUTO-TYPE and AUTO-LIC# on KEYS.
Locate AUTO-CARD# on AUTO-CARD.
Input RENTAL for AUTO-TYPE + AUTO-LIC# from RENTALS.
Set AVAILABILITY on RENTAL = "AVAILABLE".
Set AUTO-CARD, LOC & KEYS on RENTAL to AUTO-CARD, LOC & KEYS inputted.
Output RENTAL to RENTALS.
Endcase 1.

Case 2 (*Determine if any of the autos returned need maintenance*):
Input COMPL-CONTRACT from COMPL-CONTRACTS.
Search RENTALS for AUTO-TYPE + AUTO-LIC#.
Input RENTAL from RENTALS.
If (CURR-MILEAGE - OIL-MILES) > 3,000, then
Set MAINT-ORDER to "Oil change".
If (CURR-MILEAGE - TRANS-MILES) > 50,000, then
Set MAINT-ORDER to "Transmission oil change".
If (CURR-MILEAGE - BRAKE-MILES) > 35,000, then
Set MAINT-ORDER to "Brake lining change".
If (CURR-MILEAGE - TUNE-UP-MILES) > 30,000, then
Set MAINT-ORDER to "Tune-up".
If COMPL-CONTRACT has PBLMS indicated, then
Set MAINT-ORDER to PBLMS.
If any of the above maintenance is required, then
Set MAINT-ORDER to AUTO-LIC#,
Set ODOMETER-READING to CURR-MILEAGE,
Set LOC on MAINT-ORDER to LOC on COMPL-CONTRACT.
Output MAINT-ORDER to Maintenance Shop.
Else Search RENTALS for AUTO-TYPE + AUTO-LIC#,
Set MILEAGE on RENTAL entry to CURR-MILEAGE from COMPL-CONTRACT,
Set LOC on RENTAL to LOC on COMPL-CONTRACT.
Output RENTAL to RENTALS.
Output COMPL-CONTRACT to OLD-CONTRACTS.
Endcases.
Stop.

Processor Requirements: Manual process performed by the counter supervisor hourly.
Security Requirements: KEYS and AUTO-CARD must be properly placed in the RENTALS data store.
Time Requirements: N/A.
Processing Priority: Low; counter supervisor should given preference to case 1 when inputs come in from the Maintenance Shop.
Performance Requirements: COMPL-CONTRACT is processed every hour. If their are no autos available for renting, this process is done every time a COMPL-CONTRACT is ready for processing.
Safety Requirements: N/A.
Other Requirements: None.

Figure 11.7 (cont.)

MINI-SPECIFICATIONS (Cont.)

Rent Auto MINI-SPECIFICATION 1.3

```
Case 1 (Customer makes/cancels a reservation):
   Input RESERVATION from Customer.
   If RESERVATION = "CANCEL", then
      Search RESERVATIONS for CUSTM-NAME + CUSTM-ADDR + RENTAL-DATE,
      Delete CUSTM-NAME + CUSTM-ADDR + RENTAL-DATE from RESERVATIONS,
   Else Output CUSTM-NAME + CUSTM-ADDR + RENTAL-DATE + AUTO-TYPE +
      RENTAL-DAYS to RESERVATIONS.
Endcase 1.
Case 2 (Customer rents auto):
   Input CUSTM-NAME + CUSTM-ADDR from Customer.
   Set RENTAL-DATE = CURRENT-DATE.
   Search RESERVATIONS for CUSTM-NAME + CUSTM-ADDR + RENTAL-DATE.
   If there is no entry for CUSTM-NAME,
      Set APOLOGY to "no reservation",
      Output APOLOGY to Customer,
      Stop.
   Input RESERVATION from RESERVATIONS.
   Delete RESERVATION from RESERVATIONS.
   Input CONTRACT-FORM from CONTRACT-FORMS.
   Circle the RENTAL-RATE for AUTO-TYPE on RESERVATION.
   Repeat
      Search RENTALS for AUTO-TYPE.
      If RENTAL = AUTO-TYPE, then Input RENTAL from RENTALS.
      Else Set AUTO-TYPE to next larger type.
   Until RENTAL = AUTO-TYPE,
   If RENTAL not found in RENTALS then
      Set APOLOGY to "no autos available",
      Output APOLOGY to Customer,
      Stop.
   Input RENTAL for AUTO-TYPE from RENTALS.
   Set RENTAL to "RENTED".
   Keep AUTO-CARD, LOC and KEYS.
   Output RENTAL to RENTALS.
   Input DRIVER-LIC# + INSURANCE from Customer.
   Create a CONTRACT = CUSTM-NAME + CONTRACT# + CUSTM-ADDR + DRIVER-LIC#
      + DATE-RENTAL + INSURANCE + RENTAL-RATE + MILEAGE + AUTO-TYPE
      + AUTO-LIC# + AUTO-CARD#
   Output CONTRACT to CONTRACTS.
   Output AUTO-CARD + CONTRACT-COPY + LOC + KEYS to Customer.
Endcases.
Stop.
```

Processor Requirements: Manual process performed by the counter clerk.
Security Requirements: Placed KEYS + AUTO-CARD in RENTALS data store.
Time Requirements: Process should take less than 3 min.
Processing Priority: Once the process begins, no interrupts are permitted.
Performance Requirements: 75% of all customers should not wait more than 5 min and 95% of all customers should not wait more than 10 min.
Safety Requirements: N/A.
Other Requirements: None.

Figure 11.7 (cont.)

MINI-SPECIFICATIONS (Cont.)

Make Deposits MINI-SPECIFICATION 2.0

Input all but $100.00 from the CASH-REGISTER.
Set DEPOSIT to all CASH from input.
Set DEPOSIT to RENTAL-AGENCY.
Output DEPOSIT to Bank.
Stop.

Processor Requirements: Manual process performed by the manager twice a day.
Security Requirements: This process has access to all revenues collected and the daily DEPOSIT amounts must equal the TOTALS for a given day as recorded in the OLD-CONTRACT data store.
Time Requirements: N/A.
Processing Priority: N/A.
Performance Requirements: N/A.
Safety Requirements: Personal safety of the manager is of highest priority. Comply with all robbery attempts to minimize injury or harm to the manager.
Other Requirements: None.

Figure 11.7 (cont.)

MINI-SPECIFICATIONS (Cont.)

Audit Data Link MINI-SPECIFICATION 3.0

Monitor data-link signal from Telephone Company.
If data-link signal drops, then
  Disconnect phone line,
  Wait 3 sec,
  Dial data-link line,
  Establish data link.
Endif.
Wait.

Processor Requirements: Automated real-time process tied into the data port of the computer and monitors the data link from the Telephone Company.
Security Requirements: N/A.
Time Requirements: N/A.
Processing Priority: Operates at the same priority as process 6.0*.
Performance Requirements: N/A.
Safety Requirements: N/A.
Other Requirements: None.

Figure 11.7 (cont.)

MINI-SPECIFICATIONS (Cont.)

Time Share Monitor MINI-SPECIFICATION 4.0

Case 1 (*Initial start-up of the automated system causes this process to initiates a 6.0* process and two 7.0* processes. The process then waits for input data or a child process to terminate and apply the appropriate case.*):
Initiate Collect Phone Recs 6.0*.
Check ON/OFF for a process 7.1* that is "OFF".
If all processes are ON, then
Set MESSAGE = "Problem detected at " + GATE,
Output MESSAGE to Security Office.
Wait.
Else Activate a Parking Lot Controller 7.1* that is "OFF".
Endcase 1.

Case 2 (*Input arrives requesting the use of fuel pump*):
Input AUTO-CARD + PUMP# from Fuel Pump.
Initiate Fuel Autos 5.0*.
Output AUTO-CARD# + PUMP# to Fuel Autos 5.0*.
Endcase 2.

Case 3 (*Fuel Autos 5.0* completes*):
Wait for Fuel Auto 5.0* to complete.
Output UNLOCK-CARD to PUMP# Fuel Pump.
Endcase 3.

Case 4 (*Collect Phone Recs 6.0* terminated*):
* Process 6.0* should always be active.
Initiate Collect Phone Recs 6.0*.
Endcase 4.

Case 5 (*Input arrives from Parking Lot Automatic Gate*):
Input AUTO-CARD + [ENTER-REQ | EXIT-REQ] + GATE from Parking Lot Automatic Gate.
Input ON/OFF for all of the Parking Lot Controller 7.1* processes.
If all processes are ON, then
Set MESSAGE = "Problem detected at " + GATE,
Output MESSAGE to Security Office.
Wait.
Else Activate a Parking Lot Controller 7.1* that is "OFF".
Output AUTO-CARD# + GATE to Parking Lot Controller 7.1*.
Endcases.
Wait.

Processor Requirements: Automated process operating in a real-time operating environment.
Security Requirements: Process controls access to parking lot and fuel pumps by the use of AUTO-CARD; the system provides no other validation.
Time Requirements: N/A.
Processing Priority: Case 3 has highest priority, then cases 5, 4, and 2.
Performance Requirements: Case 3 must be processed within 100 millisec, cases 2 in 5 sec, and case 4 in 140 millisec. Case 1 is the initial start-up routine and does not compete with the other cases.
Safety Requirements: N/A.
Other Requirements: None.

Figure 11.7 (cont.)

MINI-SPECIFICATIONS (Cont.)

Fuel Autos MINI-SPECIFICATION 5.0

Process 5.0 is a real-time finite-state process that operates in three states according to the following state-transition table:

STATE-TRANSITION TABLE for FUEL AUTOS 5.0

INPUTS \ STATES	a	b	c
I1 : AUTO-CARD# + PUMP#	O1: START PUMP MOTOR --> a		
I2 : PUMP-ON = "ON"	O2: RATE = "FULL" --> a		
I3 : PUMP-ON = "OFF"	RETURN TO 4.0* --> a		
I4 : PUMP-REQ = "START"	O3: FUEL --> b		
I5 : PUMP-REQ = "STOP"		O4: Stop FUEL O5: RATE = "QUARTER" --> c	O4: Stop FUEL --> c
I6 : PUMP-REQ = "EMERGENCY STOP"		O4: Stop FUEL --> a	O4: Stop FUEL --> c
I7 : GAS-PRICE			O6: Output GAS-CHARGES --> a

The two state-transition diagrams below are shown as examples only. In practice we would prepare either the bubble form or the rectangular form, and, in most cases, if a state-transition table were prepared, it would suffice.

STATE-TRANSITION DIAGRAM

(BUBBLE FORM) (RECTANGULAR FORM)

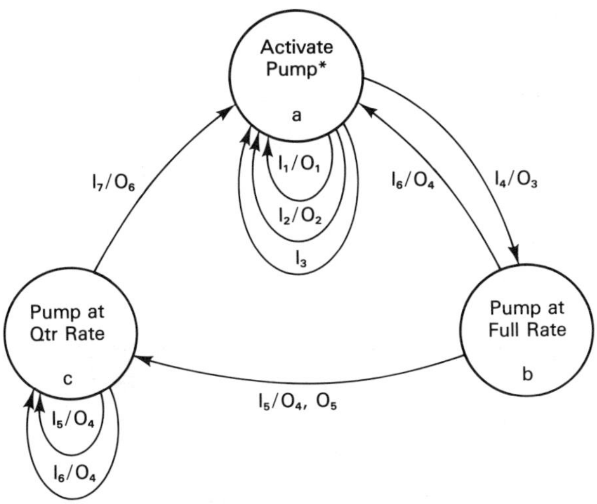

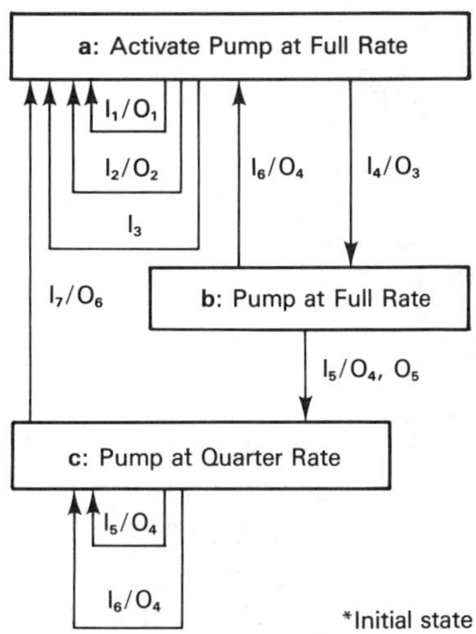

*Initial state

<u>Processor Requirements</u>: Automated process operating in a real-time operating environment.

⋮

<u>Other Requirements</u>: None.

Figure 11.7 (cont.)

MINI-SPECIFICATIONS (Cont.)

Fuel Autos MINI-SPECIFICATION 5.0a

STATE a - INITIAL STATE of the Fuel Autos finite-state process.

```
Input AUTO-CARD# + PUMP# from Time-Share Monitor 4.0*.
Repeat
    Start the pump motor for PUMP#.
    * Verify that the PUMP# pump motor is running.
    Input PUMP-ON from PUMP# Fuel Pump.
    If PUMP# = PUMP-ON, then
        Set RATE for PUMP# to "FULL",
        Output RATE to regulate FUEL-SUPPLY.
        Wait for input PUMP-REQ.
        Input PUMP-REQa from Fuel Pump,
        Change STATE-5.0 to b.
        Start PUMP# pumping at RATE.
        Output FUEL to Fuel Pump at RATE rate.
        Set START-TIME-FULL to CURRENT-TIME
        Wait.
    Else Wait 5 sec.
Until tried to start pump 5 times.
* Pump has failed to start, return to Time Share Monitor 4.0*.
Stop.
* Stop returns control to Time Share Monitor 4.0*.
```

Processor Requirements: Automated process operating in a real-time operating environment.

Security Requirements: N/A.

Time Requirements: Wait 5 sec between start pump attempts.

Processing Priority: Lowest priority among the real-time processes.

Performance Requirements: N/A.

Safety Requirements: Verify that the pump has started before changing states.

Other Requirements: None.

Figure 11.7 (cont.)

MINI-SPECIFICATIONS (Cont.)

Fuel Autos MINI-SPECIFICATION 5.0b

STATE b - Pumping fuel at full rate.

Case 1 (*Receive PUMP-REQ$_b$ = "STOP", auto's tank is almost full*):
    Input PUMP-REQ$_b$ from Fuel Pump.
    Output RATE = 0 (stop PUMP#).
    Set STOP-TIME-FULL to CURRENT-TIME.
    Set RATE for PUMP# to "QUARTER".
    Change STATE-5.0 to c.
    Start PUMP# by Output RATE = 1/4.
    Output FUEL to Fuel Pump at RATE rate.
    Set START-TIME-QTR to CURRENT-TIME.
    Wait.
Endcase 1.

Case 2 (*Receive PUMP-REQ$_b$ = "EMERGENCY STOP", pump has failed or was manually stopped*):
    Input PUMP-REQ$_b$ from Fuel Pump.
    Output RATE = 0 (stop PUMP#).
    Set STATE-5.0 to a.
    Set STOP-TIME-FULL to CURRENT-TIME
Endcases.

Stop.
* *Stop returns control to Time-Share Monitor 4.0**.

Processor Requirements: Automated process operating in a real-time operating environment.
Security Requirements: N/A.
Time Requirements: This process must begin executing within 100 millisec from the arrival of PUMP-REQ$_b$.
Processing Priority: Highest priority among the real-time processes.
Performance Requirements: N/A.
Safety Requirements: Pump must also provide for a manual "EMERGENCY STOP" (PUMP-REQ$_b$) to prevent spilling fuel.
Other Requirements: This automated process and the pumps should share the same power source, so that when power is interrupted, the pumps immediately stop. When power is resumed, a manual effort is required to restore the pumps to operating condition. All data is presumed lost.

Figure 11.7 (cont.)

MINI-SPECIFICATIONS (Cont.)

Fuel Autos MINI-SPECIFICATION 5.0c

STATE c - Pump fuel at a slow rate to "top off" the fuel tank, and, when full, compute and post the gas charges.

Case 1 (*Receive PUMP-REQ$_c$ = "STOP" - auto's tank is full*):
 Input PUMP-REQ$_c$ from Fuel Pump.
 Output RATE = 0 (stop PUMP#).
 Set STOP-TIME-QTR to CURRENT-TIME.

 ** Determine amount of charges and post to GAS-CHARGES.*
 Set GALS = ((STOP-TIME-FULL - START-TIME-FULL)/60 * 4)
 + ((STOP-TIME-QTR - START-TIME-QTR)/60 * 1)
 Set AMT-FUEL = GALS
 Input GAS-PRICE from GAS-PRICES for PUMP#.
 Set AMT$-GAS = GALS * GAS-PRICE
 Create AUTO-CARD# + AMT$-GAS + AMT-FUEL entry in GAS-CHARGES.
 Set STATE-5.0 to a.
Endcase 1.

Case 2 (*Receive PUMP-REQ$_b$ = "EMERGENCY STOP", pump has failed or was manually stopped*):
 Input PUMP-REQ$_c$ from Fuel Pump.
 Output RATE = 0 (stop PUMP#).
 Set STOP-TIME-QTR to CURRENT-TIME.

 ** Determine amount of charges and post to GAS-CHARGES.*
 Set GALS = ((STOP-TIME-FULL - START-TIME-FULL)/60 * 4)
 + ((STOP-TIME-QTR - START-TIME-QTR)/60 * 1)
 Set AMT-FUEL = GALS
 Input GAS-PRICE from GAS-PRICES for PUMP#.
 Set AMT$-GAS = GALS * GAS-PRICE
 Create AUTO-CARD# + AMT$-GAS + AMT-FUEL entry in GAS-CHARGES.
 Set STATE-5.0 to a.
Endcases.

Stop.

Processor Requirements: Automated process operating in a real-time operating environment.
Security Requirements: N/A.
Time Requirements: This process must begin executing within 100 millisec from the arrival of PUMP-REQ$_c$.
Processing Priority: Highest priority among the real-time processes.
Performance Requirements: N/A.
Safety Requirements: Pump must also provide for a manual "EMERGENCY STOP" (PUMP-REQ$_c$) to prevent spilling fuel.
Other Requirements: This automated process and the pumps should share the same power source, so that when power is interrupted, the pumps immediately stop. When power is resumed, a manual effort is required to restore the pumps to operating condition. All data is presumed lost.

Figure 11.7 (cont.)

MINI-SPECIFICATIONS (Cont.)

Collect Phone Recs MINI-SPECIFICATION 6.0

Input DETAIL-CALL-RECORD from the Telephone Company.
If DETAIL-CALL-RECORD contains an AUTO-CARD#, then
    Output DETAIL-CALL-RECORD to PHONE-CALLS.
Else Reject DETAIL-CALL-RECORD.
Endif.
Wait.

Processor Requirements: Performed by a computer linked to the Telephone Company by means of a 9600-baud data link. The Audit Data-Link process 10.0* monitors the data link between CARS and the Telephone Company and restarts the link if it "drops."
Security Requirements: N/A.
Time Requirements: The Detail-Call-Record has to be accepted and stored in 145 millisec.
Processing Priority: Third-highest priority among the real-time processes.
Performance Requirements: This process is required to meet the maximum transmission rate of 7 records per sec from the Telephone Company.
Safety Requirements: N/A.
Other Requirements: None.

Figure 11.7 (cont.)

MINI-SPECIFICATIONS (Cont.)

Parking Lot Controller MINI-SPECIFICATION 7.1

Input AUTO-CARD# + GATE from Time-Share Monitor 4.0*.
Search CONTRACTS for AUTO-CARD#.
Input CONTRACT from CONTRACTS for AUTO-CARD#.
If CONTRACT does not exist, then
    Set MESSAGE = "Unidentified auto at gate " + GATE,
    Output MESSAGE to Security Office,
    Suspend process (Output "OFF" to ON/OFF).
Endif.

* *Valid access request.*
If GATE = "1", then
    Initiate Exit Control 7.3*.
    Wait for Exit Control 7.3* to Stop.
Else (GATE = "2"), Initiate Entrance Control 7.2*,
    Output CONTRACT# to Entrance Control 7.2*.
    Wait for Entrance Control 7.2* to Stop.
Endif.
Suspend process (Output "OFF" to ON/OFF).

Processor Requirements: Automated real-time process that controls the processes that operate the entrance and exit gates to the parking lot.
Security Requirements: N/A.
Time Requirements: N/A.
Processing Priority: Second-highest priority among the real-time processes.
Performance Requirements: N/A.
Safety Requirements: If this process performs abnormally, traffic congestion and customer dissatisfaction may result.
Other Requirements: None.

Figure 11.7 (cont.)

MINI-SPECIFICATIONS (Cont.)

Entrance Control MINI-SPECIFICATION 7.2

Input CONTRACT# from Parking Lot Controller 7.1*.
Output UNLOCK to Parking Lot Automatic Gate (GATE = 2).

* *Get mileage & fuel information from returning customer.*
Initiate Obtain Mile/Fuel Details 7.4*.
Output CONTRACT# to Obtain Mile/Fuel Details 7.4*.
Wait for Obtain Mile/Fuel Details 7.4* to Stop.
* *Returns control to Parking Lot Controller 7.1*.*
Stop.

Processor Requirements: Automated real-time process that controls the processes that operate the entrance gates to the parking lot.
Security Requirements: System failure should always place the gate in a closed position.
Time Requirements: N/A.
Processing Priority: Second-highest priority among the real-time processes.
Performance Requirements: N/A.
Safety Requirements: If this process performs abnormally, traffic congestion on the main street and customer dissatisfaction may result.
Other Requirements: None.

Figure 11.7 (cont.)

MINI-SPECIFICATIONS (Cont.)

Exit Control MINI-SPECIFICATION 7.3

Output UNLOCK to Parking Lot Automatic Gate (GATE = 1).
** Returns control to Parking Lot Controller 7.1*.*
Stop.

Processor Requirements: Automated real-time process that controls the processes that operate the exit gates to the parking lot.
Security Requirements: System failure should always place the gate in a closed position.
Time Requirements: N/A.
Processing Priority: Second-highest priority among the real-time processes.
Performance Requirements: N/A.
Safety Requirements: If this process performs abnormally, traffic congestion will occur within the parking lot and result in customer dissatisfaction.
Other Requirements: None.

Figure 11.7 (cont.)

MINI-SPECIFICATIONS (Cont.)

Obtain Mile/Fuel Details MINI-SPECIFICATION 7.4

```
Input CONTRACT# from Entrance Control 7.2*.
Output CRT-MSG to Parking Lot Automatic Gate (GATE = 2).
Input CURR-MILEAGE + FUEL-LEV from Parking Lot Automatic Gate.
If Input is not received in 30 sec, then
    Set MESSAGE = "Customer not responding at CRT GATE 2",
    Output MESSAGE to Security Office.
    Output CRT-MSG to Parking Lot Automatic Gate (GATE = 2).
    Input CURR-MILEAGE + FUEL-LEV from Parking Lot Automatic Gate.
    If Input is not received in 90 sec, then
        Set MESSAGE = "Customer still not responding at Gate 2",
        Output MESSAGE to Security Office,
        Set CURR-MILEAGE to "UNKNOWN",
        Set FUEL-LEV to "UNKNOWN",
        Set GAS-NEEDED to "UNKNOWN",
        Output CURR-MILEAGE + FUEL-LEV for CONTRACT# in CONTRACTS,
        Output CONTRACT# + GAS-NEEDED to Bill Customer 8.1*.
        Stop.    * Returns control to Parking Lot Controller 7.2*.
    Endif.
Endif.
Repeat
    Input MILEAGE from CONTRACTS for CONTRACT#,
    If MILEAGE > CURR-MILEAGE, then
        Set CRT-MSG = "Mileage in error, please enter again",
        Output CRT-MSG to Parking Lot Automatic Gate,
    Else exit Repeat.
    Endif.
Until performed three times, then accept CURR-MILEAGE as inputted.
Output CURR-MILEAGE + FUEL-LEV to CONTRACTS for CONTRACT#.
If FUEL-LEV < > "FULL" then,
    Set CRT-MSG = "Fuel tank must be returned FULL, please pull to the Gas Pumps before
        coming to checkout counter.",
    Output CRT-MSG to Parking Lot Automatic Gate,
    Set GAS-NEEDED to "YES",
    Output CONTRACT# + GAS-NEEDED to Pull Customer Contract 8.1*,
Else Set CRT-MSG = "Park auto and leave the keys and AUTO-CARD in the car and remember the
        parking space number.",
    Output CRT-MSG to Parking Lot Automatic Gate,
    Set GAS-NEEDED to "NO",
    Output CONTRACT# + GAS-NEEDED to Pull Customer Contract 8.1*.
Endif.
Stop.    * Returns control to Parking Lot Controller 7.2*.
```

Processor Requirements: Automated real-time process that interacts with customer by means of a remote CRT.
Security Requirements: N/A.
Time Requirements: Customer must respond to requests for data within 30 secs (Warning message sent to Security Office), and within 2 min. (Final message sent to Security Office). If the customer fails to respond the system assumes that the customer drove past the CRT information station and indicates CURR-MILEAGE and FUEL-LEV as "UNKNOWN".
Processing Priority: Second priority among the real-time processes to prevent traffic congestion.
Performance Requirements: N/A.
Safety Requirements: N/A.
Other Requirements: None.

Figure 11.7 (cont.)

MINI-SPECIFICATIONS (Cont.)

Pull Customer Contract MINI-SPECIFICATION 8.1

Case 1 (*Customer arrival is indicated by the data flow from process 7.4**):
    Input CONTRACT# + GAS-NEEDED from Obtain Mile/Fuel Details 7.4*.
    Locate CONTRACT# in CONTRACTS.
    Input CONTRACT from CONTRACTS.
    If GAS-NEEDED = "UNKNOWN", then
        Set INSUFF-INFO to "Need auto refueled using AUTO-CARD",
        Set GAS-NEEDED to "YES",
        Output INSUFF-INFO to Customer,
        Output CONTRACT + GAS-NEEDED to HOLD-CONTRACTS,
    Endif.
    If CURR-MILEAGE = "UNKNOWN", then
        Set INSUFF-INFO to "Need current mileage",
        Output INSUFF-INFO to Customer,
        Output CONTRACT + GAS-NEEDED to HOLD-CONTRACTS,
    Endif.
    Stop.
Endcase 1.

Case 2 (*Customer has responded to the INSUFF-INFO request*):
    Input (CURR-MILEAGE) + (FUEL-LEV) + CUSTM-NAME from Customer.
    Input CONTRACT + GAS-NEEDED from HOLD-CONTRACTS for CUSTM-NAME.
    Set CURR-MILEAGE and/or FUEL-LEV on CONTRACT to CURR-MILEAGE and/or FUEL-LEV
        from input.
Endcases.

If GAS-NEEDED = "YES", then
    Output CONTRACT + GAS-NEEDED to Compute Bill 8.4*,
Else Output CONTRACT to Compute Bill 8.4*.
Endif.
Output AUTO-CARD# to Determine Phone $ 8.2*
If GAS-NEEDED = "YES", then
    Output AUTO-CARD# to Determine Gas $ 8.3*
Endif.
Stop.

Processor Requirements: Automated or manual process that retrieves CONTRACT from the data store CONTRACTS and makes sure all fuel and mileage information is present and can be used in computing the customer's bill.
Security Requirements: N/A.
Time Requirements: This process must perform its task in less than 60 sec so as to not delay customers waiting for their bills.
Processing Priority: N/A.
Performance Requirements: N/A.
Safety Requirements: N/A.
Other Requirements: None.

Figure 11.7 (cont.)

MINI-SPECIFICATIONS (Cont.)

Determine Phone$ MINI-SPECIFICATION 8.2

Input AUTO-CARD# from Pull Customer Contract 8.1*.
Search PHONE-CALLS for AUTO-CARD#.
If no entries are found, then
    Set REASON to "PHONE CALL".
    Set AMT$ to zero.
    Output REASON + AMT$ to ADD'L-CHARGES for AUTO-CARD#.
    Stop.
Endif.
For each entry in PHONE-CALLS that matches AUTO-CARD#:
    Set TIME-TALKED = FINISH-TIME - START-TIME,
    Round TIME-TALKED to the highest integer minute,
    Search PHONE-RATES for DURATION = TIME-TALKED,
    Input AMT-CALL from PHONE-RATES for DURATION,
    Add AMT-CALL to TOTAL-PHONE$,
Endfor.
Set REASON to "PHONE CALL".
Set AMT$ to TOTAL-PHONE$.
Output REASON + AMT$ to ADD'L-CHARGES for AUTO-CARD#.
Stop.

Processor Requirements: Automated or manual process with accurate arithmetic abilities to compute total charges.
Security Requirements: N/A.
Time Requirements: This process must perform its task in less than 60 sec so as to not delay customers waiting for their bills.
Processing Priority: If a manual processor is used, it should not be interrupted once the Search for PHONE-CALLS begins.
Performance Requirements: Inaccuracy in computing total charges may result in shortage of revenues or customer dissatisfaction.
Safety Requirements: N/A.
Other Requirements: None.

Figure 11.7 (cont.)

MINI-SPECIFICATIONS (Cont.)

Determine Gas$ MINI-SPECIFICATION 8.3

```
Input AUTO-CARD# from Pull Customer Contract 8.1*
Search GAS-CHARGES for AUTO-CARD#.
If no entries are found, then
    Set REASON to "GAS",
    Set AMT$ to zero,
    Output REASON + AMT$ to ADD'L-CHARGES for AUTO-CARD#,
    Stop.
Endif.
Input AMT$-GAS + AMT-FUEL from GAS-CHARGES,
Set REASON to AMT-FUEL,
Set AMT$ to AMT$-GAS,
Output REASON + AMT$ to ADD'L-CHARGES for AUTO-CARD#.
Stop.
```

Processor Requirements: Automated or manual process.
Security Requirements: N/A.
Time Requirements: This process must perform its task in less than 30 sec so as to not delay customers waiting for their bills.
Processing Priority: If manual processor is used, it should not be interrupted once the Search of GAS-CHARGES begins.
Performance Requirements: Inaccuracy in computing total charges may result in shortage of revenues or customer dissatisfaction.
Safety Requirements: N/A.
Other Requirements: None.

Figure 11.7 (cont.)

MINI-SPECIFICATIONS (Cont.)

Compute Bill MINI-SPECIFICATION 8.4

```
Input CONTRACT + (GAS-NEEDED) from Pull Customer Contract 8.1*.

* Determine gas charges.
If GAS-NEEDED, then
   Repeat
      Search ADD'L-CHARGES for AUTO-CARD# and REASON = AMT-FUEL,
      Input AMT$ + AMT-FUEL from ADD'L-CHARGES,
      Add AMT-FUEL to GALS,
      Add AMT$ to GAS$,
   Until all entries in ADD'L-CHARGES.
   If GAS$ = zero, then
      Set BILL-PBLM to "GAS NOT FOUND",
      Output BILL-PBLM to Manager,
      Wait,
      Input APPROVAL from Manager,
      Set GAS$ to APPROVAL.
   Endif.
Else Set GAS$ and GALS to zero.
Endif.

* Determine phone charges.
Repeat
   Search ADD'L-CHARGES for AUTO-CARD# and REASON = "PHONE-CALL",
   Input AMT$ from ADD'L-CHARGES,
   Add AMT$ to PHONE$,
Until all entries in ADD'L-CHARGES.

* Determine rental charges.
Set MILES = CURR-MILEAGE - MILEAGE.
If MILES is not > zero, then
   Set BILL-PBLM to "MILEAGE ERROR",
   Output BILL-PBLM to Manager,
   Wait,
   Input APPROVAL from Manager,
   Set MILES to APPROVAL.
Endif.
Set DAYS = CURRENT-DATE - DATE-RENTAL
Set TOTALS = DAYS * DAILY$ + MILES * MILE$ + PHONE$ + GAS$.
If INSURANCE = "YES", then Set TOTALS = TOTALS + INSURANCE$.
Endif.
Output TOTALS + CONTRACT to CONTRACT-STACK.
Stop.
```

Processor Requirements: Automated or manual process with accurate arithmetic abilities.
Security Requirements: N/A.
Time Requirements: Process must perform in less than 60 sec so as to not delay customers waiting for their bills.
Processing Priority: Once the process begins, it should not be interrupted.
Performance Requirements: N/A.
Safety Requirements: N/A.
Other Requirements: Manager must be available immediately for APPROVAL.

Figure 11.7 (cont.)

MINI-SPECIFICATIONS (Cont.)

Cashier MINI-SPECIFICATION 8.5

Input CONTRACT-COPY + LOC from Customer.
Input CONTRACT + TOTALS for CONTRACT# from CONTRACT-STACK.
Set AMT-DUE to TOTALS.
Output AMT-DUE to Customer.
Input PAYMENT + (PBLMS) from Customer.
If PBLMS, then Copy PBLMS onto CONTRACT.

** Collect payment and provide receipt to customer.*
Output PAYMENT to CASH-REGISTER.
Set COMPL-CONTRACT and RECEIPT to TOTALS
Mark COMPL-CONTRACT and RECEIPT as "PAID"
Set LOC in COMPL-CONTRACT to LOC as inputted from Customer
Output RECEIPT to Customer.
Output COMPL-CONTRACT to COMPL-CONTRACTS.
Stop.

Processor Requirements: Manual process required to interface with the customer on a personal basis.
Security Requirements: This processor has access to all revenues collected and disseminated to the customer.
Time Requirements: This process should be completed in 2 min.
Processing Priority: This process can be interrupted. Interruptions should be minimal, otherwise they cause customer dissatisfaction.
Performance Requirements: N/A.
Safety Requirements: N/A.
Other Requirements: Pleasant, courteous interaction with the customer is required.

Figure 11.7 (cont.)

LOGICAL ANALYSIS
. . . looking beyond the physical implementation

Input/Output Processing Chart

To obtain the essence of CARS, we can construct an input/output processing chart, as shown in Fig. 11.8. To fill in the classification for each process for the respective inputs, outputs, and data stores, consider each data flow, one by one, starting with the

INPUT/OUTPUT PROCESSING CHART

	1.1	1.2	1.3	2.0	3.0	4.0	5.0a	5.0b	5.0c	6.0	7.1	7.2	7.3	7.4	8.1	8.2	8.3	8.4	8.5	
Inputs:																				
AUTO-CARD + LOC + KEYS +MILEAGE		ECA																		
AUTO-CARD# + PUMP#						ICA	EFA													
APPROVAL																		EFA		
CONTRACT-COPY + LOC																			EFA	
CURR-MILEAGE + FUEL-LEV														ECA	ICA	EFA	EFA	EFA		
(CURR-MILEAGE) + (FUEL-LEV) +CUSTM-NAME														ECA	ICA	EFA	EFA	EFA		
CUSTM-NAME+CUSTM-ADDR & DRIVER-LIC#+INSURANCE			EFA																	
DETAIL-CALL-RECORD										ECA										
ENTER-REQ + GATE						ICA					IAA	ICA		ECA	ICA	EFA	EFA	EFA		
EXIT-REQ + GATE						ICA					IAA		EFA							
PAYMENT + (PBLMS)																			ECA	
PUMP-ON							EFA													
PUMP-REQ							EFA	EFA	EFA											
RESERVATION			ECA																	
Outputs:																				
AMT-DUE																		EFA	ICA	
APOLOGY			EFA																	
AUTO-CARD+CONTRACT-COPY +LOC+KEYS			EFA																	
BILL-PBLM																		EFA		
CAR-ORDER	EFA																			
CRT-MSG														ECA						
DEPOSIT				EFA																
FUEL							EFA	EFA												
INSUFF-INFO															ECA					
MESSAGE						EFA					EFA			EFA						
MAINT-ORDER		EFA																		
RECEIPT																			EFA	
UNLOCK												EFA	EFA							
UNLOCK-CARD						EFA														
PROCESS SUMMARY:	EFA	EFA	EFA	EFA	??? (IAA)	EFA	EFA	EFA	EFA	ECA	EFA	EFA	EFA	EFA	ECA	EFA	EFA	EFA	EFA	
Data Stores:																				
ADD'L-CHARGES																ICM	ICM	ICM		= ICM
CASH-REGISTER				EM															EM	= EM
COMPL-CONTRACTS		ICM																	ICM	= ICM
CONTRACT-FORMS			IAM																	= IAM
CONTRACT-STACK																		ICM	ICM	= ICM
CONTRACTS														EM	EM					= EM
FUEL-SUPPLY							EM	EM												= EM
GAS-CHARGES									ICM								ICM			= ICM
GAS-PRICE									EM											= EM
HOLD-CONTRACTS															IAM					= IAM
OLD-CONTRACTS		IAM																		= IAM
PHONE-CALLS										EM						EM				= EM
PHONE-RATES																EM				= EM
RENTALS	EM	EM	EM																	= EM
RESERVATIONS	EM		EM																	= EM

Figure 11.8

Inputs section. Examine the data flow diagrams and mini-specs to determine the type of function being performed. What follows is a summary of each data flow and the justification for the classifications that appear in Fig. 11.8.

Input Section

AUTO-CARD + LOC + KEYS + MILEAGE is supplied by the Maintenance Shop to process 1.2*, which updates the RENTALS data store to show that the auto is now available for renting. RENTALS contains all of the information that the system must remember about autos and is thus an essential memory (EM). Process 1.2* keeps the data store RENTALS up to date and is therefore an essential custodial activity (ECA).

So we have:

Process 1.2* (ECA), using RENTALS (EM)

AUTO-CARD# + PUMP# is inputted into the system when a customer or attendant wishes to fill an auto with fuel, and inserts the AUTO-CARD into the Automatic Pump. The data AUTO-CARD + PUMP# is sent to process 4.0* which inputs the data, initiates process 5.0a*, and passes the data on to process 5.0a*. Thus, process 4.0* for this input, performs an internal communications activity (ICA). Process 5.0a* sets up to receive the "START" signal (PUMP-REQ), which will cause the output of FUEL from the FUEL-SUPPLY, an essential memory (EM). Process 5.0a* performs an essential fundamental activity (EFA) of supplying FUEL.

So we have:

Process 4.0* (ICA)
Process 5.0a, (EFA), using FUEL-SUPPLY (EM)

APPROVAL is needed as input when the system detects a billing problem from the input data found in the internal communications memory (ICM) ADD'L-CHARGES. Process 8.4* outputs BILL-PBLM to the manager and waits for a response (APPROVAL) on how to handle the problem. Process 8.4* then continues to prepare the customer's bill. It performs an essential fundamental activity (EFA) and outputs the results into the data store CONTRACT-STACK. Looking ahead, process 8.5* reads all of the data from CONTRACT-STACK; thus, CONTRACT-STACK serves only as an internal communications memory (ICM).

So we have:

Process 8.4* (EFA), using ADD'L-CHARGES (ICM)
CONTRACT-STACK (ICM)

CONTRACT-COPY + LOC is received by process 8.5* as input from the customer who is returning the rented auto. It matches the CONTRACT-COPY with the original CONTRACT that it obtains from the CONTRACT-STACK (ICM). In matching the CONTRACT with CONTRACT-COPY, process 8.5* performs an internal administrative activity (IAA). It then proceeds to pass the TOTALS amount found in CONTRACT-STACK to the customer as AMT-DUE. At this point, process 8.5* performs an essential fundamental activity (EFA) by providing the customer with one of the outputs of the system, AMT-DUE.

So we have:

Process 8.5* (EFA), using CONTRACT-STACK (ICM)

CURR-MILEAGE + FUEL-LEV is received by process 7.4* and is stored in the essential memory (EM) CONTRACTS. CONTRACTS is an essential memory because it contains all of the information that the system must remember about the customer and the rented auto. Thus, process 7.4*, at this point, performs an essential custodial activity (ECA). It also passes data to process 8.1*, which sends the appropriate information to three processes: 8.2*, 8.3*, and 8.4*, which are all essential fundamental activities (EFA) because they are responsible for computing the customer's bill.

So we have:

Process 7.4* (ECA), using CONTRACTS (EM)
Process 8.1* (ICA), using CONTRACTS (EM)
Processes: 8.2*, 8.3*, and 8.4* (EFA)

(CURR-MILEAGE) + (FUEL-LEV) + CUSTM-NAME enters the system at the CRT information station. If a customer fails to provide the current mileage (CURR-MILEAGE) and fuel level (FUEL-LEV) of the auto, process 7.4* passes the CONTRACT# + GAS-NEEDED to process 8.1* and updates the essential memory (EM) CONTRACTS, indicating mileage and fuel level is "UNKNOWN." Process 7.4* performs an essential custodial activity (ECA). While waiting for the customer to provide the information, process 8.1* holds the rest of the information about the customer in the internal administrative memory (IAM) HOLD-CONTRACTS. After receiving the missing information process 8.1* locates the customer's original CONTRACT in the essential memory (EM) CONTRACTS. Process 8.1* passes on the information about any phone calls made from the auto and information about any gas purchased in the parking lot to processes 8.2* and 8.3*, respectively. At this point, process 8.1* performs an internal communications activity (ICA). Processes 8.2* and 8.3* both are essential fundamental activities that compute part of the customer's bill, and send their results to process 8.4*. Process 8.4* is an Essential Fundamental Activity (EFA) and is used to compile the customer's bill.

So we have:

Process 7.4* (ECA), using CONTRACTS (EM)
Process 8.1* (ICA), using HOLD-CONTRACTS (IAM)
CONTRACTS (EM)
Processes: 8.2*, 8.3*, and 8.4* (EFA)

CUSTM-NAME + CUSTM-ADDR and DRIVER-LIC# + INSURANCE is received by process 1.3* from a customer when he or she arrives to rent a reserved auto. Process 1.3* performs one of the essential fundamental activities of the system, renting autos, and is an EFA. To rent an auto, process 1.3* has to select a blank contract form from the internal administrative memory (IAA) CONTRACT-FORMS, reference the essential memory (EM) RENTALS to obtain information about the auto, and input information from the essential memory (EM) RESERVATIONS about the car that the customer reserved. If the system finds the customer's reservation, the process outputs AUTO-CARD + CONTRACT-COPY + LOC + KEYS to the customer. If there is no reservation for the customer, the system sends an appropriate APOLOGY to the customer. If the system finds the reservation but there are no autos available, the process also outputs an APOLOGY to the customer.

So we have:

Process 1.3* (EFA), using CONTRACT-FORMS (IAM)
RENTALS (EM)
RESERVATIONS (EM)

DETAIL-CALL-RECORD is input from the Telephone Company directly to process 6.0*, which is always waiting for the input data. The Telephone Company provides a record every time a customer makes a phone call from a rented auto. The process examines the record to see if the call was charged to the AUTO-CARD#. If so, the DETAIL-CALL-RECORD is put into the essential memory (EM) PHONE-CALLS. If the customer charged the phone call to a credit card, the process rejects the record. PHONE-CALLS is an essential memory because the system must remember all of the calls the customer charges to the AUTO-CARD during the rental period, so the system can add the cost of the calls to the customer's bill. A process that updates an essential memory is an essential custodial activity (ECA).

So we have:

Process 6.0* (ECA), using PHONE-CALLS (EM)

ENTER-REQ + GATE is received by process 4.0* when a customer wishes to enter the parking lot to return a rented auto. Under normal conditions, process 4.0* passes on the information after activating process 7.1*; it is an internal communications activity (ICA). If process 4.0* cannot activate process 7.1*, another action is taken, resulting in output, however, we will discuss this when we consider output data flows.

Following the data further through the system, we see that process 7.1* not only initiates and passes data onto process 7.2*, but it also makes sure that the ENTER-REQ is being made by a customer that has rented an auto. Such a verification activity is an administrative task and process 7.1* is an internal administrative activity (IAA). However, if the system cannot identify the person seeking access to the parking lot, process 7.1* provides an output data flow that will be discussed later in the output section.

When process 7.2* receives the input, it initiates process 7.4* and passes data to it. At this point, process 7.2* performs an internal communications activity (ICA).

Process 7.2* also outputs the UNLOCK data flow, which will also be covered in the output section. Process 7.4* is responsible for collecting the current mileage (CURR-MILEAGE) and fuel level (FUEL-LEV) information from the customer at the CRT information station. The process gathers this data, updates the essential memory (EM) CONTRACTS, and passes information onto process 8.1*. The highest-priority function which process 7.4* performs is updating the essential memory (EM) CONTRACTS, and thus is an essential custodial activity (ECA). Process 7.4* is also responsible for outputting the CRT-MSG to the customer at the CRT information station. CRT-MSG is a dialogue data flow used to obtain information from the returning customer to update the essential memory (EM) CONTRACTS. Process 7.4* is really performing an essential custodial activity (ECA) when issuing the CRT-MSG. If any unusual event occurs during the dialogue, the process produces an output data flow, which will be discussed later.

Upon receiving CONTRACT# and GAS-NEEDED, process 8.1* matches the data with the information found in the essential memory (EM) CONTRACTS. It then passes information to processes 8.2* and/or 8.3*, as well as process 8.4*. Process 8.1* performs an internal communications activity (ICA) at this point. Process 8.2* searches the essential memory (EM) PHONE-CALLS for any phone calls made by the customer and charged to the AUTO-CARD#. If such phone calls were made, the process determines the cost by using the essential memory (EM) PHONE-RATES, and creates a portion of the customer's bill by placing the information into the internal communications memory (ICM) ADD'L-CHARGES.

Likewise, process 8.3* performs a similar activity, but for any charges for refueling the auto upon its return. The system stores the information about refueling in the internal communications memory (ICM) GAS-CHARGES. Process 8.3* also passes its portion of the customer's bill to process 8.4* by using the internal communications memory (ICM) ADD'L-CHARGES.

Process 8.4* computes the amount the customer owes for the basic rental of the auto, and adds to the bill any charges computed by processes 8.2* and 8.3*. The billing activity is an essential fundamental activity (EFA) for CARS and therefore all three processes, 8.2*, 8.3*, and 8.4*, are EFAs. When the system completes the customer's bill, it passes the bill onto the cashier process by means of the internal communications memory (ICM) CONTRACT-STACK. Finally, process 8.5* takes the customer's bill from CONTRACT-STACK and presents it to the customer.

So we have:

Process 4.0* (ICA),
Process 7.1* (IAA),
Process 7.2* (ICA),
Process 7.4* (ECA), using CONTRACTS (EM)
Process 8.1* (ICA), using CONTRACTS (EM)
Process 8.2* (EFA), using PHONE-CALLS (EM)
PHONE-RATES (EM)
ADD'L-CHARGES (ICM)

Process 8.3* (EFA), using GAS-CHARGES (ICM)
ADD'L-CHARGES (ICM)
Process 8.4* (EFA), using ADD'L-CHARGES (ICM)
CONTRACT-STACK (ICM)

EXIT-REQ + GATE comes into process 4.0* when a customer wishes to exit the parking lot with a rented auto. Process 4.0* passes on the information after activating process 7.1*; it is an internal communications activity (ICA). If process 4.0* cannot activate process 7.1*, it provides an output that will be discussed later.

Following the EXIT-REQ+GATE data flow, process 7.1* not only initiates and passes data onto process 7.3*, but it also makes sure that the EXIT-REQ is being made by a customer that has rented an auto. This activity is an internal administrative activity (IAA) and verifies the EXIT-REQ. However, if the person seeking to leave the parking lot cannot be properly identified, process 7.1* provides additional output, which is discussed in the output section. When process 7.3* receives control, it performs the essential fundamental activity (EFA) of outputting the UNLOCK data flow to the exit gate. (Normally, we would discuss the processes that output data in the next section, however, we have to trace all inputs through the system until they reach an essential activity and, in this case, that essential activity is an EFA that outputs the data flow UNLOCK.)

So we have:

Process 4.0* (ICA)
Process 7.1* (IAA)
Process 7.3* (EFA)

PAYMENT + (PBLMS) is received by process 8.5* after it has outputted the AMT-DUE to the customer. The process performs an internal communications activity (ICA) by presenting the customer with the AMT-DUE. When process 8.5* receives PAYMENT and any problem (PBLMS) from the customer, it updates the essential memory (EM) CASH-REGISTER. This qualifies the process as an essential custodial activity (ECA). It also records any problem (PBLMS) on the contract and communicates it to process 1.2* by way of the internal communications memory (ICM) COMPL-CONTRACTS. Finally, process 8.5* is responsible for preparing a RECEIPT and outputting it to the customer. This is covered in the output section.

So we have:

Process 8.5* (ECA), using CASH-REGISTER (EM)
COMPL-CONTRACTS (ICM)

PUMP-ON is a signal received from a fuel pump (PUMP#) by process 5.0a, which makes sure that the pump is "ON" (running at the RATE of zero gallons per minute) before it sends the signal to begin pumping the fuel. If the pump is "ON", then process continues, else the process makes five attempts to turn the pump "ON". If the pump fails to indicate it is "ON", then process 5.0a ends and returns control to process 4.0*, as if the refueling were completed. Process 5.0a performs an essential fundamental activity (EFA).

So we have:

Process 5.0a* (EFA)

PUMP-REQ is a set of start and stop signals received from the fuel pump by process 5.0*. Processes 5.0a and 5.0b control the flow of FUEL from the essential memory (EM) FUEL-SUPPLY to the fuel auto sink, and as such are essential fundamental activities (EFA). In addition, process 5.0c computes the amount owed for the refueling of the auto and creates a portion of the customer's bill by accessing the essential memory (EM) GAS-PRICE. The process sends the gas portion of the customer's bill to process 8.3* by means of the internal communications memory (ICM) GAS-CHARGES. As a result process 5.0c is an essential fundamental activity (EFA). (See also AUTO-CARD + PUMP#.)

So we have:

Process 5.0a* (EFA), using FUEL-SUPPLY (EM)
Process 5.0b* (EFA), using FUEL-SUPPLY (EM)
Process 5.0c* (EFA), using GAS-PRICE (EM)
GAS-CHARGES (ICM)

RESERVATION enters process 1.3* when a customer wishes to reserve an auto. The process updates the essential memory (EM) RESERVATIONS. Thus, process 1.3* is an essential custodial activity (ECA).

So we have:

Process 1.3* (ECA), using RESERVATIONS (EM)

Output Section

AMT-DUE (see also input PAYMENT) is part of a dialogue data flow, where process 8.5* passes this output data flow to the customer and expects to receive PAYMENT in return. Process 8.5, in this case, performs an internal communications activity (ICA). Tracing the data flow back further into the system, we find that the bill (AMT-DUE) is computed by the essential fundamental activity (EFA) Compute Bill 8.4*. The data flow is passed between these processes by the internal communications memory (ICM) CONTRACT-STACK.

So we have:

Process 8.5* (ICA), using CONTRACT-STACK (ICM)
Process 8.4* (EFA), using CONTRACT-STACK (ICM)

APOLOGY (see also input CUSTM-NAME + CUSTM-ADDR) is provided to the customer in the event that there are no autos to rent as determined from the data in the essential memories (EM) RENTALS and RESERVATIONS by process 1.3*. The process is responsible for determining if an auto is available and presenting the output to the customer. Thus, process 1.3* is an essential fundamental activity (EFA).

So we have:

Process 1.3* (EFA), using RENTALS (EM)
RESERVATIONS (EM)

AUTO-CARD + CONTRACT-COPY + LOC + KEYS (see also input CUSTM-NAME + CUSTM-ADDR) is provided by process 1.3* if there are autos to rent, as determined from the data in the essential memories (EM) RENTALS and RESERVATIONS. Process 1.3* is responsible for the following: obtaining a CONTRACT-FORM from the internal administrative memory (IAM) CONTRACT-FORMS, completing the CONTRACT, and presenting the customer with this output. Process 1.3* performs an essential fundamental activity (EFA).

So we have:

Process 1.3* (EFA), using CONTRACT-FORM (IAM)
RENTALS (EM)
RESERVATIONS (EM)

BILL-PBLM (see also input APPROVAL) is part of a dialogue data flow used in the event that there is a discrepancy in determining the charges, found in the internal communications memory (ICM) ADD'L-CHARGES, that comprise the customer's bill. Process 8.4* is responsible for informing the manager (BILL-PBLM) and expects a response (APPROVAL) on how to resolve the discrepancy. Thus, process 8.4* fulfills one of the system's requirements by performing this essential fundamental activity (EFA), and outputs its results to the internal communications memory (ICM) CONTRACT-STACK.

So we have:

Process 8.4* (EFA), using ADD'L-CHARGES (ICM)
CONTRACT-STACK (ICM)

CAR-ORDER is outputted at the appointed time by process 1.1* after it determines that there are not enough autos on hand to meet the demand for the remaining day and the next day. To determine the demand, the process retrieves information from two essential memories (EM): RENTALS and RESERVATIONS. If the rental agency needs additional autos, the process, which performs an essential fundamental activity (EFA), sends this output to the Central Auto Pool.

So we have:

Process 1.1* (EFA), using RENTALS (EM)
RESERVATIONS (EM)

CRT-MSG (see also input ENTER-REQ) is part of a dialogue data flow used when a customer returns a rented auto. Process 7.4* is responsible for acquiring CURR-MILEAGE and FUEL-LEV information from the customer at the CRT information station by sending the customer this output. When the customer responds, the process updates the essential memory (EM) CONTRACTS. Thus, process 7.4* basically performs an essential custodial activity (ECA) by obtaining the information for the purpose of updating an essential memory in the system.

So we have:

Process 7.4* (ECA), using CONTRACTS (EM)

DEPOSIT is outputted at the appointed time by process 2.0*. The process inputs CASH from the essential memory (EM) CASH-REGISTER, creates a bank DEPOSIT, and sends it to the bank. As a result, process 2.0* performs one of the system's essential fundamental activities (EFAs), making bank deposits.

So we have:

Process 2.0* (EFA), using CASH-REGISTER (EM)

FUEL (see also input PUMP-REQ) is part of a dialogue data flow used by processes 5.0a* and 5.0b*. Both provide this output by obtaining it from the essential memory (EM) FUEL-SUPPLY and regulating the rate of delivery into the customer's auto. Both of these processes perform the essential fundamental activity (EFA) of providing one of the required outputs of the system in the required manner (RATE).

So we have:

Processes 5.0a* (EFA) and 5.0b* (EFA), using FUEL-SUPPLY (EM)

INSUFF-INFO (see also (CURR-MILEAGE) + (FUEL-LEV) + CUSTM-NAME) is part of a dialogue data flow from process 8.1* used to obtain information from the customer about the mileage and fuel level of the returned auto. If any of the information is missing in the CONTRACTS data store, the process sends this output to the customer, requesting the missing information, and places the customer's CONTRACT in the internal administrative memory (IAM) HOLD-CONTRACTS. When the process receives the customer's reply, it updates the CONTRACT in the essential memory (EM) CONTRACTS. Process 8.1* performs the administrative activity of making sure that the information on the customer's contract is correct, and ultimately performs an essential custodial activity (EFA) of obtaining the missing information from the customer and updating the essential memory CONTRACTS. The highest-priority activity applies, which is an essential custodial activity (ECA).

So we have:

Process 8.1* (ECA), using CONTRACTS (EM)
HOLD-CONTRACTS (IAM)

MESSAGE (see also inputs ENTER-REQ + GATE and EXIT-REQ + GATE) is an output produced by several processes responsible for notifying the Security Office when certain situations occur. The processes that perform this essential fundamental activity (EFA) of outputting data to a sink are 4.0*, 7.1*, and 7.4*. *(Note: If the Security force was within the system's boundary, these processes would only be performing an internal communications activity.)*

So we have:

Processes 4.0* (EFA), 7.1* (EFA) and 7.4* (EFA)

MAINT-ORDER is an output of process 1.2*, which is responsible for examining all of the entries in the internal communications memory (ICM) COMPL-CONTRACTS and the information about the auto in the essential memory (EM) RENTALS to determine if the returned auto needs maintenance. If the auto does process 1.2* outputs a MAINT-ORDER to the Maintenance Shop, an essential fundamental activity (EFA). If no maintenance is required, the process updates the RENTALS data store, which is an essential custodial activity (ECA) and files COMPL-CONTRACT in OLD-CONTRACTS (IAM). The classification of process 1.2* is the highest-priority activity it performs, namely EFA.

So we have:

Process 1.2* (EFA), using COMPL-CONTRACTS (ICM)
OLD-CONTRACTS (IAM)
RENTALS (EM)

RECEIPT (see also input PAYMENT + (PBLMS)) is provided by the essential fundamental activity (EFA) process 8.5*. It is responsible for creating and outputting the RECEIPT to the customer after receiving the PAYMENT. The completed copy of the customer's contract is also outputted to the internal communications memory (ICM) COMPL-CONTRACTS.

So we have:

Process 8.5* (EFA), using COMPL-CONTRACTS (ICM)

UNLOCK (see also inputs ENTER-REQ + GATE and EXIT-REQ + GATE) is provided by both processes 7.2* and 7.3*. These perform the essential fundamental activity (EFA) of outputting this data flow, which controls the entrance and exit gates of the parking lot. This output controls access to the parking lot and is one of the reasons this system exists.

So we have:

Processes 7.2* (EFA) and 7.3* (EFA)

UNLOCK-CARD (see also input AUTO-CARD + PUMP#) is issued by process 4.0* when a customer or attendant finishes refueling an auto. Process 4.0* is responsible for returning the AUTO-CARD by outputting this data flow to the automatic pump at the Fuel Auto station. Process 4.0* performs an essential fundamental activity (EFA) when it provides this output.

So we have:

Process 4.0* (EFA)

PROCESS SUMMARY is completed after all of the system's inputs and outputs have been analyzed and classified in the input/output processing chart. The final classification for each process is determined by applying the highest-priority activity in each column to the functional primitive process. The highest-priority classification is EFA, followed by ECA, IAA, and finally ICA.

Likewise, in each row in the data store section, we summarize the classifications of each data store with the highest-priority classification at the extreme right column beyond the equal sign. The highest priority classification is EM, followed by IAM, and finally ICM. We can use the results from the chart to identify the essence of the system which contains the essential memories needed to perform object analysis.

Notice in this example that after the process-summary entries are completed, there is no activity classification for the functional primitive 3.0*. In examining the mini-spec for this process, we find that it restores the telecommunications link to the Telephone Company when an abnormal condition exists, causing the lines to be disconnected. This task is an internal administrative activity (IAA) and the input/output processing chart has pointed out an internal event in the system. We will discuss this again during event analysis.

OBJECT ANALYSIS
. . . determine the "things" about which the system stores the data

From the input/output processing chart, we can identify all of the essential memories in the system. To identify the objects of the system and separate all of the data elements according to those objects, we use the data-base design tool of normalization. By placing the essential memory data stores in Boyce/Codd normal form, we can immediately identify the objects of the system. After we define the objects of the system, we construct object data stores, which are used to construct the event diagram for the system.

First Normal Form

From the logical analysis, we have determined that the following data stores are essential memories:

```
CASH-REGISTER = {CASH}
CONTRACTS     = {CONTRACT# + CUSTM-NAME + CUSTM-ADDR + DRIVER-LIC#
                 + DATE-RENTAL + INSURANCE + RENTAL-RATE + MILEAGE
                 + FUEL-LEV + AUTO-TYPE + AUTO-LIC# + AUTO-CARD#
                 + CURR-MILEAGE}
FUEL-SUPPLY   = {FUEL}
GAS-PRICES    = {PUMP# + GAS-PRICE}
PHONE-CALLS   = {AUTO-CARD# + DATE-CALL + START-TIME + FINISH-TIME
                 + MOBILE-PHONE# + CALLED-PHONE#}
PHONE-RATES   = {DURATION + AMT-CALL}
RENTALS       = {AUTO-TYPE + AUTO-LIC# + MILEAGE + OIL-DATE + OIL-MILES
                 + TRANS-DATE + TRANS-MILES + BRAKE-DATE + BRAKE-MILES
                 + TUNE-UP-DATE + TUNE-UP-MILES + LOC + CAR-KEYS
                 + RENTAL-AGENCY + AUTO-CARD# + AVAILABILITY}
```

RESERVATIONS = {CUSTM-NAME + CUSTM-ADDR + AUTO-TYPE + RENTAL-DATE + RENTAL-DAYS}

Note: In the RENTALS data store, we replaced the group data element KEYS with its elementary data elements: AUTO-TYPE, AUTO-LIC#, CAR-KEYS, and RENTAL-AGENCY unless the data element already existed in the data store definition.

Because none of these data stores contains repeating group data elements and all of the elementary data elements are single-valued, the data stores are in first normal form.

Second Normal Form

To place the data stores into second normal form, we have to consider the following data stores that have compound primary keys:

RENTALS = {AUTO-TYPE + AUTO-LIC# + MILEAGE + OIL-DATE + OIL-MILES + TRANS-DATE + TRANS-MILES + BRAKE-DATE + BRAKE-MILES + TUNE-UP-DATE + TUNE-UP-MILES + LOC + CAR-KEYS + RENTAL-AGENCY + AUTO-CARD# + AVAILABILITY}

RESERVATIONS = {CUSTM-NAME + CUSTM-ADDR + AUTO-TYPE + RENTAL-DATE + RENTAL-DAYS}

To place these data stores into second normal form, we have to examine every nonkey attribute and determine if it describes or is a fact about the *whole* primary key. If it is a description or a fact about just some portion of the primary key, we will form a new data store. All of the nonkey attributes in RENTALS describe the auto designated by the key AUTO-TYPE + AUTO-LIC#. In the RESERVATIONS data store, AUTO-TYPE describes the type of auto the particular customer (CUSTM-NAME + CUSTM-ADDR) has reserved for the RENTAL-DATE. The nonkey attribute RENTAL-DAYS describes the planned length of time the customer expects to keep the rented vehicle (AUTO-TYPE) beginning on RENTAL-DATE. Therefore, both of these data stores pass the test for second normal form. Because the other data stores do not have composite keys, they are automatically in second normal form.

Third Formal Form

To place the data stores into third normal form, we see if any nonkey attribute in second-normal-form data stores describes or is a fact about another nonkey attribute. Let's consider the data stores one at a time, place them into third normal form, and try to identify an object that pertains to a data store.

CASH-REGISTER = {CASH}: The cash register is an essential memory because a process uses it to provide the output data flow DEPOSIT. There are no other attributes than the data element CASH. The object that this data store describes is *money.*

CONTRACTS = {CONTRACT# + CUSTM-NAME + CUSTM-ADDR + DRIVER-LIC# + DATE-RENTAL + INSURANCE + RENTAL-RATE + MILEAGE + FUEL-LEV + AUTO-TYPE + AUTO-LIC# + AUTO-CARD# + CURR-MILEAGE}: The table lists all of the nonkey attributes from the data store CONTRACTS, the attribute it describes and the object to which it relates.

ATTRIBUTE-NAME	*fact about or describes*	ATTRIBUTE-NAME	*relates to*	OBJECT
CUSTM-NAME		CUSTM-NAME+CUSTM-ADDR		Customer
CUSTM-ADDR		CUSTM-NAME+CUSTM-ADDR		Customer
DATE-RENTAL		CONTRACT#		Customer & Auto
DRIVER-LIC#		CUSTM-NAME+CUSTM-ADDR		Customer
INSURANCE		CUSTM-NAME+CUSTM-ADDR		Customer
RENTAL-RATE		AUTO-TYPE		Auto Rate
MILEAGE		AUTO-LIC#		Auto
FUEL-LEV		AUTO-LIC#		Auto
AUTO-TYPE		AUTO-LIC#		Auto
AUTO-LIC#		AUTO-LIC#		Auto
AUTO-CARD#		AUTO-LIC#		Auto
CURR-MILEAGE		AUTO-LIC#		Auto

Notice that DATE-RENTAL is the only nonkey attribute that does describe CONTRACT#. It also describes a relationship between a *customer* and an *auto.* From this analysis, we find that we have to define four separate data stores out of CONTRACTS, one for each of the following objects: *contract*, *customer*, *auto*, and *auto rate*. We will again leave the two relationship descriptors DATE-RENTAL and CONTRACT# in the data store of the contract, however, we will use the above information obtained in our analysis when we construct the E-R diagram.

CONTRACT = {CONTRACT# + DATE-RENTAL}
CUSTOMER = {CUSTM-NAME + CUSTM-ADDR + DRIVER-LIC# + INSURANCE}
AUTO = {AUTO-LIC# + MILEAGE + FUEL-LEV + AUTO-CARD# + CURR-MILEAGE}
AUTO-RATE = {AUTO-TYPE + RENTAL-RATE}

RELATIONSHIP

R1[CUSTOMER/AUTO, 1-1]

FUEL-SUPPLY = {FUEL} and **GAS-PRICES = {PUMP# + GAS-PRICE}**: The system has many pumps that pump from the same fuel supply. The price of the fuel is related to the PUMP# even though all fuel is the same. We will define a data store with the key the physical item FUEL, which comes from a specific pump (PUMP#), and costs GAS-PRICE per gallon. The object of the data store is *fuel.*

FUEL = {FUEL + PUMP# + GAS-PRICE}

PHONE-CALLS = {AUTO-CARD# + DATE-CALL + START-TIME + FINISH-TIME + MOBILE-PHONE# + CALLED-PHONE#}: All of the attributes listed in the data store PHONE-CALLS do in fact describe a phone call, but the primary key AUTO-CARD# is not the unique identifier. Instead AUTO-CARD# is the means used to pay for the phone call and is considered a nonkey attribute. A *phone call* that is the object can be identified by the originating number (MOBILE-PHONE#) and the date and time of the call (DATE-CALL + START-TIME). Thus, the following data store is in third normal form:

PHONE-CALL = {MOBILE-PHONE# + DATE-CALL + START-TIME + FINISH-TIME
+ CALLED-PHONE# + AUTO-CARD#}

PHONE-RATES = {DURATION + AMT-CALL}: PHONE-RATE is a table of duration of the phone calls and the associated price is in the *phone rate* data store.

PHONE-RATES = {DURATION + AMT-CALL}

RENTALS = {AUTO-TYPE + AUTO-LIC# + MILEAGE + OIL-DATE + OIL-MILES + TRANS-DATE + TRANS-MILES + BRAKE-DATE + BRAKE-MILES + TUNE-UP-DATE + TUNE-UP-MILES + LOC + CAR-KEYS + RENTAL-AGENCY + AUTO-CARD# + AVAILABILITY}: All of the attributes in the RENTALS data store are descriptions of the auto. Each auto is identified uniquely by its license number (AUTO-LIC#). Therefore, we can combine all of the attributes in the RENTALS data store with the AUTOS data store defined earlier. The following data store in third normal form describes the object *auto*:

AUTO = {AUTO-LIC# + MILEAGE + FUEL-LEV + AUTO-CARD# + CURR-MILEAGE
+ AUTO-TYPE + OIL-DATE + OIL-MILES + TRANS-DATE + TRANS-MILES
+ BRAKE-DATE + BRAKE-MILES + TUNE-UP-DATE + TUNE-UP-MILES + LOC
+ CAR-KEYS + RENTAL-AGENCY + AUTO-CARD# + AVAILABILITY}

RESERVATIONS = {CUSTM-NAME + CUSTM-ADDR + AUTO-TYPE + RENTAL-DATE + RENTAL-DAYS}: The only attribute not already accounted for in one of the above data stores is RENTAL-DAYS. It is an attribute of contract and is placed in the following data store that describes the object *contract*:

CONTRACT = {CONTRACT# + DATE-RENTAL + RENTAL-DAYS}

In searching for the objects and attributes in a system, we may find it sufficient to take the essential memories to third normal form. Usually, little is gained by putting the data stores into Boyce/Codd normal form. From this point, we combine any data stores in third normal form that pertain to the same object. However, in this case, all of the objects are unique, therefore, the above data stores in third normal form are also the object data stores of the system. From the analysis, the following table lists the data stores in third normal form.

OBJECTS	DATA STORES IN THIRD NORMAL FORM
MONEY	= {CASH}
CUSTOMER	= {CUSTM-NAME + CUSTM-ADDR + DRIVER-LIC# + INSURANCE}
AUTO-RATE	= {AUTO-TYPE + RENTAL-RATE}
FUEL	= {FUEL + PUMP# + GAS-PRICE}
PHONE-CALL	= {MOBILE-PHONE# + DATE-CALL + START-TIME + FINISH-TIME + CALLED-PHONE# + AUTO-CARD#}
PHONE-RATE	= {DURATION + AMT-CALL}
AUTO	= {AUTO-LIC# + MILEAGE + FUEL-LEV + AUTO-CARD# + CURR-MILEAGE + AUTO-TYPE + OIL-DATE + OIL-MILES + TRANS-DATE + TRANS-MILES + BRAKE-DATE + BRAKE-MILES + TUNE-UP-DATE + TUNE-UP-MILES + LOC + CAR-KEYS + RENTAL-AGENCY + AUTO-CARD# + AVAILABILITY}
CONTRACT	= {CONTRACT# + DATE-RENTAL + RENTAL-DAYS}

Now that we have identified all of the objects in the system, let's go back and apply the information derived during normalization and define any relationships that exist among the object data stores of the system.

One *customer* can rent one *auto* for any period of time (relationship R1). When the relationship of a *customer* and an *auto* is formed, a single *contract* exists (R2). There is also a relationship between a *contract* and a single amount of *money* (R3). For every *auto* that is rented there exists a *rate* that is charged. However, there can be many *autos* to which a single *auto rate* applies (R4). Thus, we have the following relationships:

R1[CUSTOMER/AUTO, 1-1]

R2[(CUSTOMER/AUTO)/CONTRACT, 1-1-1]

R3[CONTRACT/MONEY, 1-1]

R4[AUTO/AUTO-RATE, M-1]

Let's go another step further. When a contract exists, the customer is able to make many *phone calls* from the mobile phone and the charges apply to a *contract* (R5). Each *phone call* has a specific *phone rate* that applies; however, many *phone calls* can have the same *phone rate* (R6). Also, the customer can add *fuel* to the auto on its return and have the charge apply to the *contract* (R7). Thus, the additional three relationships exist:

R5[CONTRACT/PHONE-CALL, 1-M]

R6[PHONE-CALL/PHONE-RATE, M-1]

R7[CONTRACT/FUEL, 1-1]

The above seven relationships can be illustrated by means of a E-R diagram, as shown in Fig. 11.9. As always other relationships can exists among the objects of this system, however the ones illustrated in Fig. 11.9 are the ones that are used by the system.

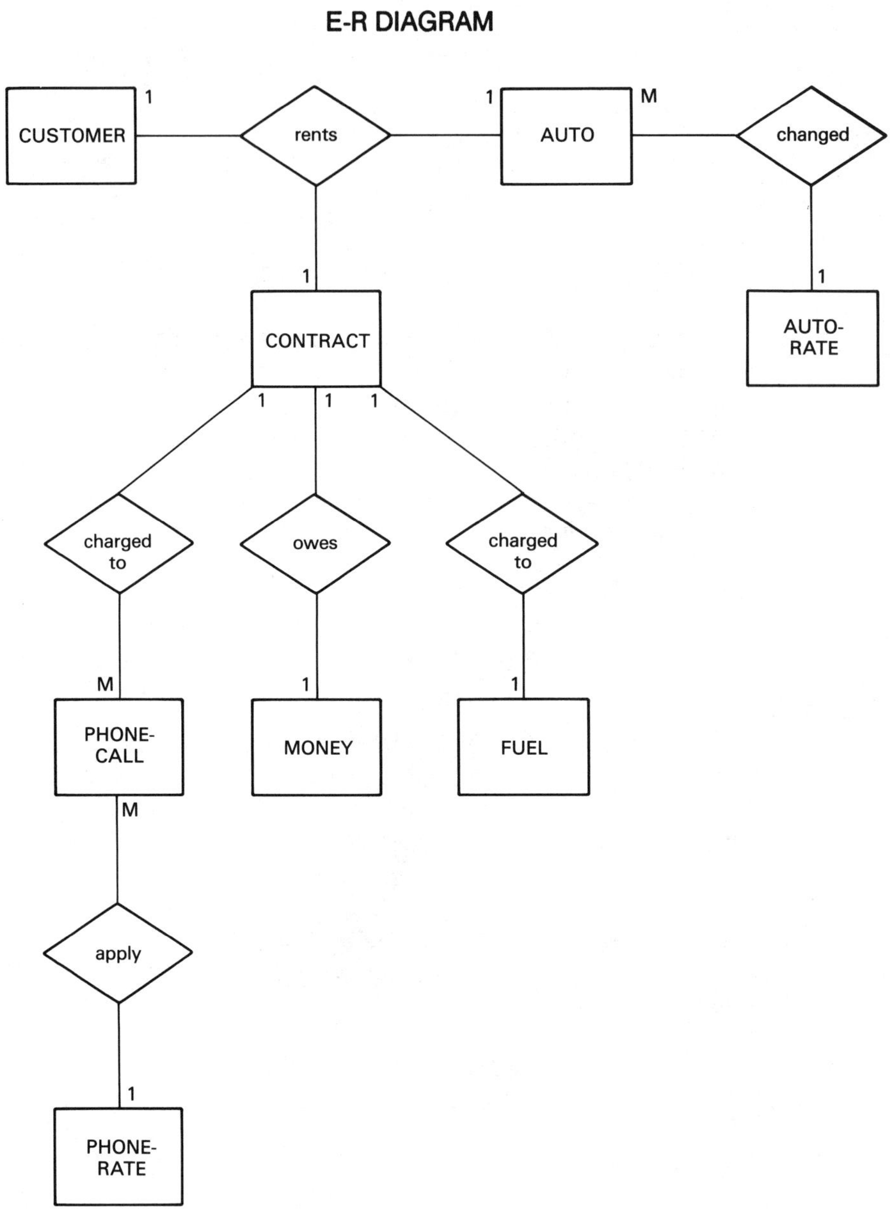

Figure 11.9

EVENT ANALYSIS
. . . determine what causes this system to act

To analyze the events to which the system responds, we begin with the input/output processing chart (see Fig. 11.8). Consider each input and each output and determine what event caused the data flow and the type of event (external, temporal, or internal).

Input Section

AUTO-CARD + LOC + KEYS + MILEAGE input arrives when the Maintenance Shop has completed the repairs of an auto. This is an external event that causes the input data flow to come into the system. Thus, we expand the chart and indicate the event number, event name, and the event type, as shown in Fig. 11.10.

EVENT #1, Maintenance Shop has completed repairing auto, External

AUTO-CARD# + PUMP# input comes into the system when a customer or an attendant wants to refuel a rented auto. Because refueling is not always necessary when a customer returns an auto, a separate external event is identified.

EVENT #2, Customer/Attendant wants to refuel a rented auto, External

APPROVAL input comes into the system if there is a problem in computing the customer's bill after the customer returns a rented auto.

EVENT #3, Customer comes in to return a rented auto, External

CONTRACT + LOC, CURR-MILEAGE + FUEL-LEV, and (CURR-MILEAGE) + (FUEL-LEV) + CUSTM-NAME are the inputs that provide information to the system when the customer returns an auto. These all apply to event number 3, which was previously identified.

EVENT #3, Customer returns a rented auto, External

CUSTM-NAME + CUSTM-ADDR and DRIVER-LIC# + INSURANCE are the inputs that provide information to the system when the customer rents an auto.

EVENT #4, Customer rents a reserved auto, External

DETAIL-CALL-RECORD is received by the system from the Telephone Company every time a customer makes a phone call from a rented auto.

EVENT #5, Customer completes a mobile phone call from an auto, External

ENTER-REQ + GATE comes into the system when a customer returns an auto and wishes to enter the parking lot.

EVENT #3, Customer returns a rented auto, External

EXIT-REQ + GATE comes into the system when a customer wishes to leave the parking lot after renting an auto. This data flow is a consequence of a set of activities that begins when the customer rents an auto.

EVENT #4, Customer rents a reserved auto, External

PAYMENT + (PBLMS) comes into the system during a dialogue with the customer after the system requests the AMT-DUE. This dialogue occurs when the customer returns the rented auto.

EVENT #3, Customer returns a rented auto, External

PUMP-ON and PUMP-REQ comes into the system when a customer or an attendant refuels a rented auto. Because refueling is not always necessary when an auto is returned and refueling was identified earlier as an event different from the event "a customer returns an auto," this input is related to event #2.

EVENT #2, Customer/Attendant wants to refuel a rented auto, External

RESERVATIONS comes into the system when a customer wishes to reserve an auto to rent.

EVENT #6, Customer wants to reserve an auto to rent, External

Output Section

AMT-DUE informs the customer of the amount due after they have returned a rented auto.

EVENT #3, Customer returns a rented auto, External

APOLOGY or **AUTO-CARD + CONTRACT-COPY + LOC + KEYS** occur when a customer comes in to rent an auto and there are autos available, or there are no autos available.

EVENT #4, Customer rents a reserved auto, External

BILL-PBLM is sent to the manager when there is a problem in computing the customer's bill. This is a consequence of the customer returning a rented auto.

EVENT #3, Customer returns a rented auto, External

CAR-ORDER is generated by the system at the end of the day when it is determined that more cars are needed to meet the reservations for the next day. In order for the system to know when to perform this event, the system must keep track of the time and initiate this activity at approximately 6:00 P.M. every day.

EVENT #7, End of day, order cars for the next day, Temporal

CRT-MSG is a message sent to the customer at the CRT information station when he or she returns a rented auto.

EVENT #3, Customer returns a rented auto, External

INPUT/OUTPUT PROCESSING CHART WITH EVENTS

	1.1	1.2	1.3...	8.5		EVENTS	TYPE
Inputs:							
AUTO-CARD + LOC + KEYS + MILEAGE			ECA		1.	Maintenance Shop completed repairing auto.	Ext.
AUTO-CARD# + PUMP#					2.	Customer/Attendant wants to refuel auto.	Ext.
APPROVAL					3.	Customer returns rented auto.	Ext.
CONTRACT-COPY + LOC				IAA	3.	(See event above.)	
CURR-MILEAGE + FUEL-LEV					3.	(See event above.)	
(CURR-MILEAGE) + (FUEL-LEV) + CUSTM-NAME					3.	(See event above.)	
CUSTM-NAME + CUSTM-ADDR & INSURANCE			EFA		4.	Customer rents a reserved auto.	Ext.
DETAIL-CALL-RECORD					5.	Customer completes a phone call from auto.	Ext.
ENTER-REQ + GATE				ICA	3.	(See event above.)	
EXIT-REQ + GATE					4.	(See event above.)	
PAYMENT + (PBLMS)				ECA	3.	(See event above.)	
PUMP-ON					2.	(See event above.)	
PUMP-REQ					2.	(See event above.)	
RESERVATION			ECA		6.	Customer reserves an auto to rent.	Ext.
Outputs:							
AMT-DUE				ICA	3.	(See event above.)	
APOLOGY			EFA		4.	(See event above.)	
AUTO-CARD + LOC + KEYS + CONTRACT-COPY			EFA		4.	(See event above.)	
BILL-PBLM					3.	(See event above.)	
CAR-ORDER	EFA				7.	End of the day, order cars for the next day.	Temp.
CRT-MSG					3.	(See event above.)	
DEPOSIT					8.	Twice a day, deposit cash in the Bank.	Temp.
FUEL					2.	(See event above.)	
INSUFF-INFO					3.	(See event above.)	
MESSAGE					3,4.	(See event above.)	
MAINT-ORDER		EFA			9.	Hourly, check if cars need maintenance.	Temp.
RECEIPT				EFA	3.	(See event above.)	
UNLOCK					3,4.	(See events above.)	
UNLOCK-CARD					2.	(See event above.)	
PROCESS SUMMARY:	EFA	EFA	EFA...	EFA			

Figure 11.10

DEPOSIT occurs twice a day when the manager takes the excess cash from the cash register to the bank. The manager within the system is responsible for remembering the time to initiate this activity.

EVENT #8, Twice a day, deposit cash in Bank, Temporal

FUEL is a consequence of a customer or an attendant refueling an auto.

EVENT #2, Customer/Attendant wants to refuel a rented auto, External

INSUFF-INFO and **MESSAGE** are generated by the system when a customer returns an auto and neglects to provide the necessary information to generate a bill, when there is a problem at the entrance gate, or when there is a problem at the gas pump. **MESSAGE** can also be generated when there is a problem at the exit gate after a customer comes in to rent an auto and wishes to leave the parking lot.

EVENT #3, Customer returns a rented auto, External
EVENT #4, Customer rents a reserved auto, External

MAINT-ORDER is produced every hour when the system examines the COMPL-CONTRACTS data store to see if any returned autos need maintenance before they can be rented again. The system is responsible for keeping track of the time and performing this activity.

EVENT #9, Hourly, check if any returned car needs maintenance, Temporal

RECEIPT and **UNLOCK** are generated by the system when a customer wishes to enter the parking lot to return a rented auto and after the customer pays the bill. UNLOCK can also be generated when a customer wishes to exit the parking lot after he/she has rented an auto.

EVENT #3, Customer returns a rented auto, External
EVENT #4, Customer rents a reserved auto, External

UNLOCK-CARD is generated by the system when a customer or attendant has completed refueling an auto.

EVENT #2, Customer/Attendant wants to refuel a rented auto, External

From the event analysis, we determined there are nine events to which this system responds. Six events are external and three are temporal. There is a tenth event hidden in this system. Because there are real-time aspects involved in CARS, we must look for internal events. The audit process 3.0* monitors the real-time processes. If any of the processes that collect DETAIL-CALL-RECORDs from the Telephone Company lose the data-link connection to the Telephone Company, audit process 3.0* will properly reestablish it. This internal event is an example that does not produce any external data flow and is not detected during the event analysis using the input/output processing chart. However, we detect internal events during logical analysis (see Fig. 11.8, column 3.0, row PROCESS SUMMARY). If we complete a chart and there are no classifications assigned to a functional primitive, we have either made an error or there is an anomalous event to which that process responds.

EVENT #10, System loses the data link to the Telephone Company, Internal

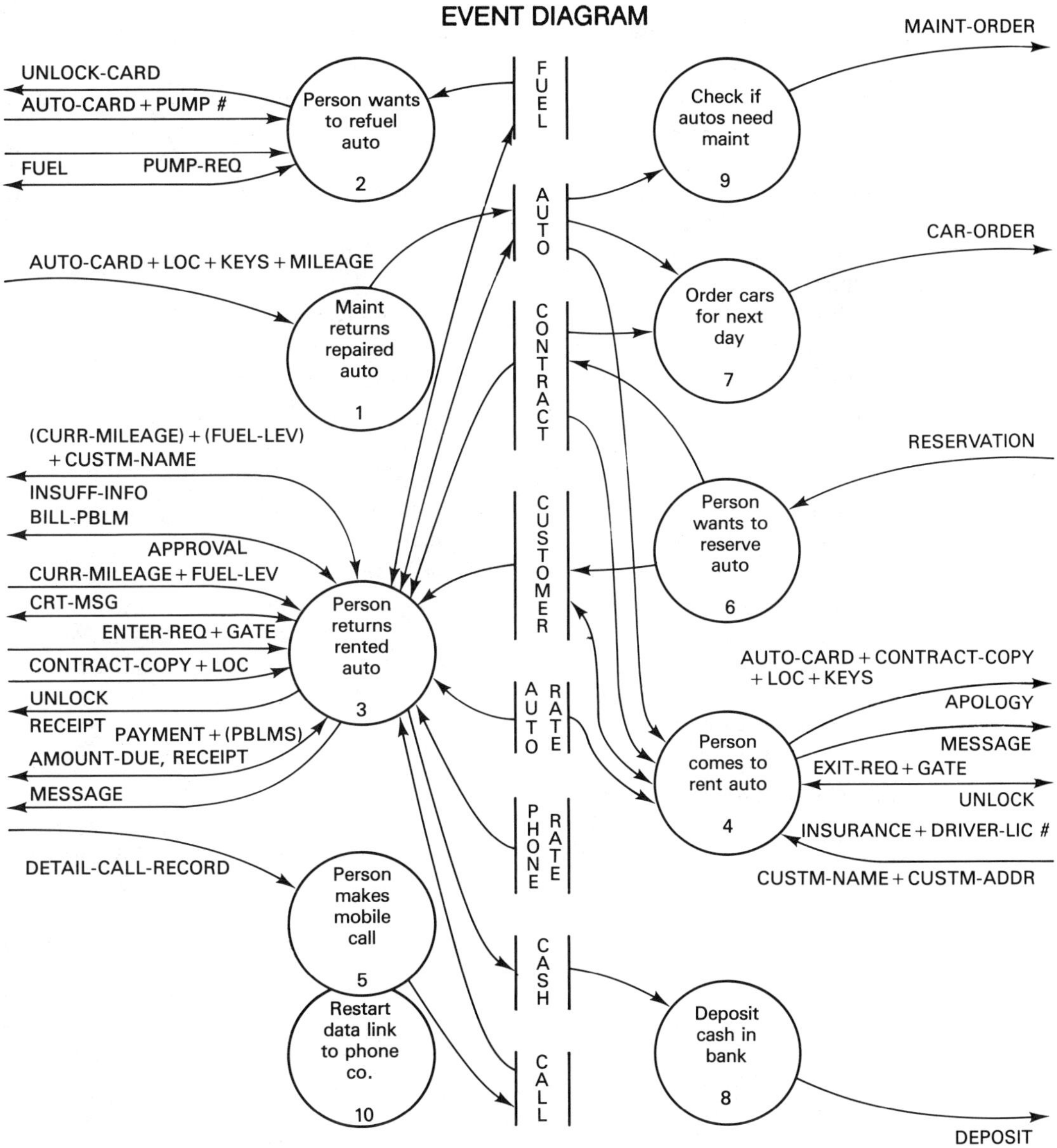

Figure 11.11

Given the events and object data stores found earlier, we are now in a position to construct the event diagram for CARS, as shown in Fig. 11.11.

SUMMARY
. . . you are now an expert on the CARS system

After we complete the event analysis, all of the information is available to assess the impact of any change to the system that the users may request. Given a request to change the system, the analyst merely relates to one of the events in the system. Then, from the input/output processing charts and event diagrams, the analyst can identify all of the inputs and outputs that relate to that event, all of the functional primitive processes involved in processing that event, and all of the data stores involved when the event occurs. With this tool, the analyst can estimate the impact of the change, and, with some experience, can estimate the time required to implement the request.

Even after the change is implemented, the chart is useful. It tells what events may need testing to guarantee that the implemented change has not inadvertently changed the processing when an unrelated event occurs. This is done for those events that share the same functional primitive or data store used to service the event that changed.

A third advantage we can draw from the structured analysis work applied to this system is that the event diagram explains the system from a very high-level functional view. In dealing with the users' management or the owners of the system, it is important to be able to explain the purpose and operations of a system without discussing unnecessary details. The event diagrams provide a very high-level logical view of the system's inputs, outputs, and the events that initiate activity in the system. From these diagrams, we can proceed to discuss the aspects of the system that are of the most interest to the owners of the system, the services provided to the users when an event occurs.

The event diagrams can also be used to explain alternative solutions or changes to the system from the user's perspective, while relating costs to the benefits of each alternative solution. It is tools such as those presented in this text that enable a systems analyst to provide the information and explanations expected by the user community.

QUESTIONS

1. a. What is the domain of the data element GATE?

   b. What is the cardinality of the data store GATE-REQ?

   c. What is the degree of the data store GATE-REQ?

2. Management suspects workers are using AUTO-CARDs of autos assigned to maintenance to refuel their personal vehicles. What changes should be made to the system to avoid this possibility?

3. a. How can we pump gas and not have it recorded or charged to anyone when refueling an auto?

   b. What changes should be made to the system to avoid the possibility described in Question 3a?

4. What happens when a customer enters the Parking Lot and neglects to provide CURR-MILEAGE and FUEL-LEV at the CRT station?

5. a. Is AUTO-CARD# a valid candidate key for the data store CONTRACTS?

   b. Is AUTO-CARD# a valid candidate key for OLD-CONTRACTS?

6. Explain the relative priorities the system uses for the following real-time functions: Start Fuel Pump, Stop Fuel Pump, Open Gate, and Accept/Store Phone Call Information.

PROBLEMS

1. Specify in the system requirements an override that can be invoked by the manager of the rental agency (protected by a password) that permits the fuel to be estimated to refuel the auto so the system can provide the customer a bill if the fuel pumps are not working.

2. Modify the structured specification in this chapter to include a PARKING-LOT-ACCESS log data store. The log would be posted every time someone requested to enter or exit the parking lot. It is organized by CURRENT-DATE and CURRENT-TIME and stores the AUTO-CARD# and the GATE at which access took place. Which of the following structured specifications would require modification?

Context diagram	Input/output processing chart
Data flow diagrams	Object data-store definitions
Data dictionary	E-R diagram
Mini-specifications	Event diagram

3. Modify the system so that if a customer uses the mobile phone but charges the call to a credit card, a minimum service charge of $.50 is billed to the customer by the rental agency for use of the mobile phone.

4. Figure 11.12a is an aerial view of the original rental agency location, showing the single entrance and exit gates. Figure 11.12b shows the expansion plans that add

additional entrance and exit gates. Modify the system requirements to control several entrance and exit gates simultaneously.

AERIAL VIEW OF CAR'S PARKING LOT

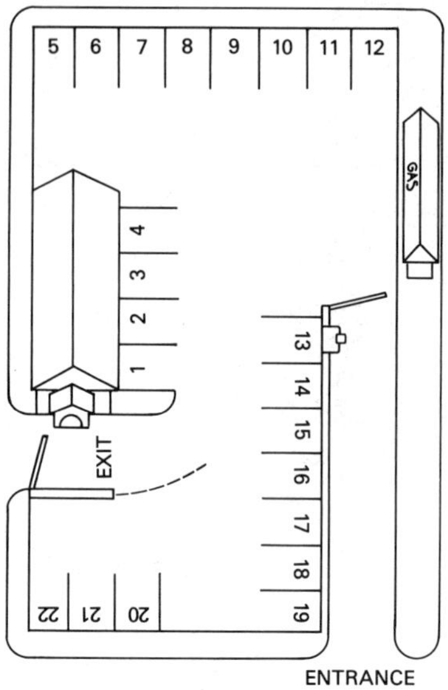

Figure 11.12a

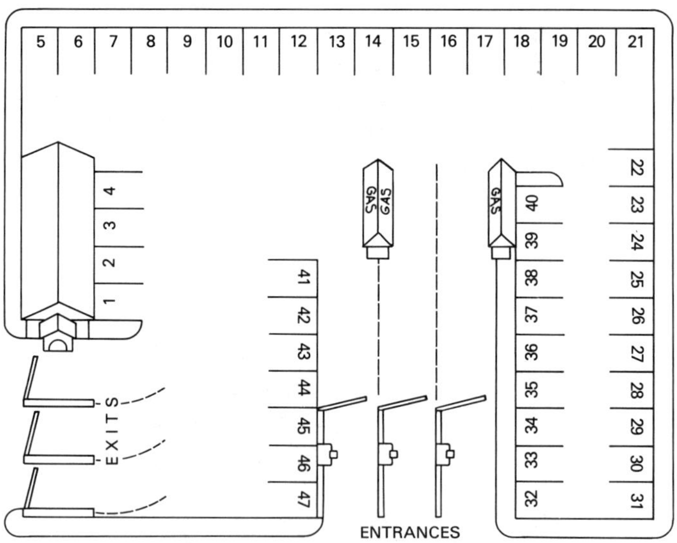

Figure 11.12b

12

SYSTEMS ARCHITECTURE

This text presents methods for modeling the *processes* as well as the *data* used in systems. We use *process-oriented* methods to document the processes and procedures used in developing and maintaining systems. Structured specifications, as originally proposed by Tom DeMarco,[D-2] make a distinction between the physical implementation and the logical aspects of the system. First, the analyst models the system as it physically exists. The users make sure the model is a complete and accurate representation of the current system. Second, the analyst derives the logical model from the physical model of the current system. The logical model concentrates on what is being done as opposed to the physical model, which includes more details (infrastructure) relating to the technology of *how* the system operates. Third, when new constraints, requirements, and/or changes are included in the system, the analyst makes the modifications and/or enhancements to the structured specification, producing a fourth model known as the *new* logical model. Once there is concurrence on *what* needs to be done, the analyst produces a *new* physical model, which proposes the necessary infrastructure that defines *how* the new system will operate.

There are those who have studied the process-oriented approach, or one similar to it, and have adopted *process-oriented* methods as their standard for developing systems, as shown in Fig. 12.1.

A second approach to systems development is strictly *data-oriented*. The data approach is fundamentally different. Instead of examining processes, we examine the data at a high level and organize it to meet the goals and objectives of the enterprise. There are basically two types of data elements: those that uniquely identify the *things* for which the system must store data about, (i.e., objects) and those that describe or contain a fact about one of the *things* (i.e., attributes). The data-oriented approach requires interviews with the top management to understand the data requirements of the enterprise and the relationships among the data elements. The analyst then constructs models that illustrate the logical relationships among the data elements.

The underlying principle here is that data structures are more stable than the procedures and activities that process the data. After the analyst defines the data structures, much of the work in defining a system solution is already done. The

PROCESS-ORIENTED SYSTEM-DEVELOPMENT DOCUMENTATION

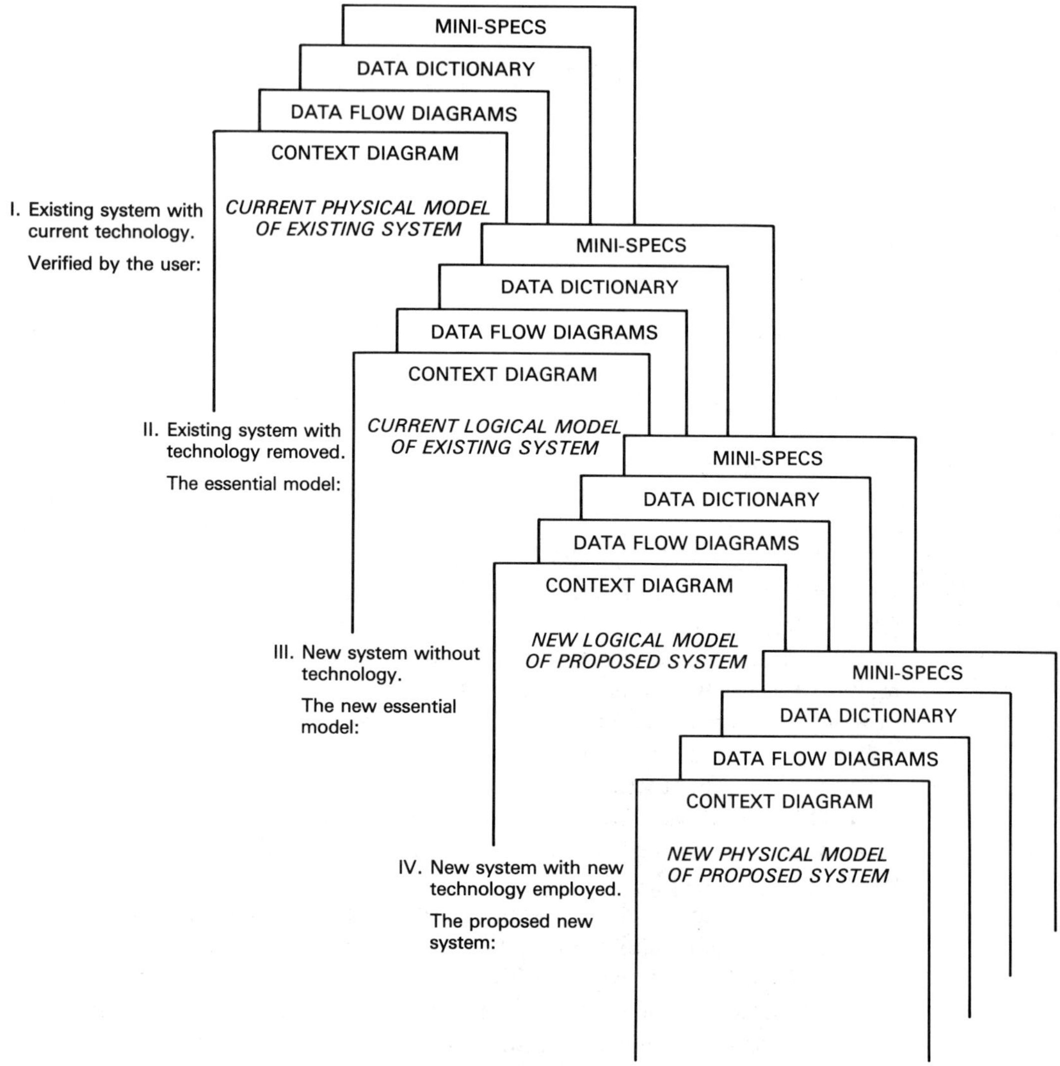

The above illustrates the documentation used in structured specifications as originally proposed by Tom DeMarco.[D-2]

Figure 12.1

analyst has to determine what *events* change or use the data. Powerful reporting tools can be applied to minimize the development effort involved in displaying data or information on retrieval screens or printed reports. In addition, software packages are available to provide decision support and transaction processing to further minimize the costs of developing, implementing, and maintaining information-systems software.

Interactive *decision-support system* software packages can provide a user with direct access to data and provide update and retrieval processes that operate on information for the classical *what if* situations. **A DECISION-SUPPORT SYSTEM is an interactive computer-based system that provides managers with information to quantify and/or reduce the risks of decision making.** Decision-support systems do not attempt to automate the decision-making process, but rather support the business executive by providing information to minimize or quantify the risks involved in a business decision.

To interactively input, validate, and retrieve data from a data bank, we can use a *transaction-support system*. **A TRANSACTION-SUPPORT SYSTEM is an interactive computer-based system that provides the services of collecting, validating, editing, and displaying data upon a request from a user.** Transaction-support systems are tools that interface with data base management system software to provide quick and convenient access to the data stored in a system.

The data approach shown in Fig. 12.2 begins by involving top-level management in the first step, Strategic Requirements Planning. From there, analysis of the data needs of the enterprise continues until we can propose a conceptual (logical) data model along the left side, and a process (or function) model along the right side of Fig. 12.2. The analyst's attention is focused on creating a data model and determining the usage requirements of the data so that he or she can design the applications model of the data base. Ultimately, this method will yield a data-base design from which we can specify programs, produce reports by means of report generators, and/or apply other available software to provide the users with the information they need.

There is much controversy and competition between the *process-oriented* and *data-oriented* approaches. Some groups have become fanatics over a particular approach and apply the one method they've chosen to all data-processing problems.

There are those problems that are well suited for the *process-oriented* methods, and there are those that are solved much more elegantly and efficiently by the *data-oriented* methods. The fact of the matter is that no one method is the best to solve all of the problems facing the systems professional. To be a highly valued *problem solver*, we have to know several methods and have the knowledge to apply the appropriate method to the problem at hand.

Technology has provided us with faster and faster processors, larger and larger memories, and highly sophisticated input and output devices. Likewise, the problems we are able to address have grown larger and larger in scope. Let's consider the overall architecture of the *information system*. Originally, we could easily identify three aspects of a system solution: the capturing and converting of source data into

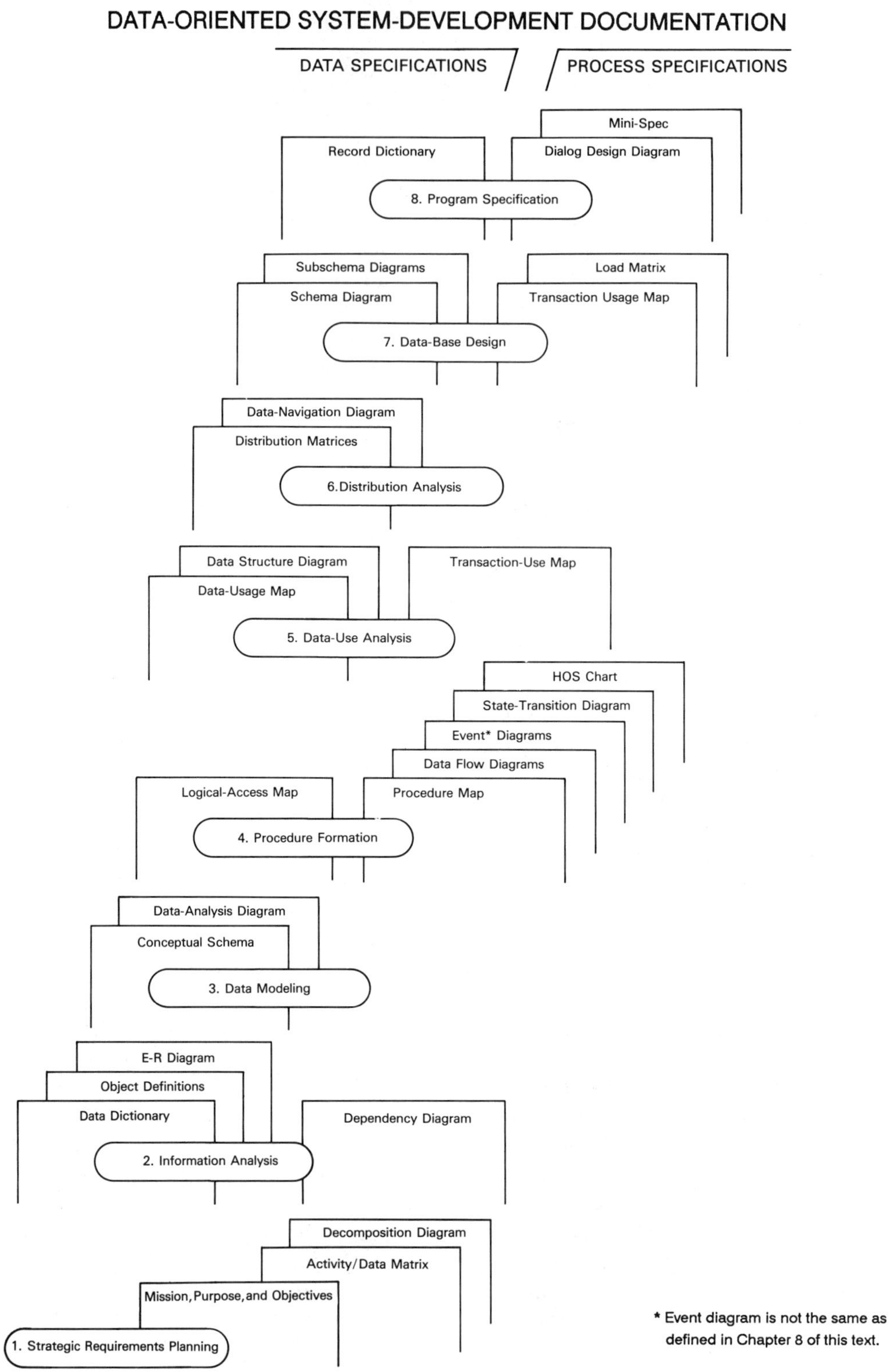

The above illustrates the information engineering approach to systems developments.[M-3]

Figure 12.2

machine-readable form, the processing of raw data into information, and disseminating data and information. Let's look at how we can engineer systems and combine the process and data methods to best address the needs of the enterprise.

GENERAL BUSINESS INFORMATION SYSTEM
. . . great software isn't written, it's engineered

The costs of developing and maintaining *custom-built* software for business applications have exceeded the tolerance threshold of the management of many major corporations. As a result, management has implemented somewhat radical policies regarding new systems developments, such as:

"No new developments shall use the COBOL language."

"All new information systems shall use purchased software packages."

These seem to be rather extreme reactions to the high cost of developing business systems. However, it was inevitable. Software developers kept reinventing software to do the same business functions over and over again at research and development rates instead of engineering systems by reusing existing routines, modules, and subsystems. Many developers build *new* systems by redeveloping all of the software from scratch every time, even if there are programs that perform the same functions in another system. This is due in part to the rapid changes in technology as well as improved ideas and techniques known for processing data. Advances in hardware and software often provide cost justifications in replacing existing automated systems with new ones. Lower maintenance costs, greater productivity, and higher reliability are often claimed as the benefits that justify these reoccurring corporate investments. However, upper management is taking a hard look at the costs and expenses in running the enterprise and is no longer willing to underwrite another payroll or accounts payable system — especially one that costs several millions of dollars because it will be totally redesigned from input to final output. Why does it have to be *totally* redesigned and rebuilt from initial inputs to the final outputs? During the past five years, changes in hardware and software have brought us more reliable technology with greater capacity for work. Furthermore, it is unlikely there will be any change in this trend over the next decade.

Let's look at the original high-level architecture of a business-information system, as shown in Fig. 12.3. The figure is a high-level system diagram of a business-information system based on the input, process, output subsystem architecture. The system also centers around the concept of a *data bank*. **A DATA BANK is any collection of data, whether in the form of files, data-bases, or in an information retrieval system.**[M-4] The term data-base is a more precise term and implies the data has a formal structure and support software to maintain the data. Likewise, an **INFORMATION RETRIEVAL SYSTEM is a collection or network of data and data relationships under the control of a data-management system that maintains the**

data and data relationships as well as monitoring and controlling access to the data. However, the input and output subsystems service the same hardware devices in many systems. Unfortunately, in most systems architecture, the structure of the data bank has to be different, depending on the requirements for processing the data and presenting it to the user. If the data bank is defined in terms of data objects and data relationships, there is basically only one data structure, the structure that is inherent in the data itself. Given a stable data structure with definable data formats (stored in some convenient place such as a mechanized data dictionary), input and output subsystems would have a well-defined interface to the data bank. This portion of the overall system can best be analyzed and designed by data-oriented methods and tools.

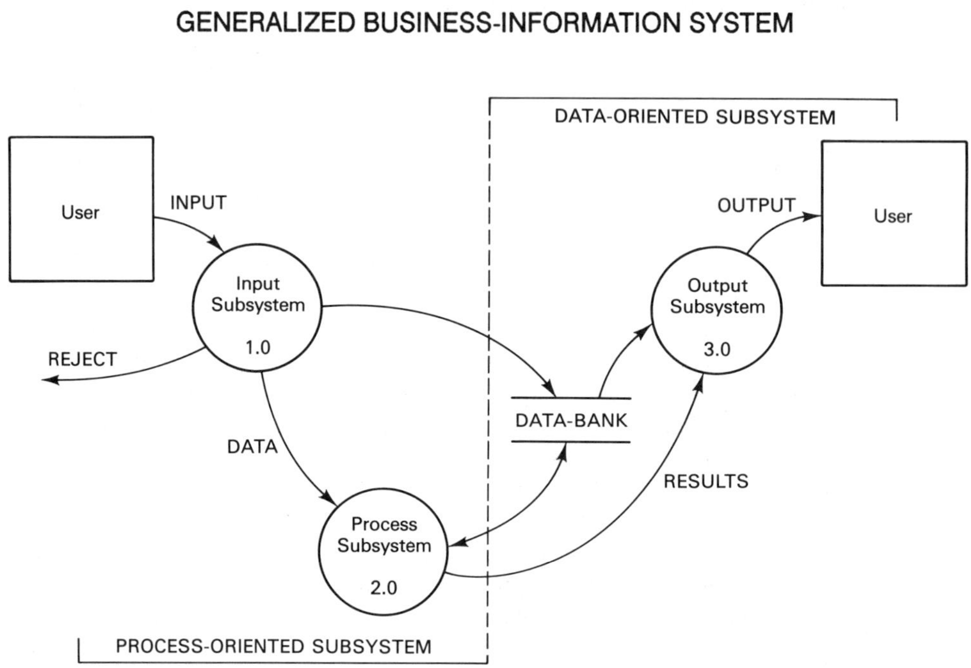

Figure 12.3

The interface between the end user and the input and output subsystem will constantly change due to advances in technology. However, the I/O subsystem does not need to be *custom-built* application software that is reinvented for each new system.

The task of the analyst is one of separating the functions of inputting and processing the data from the functions of reporting the data. Systems-modeling methods have addressed these two functions of processing and reporting data somewhat differently. Process-oriented methods do well in decomposing and modeling input processes and the more complex processes that transform the input into the desired results. Data-oriented methods are very productive when applied to problems that must report data and information in a variety of ways. The analyst can derive the benefits of both approaches by factoring the system into the input/process functions and the update/output functions. The analyst can bring the power of automatic reporting and information retrieval to the output function of the problem by using *data-oriented* methods. Decision-support software, transaction-support software, and other existing software packages can interface with the system's data bank once it is organized around the inherent structure of the data itself, and not built exclusively to support a unique application. Meanwhile, the input processes and the complexities of data transformation used in process control can be more effectively handled by applying the *process-oriented* methods of functional decomposition.

The benefits of organizing the system in a manner that permits the analyst to take advantage of the best aspects of a variety of methods are greater than applying a single method to all system problems. Furthermore, we can gain many benefits from engineering a solution by reusing *component parts*, such as using standard communications-interface software, existing information-retrieval software, decision-support software, or report generators.

Specifically the benefits are:

1. Lower development costs by using existing software.
2. Lower maintenance cost by reusing *working* software.
3. Higher quality software by using existing software with proven reliability.
4. More rapid access to information to satisfy the users' needs by building an environment where the users can retrieve their own information by using report-generation software.
5. Easier capability to incorporate new data collection/retrieval technology.
6. Shorter development intervals because input processes, transformation processes, and information retrieval and reporting processes can be done in parallel and are often provided by software venders.

In the overall systems architecture shown in Fig. 12.4, batch, on-line, or real-time processing methods can be employed to gather the data at the point closest to the source. Likewise, output information can be provided to support routines, process control, accounting, and manufacturing operations. We can use the data flow and *process-oriented* methods to define the functions that transform the inputs into the stored data elements. The *data-oriented* methods are useful in building a user-accessible data bank of processed data. The methods of *software engineering* complete the picture by integrating existing software into the system. The user is free to use information-retrieval system and report generators to provide information on demand. We should apply the tools and methods that best fit the problem. No one approach or technique can work best on all problems. The most valuable problem-solver is one who knows many approaches and techniques, and when to apply them

GENERAL ARCHITECTURE OF BUSINESS-INFORMATION SYSTEMS

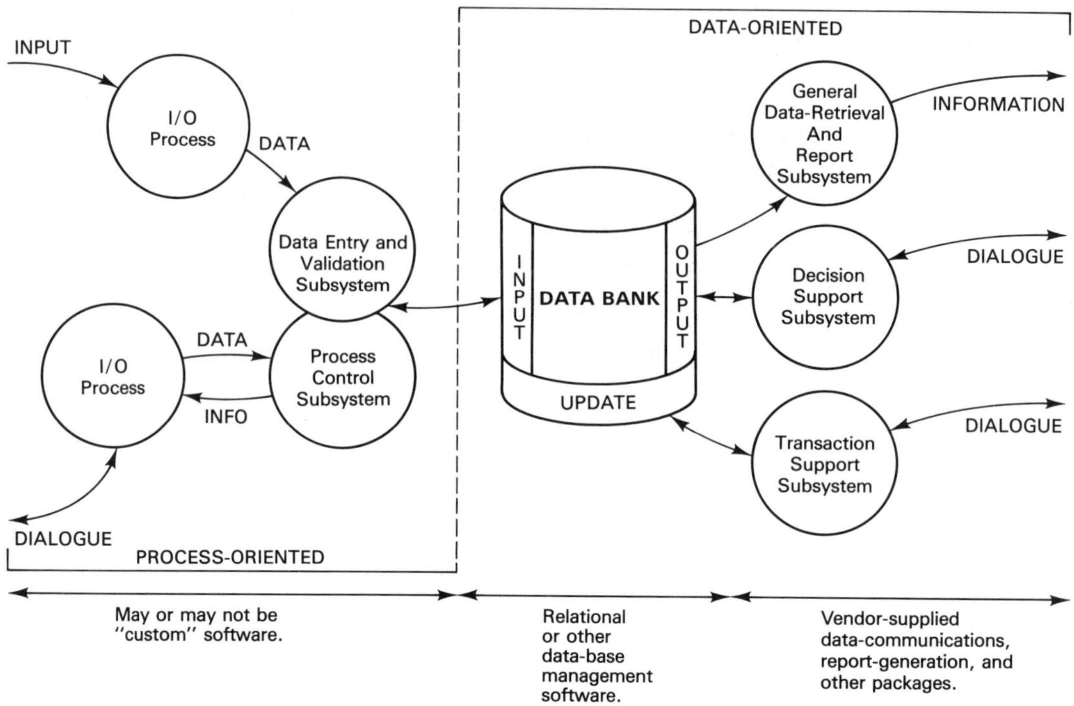

Figure 12.4

SYSTEMS DEVELOPMENT DOCUMENTATION
. . . documenting the requirements for a system

The basic systems-development documentation covered in this text and used in both the process-oriented and data-oriented development methods is shown in Fig. 12.5.

The theory and methods used in Chapters 6 and 7 (normalization and E-R diagrams) for modeling objects and relationships are applicable to all popular data-base models. From an E-R diagram, we can map the data model into an inverted list, hierarchical, network, or relational data base management system.

The systems analyst's experience and knowledge, combined with the appropriate *tools of the trade*, can yield the best, least-costly solution to the problem. In determining the least-cost solution, we must consider not only the development costs but the maintenance costs over the entire system life cycle.

In most information systems, the data-transformation and process-control subsystems often are unique. It is in this area that the input data elements are transformed and saved in the system's data bank. Depending on the business needs, data transformations will change and/or new ones will be necessary. The most volatile area in the system is the information system's report-generation and transaction subsystems, which usually have the highest number of changes required. The software that performs these functions must accommodate changes very readily because of two phenomena: the *uncertainty principle of data processing* and the *spotlight effect*.

In addition several other reasons have contributed to the need to accommodate changes. The economics of using computers, high-tech input and output devices have

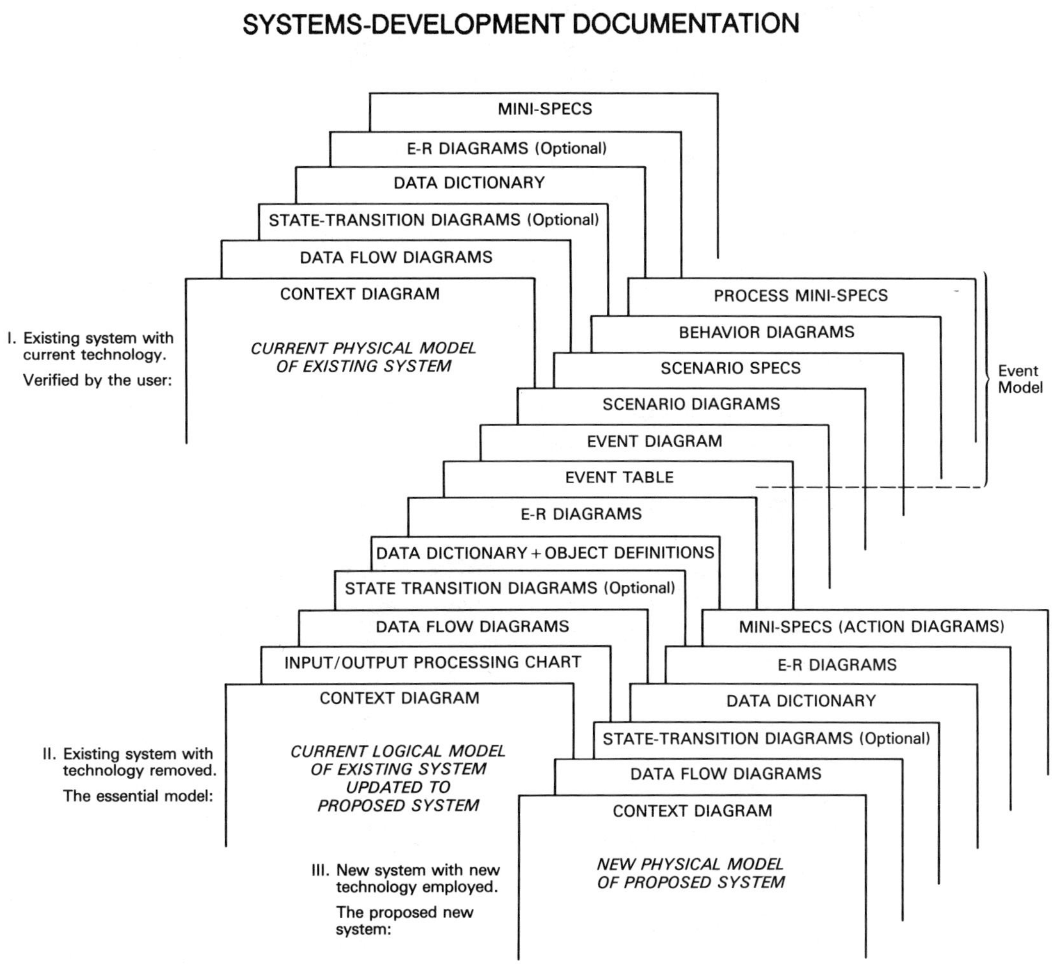

The above illustrates the revised structured-specifications development documentation as presented in this text.

Figure 12.5

increased the rate of change in systems particularly with in personal computer technology. Other benefits such as faster and more accurate processing, new services and conveniences, and providing new (or changed) governmental and business reports have made it necessary to turn to automated activities and tools.

FOURTH-GENERATION LANGUAGES
. . . systems need to be able to respond to change

A system's longevity is dependent on its ability to respond to changes. There are many reasons why information systems are changed frequently. Among them are the *uncertainty principle of data processing*, the *spotlight effect*, and the unanticipated needs of the business.

Uncertainty Principle of Data Processing

There is an uncertainty principle in atomic physics that states:

The act of observing subatomic events changes those events.

Similarly, there is an **UNCERTAINTY PRINCIPLE IN DATA PROCESSING that states:**

The act of providing what an end user says he needs changes his perception of those needs.[M-6A]

The mere act of implementing a user-driven system changes the requirements for that system. Another way of saying it: the solution to a problem changes the problem. The uncertainty principle of data processing was likely to have been the underlying motivational factor for such as E.J. (Ted) Codd when he began formulating relational data-base theory. Ted sought for a way to eliminate the need to restructure the data in a data base because the needs of the business have changed. Later, other developments led to *user-friendly* information-retrieval systems and fourth generation nonprocedural languages, which I will define shortly. Ignoring the uncertainty principle of data processing soon creates a large backlog of work for the information systems development organization. With a large backlog of work, the development organization loses its ability to be responsive to the needs of the business and management.

Spotlight Effect

Many management-information systems provide information to managers to measure the output or performance of individuals or organizations. There are those in management who believe in the *spotlight effect*. **The SPOTLIGHT EFFECT focuses management's attention on a specific aspect of the organization until the results or**

expectations are improved. The spotlight effect is characterized by a manager who receives data that measures the specific output or performance of individuals or organizations. Management in its attempt to improve the output, performance, or the profitability of the enterprise looks for inefficiencies and uses the data to provide factual and objective feedback to employees. The response by the employees usually causes the measured results to come closer to the manager's expectations. The manager, however, believes that the pervasive inefficiencies have just moved elsewhere in the system and now needs more or different information to determine where the inefficiencies now reside. The need for new information in effect places the *spotlight* on other areas or individuals in the organization. Once again, the organization begins to optimize to achieve management's objectives. Keeping the *spotlight* moving requires frequent changes to the management-information system. Requests for new and/or additional information continues as long as management is intent on reducing waste, improving efficiency, or increasing the profitability of the enterprise.

Anticipating the Unanticipated Requests

Requests for information that relates data elements in a way no one anticipated often creates a problem in some data-base systems. We usually store data elements in a structure designed to accommodate requests that were either in the system requirements or anticipated during the development of the system. When we present a request for information to the systems-development organization that was not anticipated, it usually is not able to respond quickly to such changes or requests, and a backlog of work develops.

In looking at the demands placed on the typical information-management system, we find that to accommodate the changes often only requires modifications to the output subsystem. This is because most of the data elements already exist in the data bank. In cases where new data elements have to be created, the development organization must develop a new subroutine that will add the data to the data bank. In such cases, the data bank itself needs modifications to accept and store the new data elements. Once we have all the necessary data elements present, the problem becomes simply one of reporting the data, which is a problem we have solved many times. Furthermore, users can even solve the problem with purchased software products. With the proper software, such as a *fourth-generation nonprocedural query language*, the data and information can be provided by the users themselves. **A FOURTH-GENERATION LANGUAGE (4GL) is one that is a very high-level and "user-friendly" way of specifying the results to be obtained from data usually stored in a data base management system. Elements of the language do not necessarily have to be specified in any logical order and the results can be obtained in an order-of-magnitude less time than with conventional programming languages.**[M-6A] Examples of fourth-generation languages are IBM's Query-By-Example, National CSS's NOMAD, Information Builders Inc.'s FOCUS, Unisys' MAPPER, etc.

The term *fourth generation* refers to the stages through which computer languages have evolved. The *first generation* was machine language, which came into use during the introduction of stored-program machines in the late 1940s and 1950s. The *second generation* was introduced in the 1950s when assemblers were developed that greatly eased the drudgery of programmers in dealing with the binary language of computers. We entered the *third generation* during the 1960s with the development of high-level languages such as COBOL, FORTRAN, ALGOL, BASIC, etc.

The level of a language relates to the number of machine instructions generated with one instruction of the command language. In machine language, there is a one-to-one correspondence between the machine-language instruction and the binary-coded instruction performed by the computer. In assembly language, most instructions have a one-to-one correspondence to computer instructions; however input/output commands often generate several computer instructions. (Many vendors' assembly languages have a macro capability that raises the correspondence between the assembly-language instruction and the binary instructions executed by the computer.) In higher-level languages, such as COBOL, FORTRAN, PASCAL, etc., there is a much higher ratio between the commands used in the language and the machine's executable instructions. However, in all of these programming languages, the logical ordering of the instructions is crucial to the outcome.

In fourth-generation languages, not only is there an even higher ratio of machine-language instructions to the commands of the language, but much of the logical ordering of the commands is relaxed. Vendors relieve the requirement of the logical ordering of the command statements by having the users format a screen according to the way they wish to see the data displayed. Selection criteria, organization, accumulated totals, and other formatting characteristics can be entered at any time onto the screen and in any order. Once a user is satisfied with the screen format, it is released to the system. The system interrupts the commands on the screen, compiles them into a data-retrieval commands, executes the retrieval commands, and populates the screen with the resultant data selected from the data bank. As a result, fourth-generation languages are very well suited for nontechnical users to retrieve data from a data bank. These languages are particularly appropriate for the output subsystem in the general architecture of business-information systems shown in Fig. 12.4.

If the output subsystem has access to a common data bank, then the request for new or additional data becomes one of serving the *unanticipated query*. By use of a fourth-generation language and an information-retrieval system, much of the development to produce information displays and reports is under the control of the users themselves. By apply this type of technology we begin to realize what was predicted in the opening chapter, "everyone will become a programmer, but programming will not be what it is today." To take advantage of the type of technology that puts information in the hands of the users, we must plan our information systems so that their underlying structure does not preclude our ability to incorporate new tools for managing and disseminating information. Lets look are the architecture for information systems that support these ideas.

BASIC INFORMATION SYSTEMS ARCHITECTURE
. . . the ultimate in information retrieval is the ability to service the unanticipated query

As stated in an earlier chapter, a *query* is a request for information. When the data contents of the request are known at the time the system is developed, or are otherwise anticipated by the system developers, the information system can usually provide the desired results easily. The problem occurs when a request for information was not anticipated and the structure of the data bank does not lend itself to a straightforward retrieval of the information. The attempts to satisfy such *unanticipated queries* are responsible for a significant part of the data processing backlog; but, why should they be? In most cases, the data is in the data bank. If not, a routine has to be added or changed to collect the data element and store it in the data bank. Adding a new data element should in no way require changes to existing programs merely because the data bank now contains additional data. The issue here is the independence of the data and the organization of the data bank to the applications programs. One way to reduce the backlog for data-processing efforts and reduce the maintenance cost of information systems is to separate the output function from the information-processing function in the system. Once we achieve this architecture, the users themselves can provide their own information through the use of independent routines or vendor-supplied packages.

The methods and tools used to specify the requirements of a system that employs both the process-oriented and data-oriented methods are described in this text. To specify the input, update, and transformation processes, the *process-oriented* methods are usually very useful. The methods used to specify the data bank, which serves as the interface between the data-transformation functions of the system and the retrieval functions, have to express data elements and their relationships. The tools to model data and relationships among data are available by using the *data-oriented* methods.

To service the information needs of the enterprise, the following components of a computer-based information system are needed:

1. Data banks
    - whose structure is not application-dependent
    - that can independently organize data around the objects of the system
    - that can associate data elements of one object to those of another without application software modifications
    - where the format of the reports is defined separately from the structure used to store the data
    - where input, output, and update software are available to maintain the stored data

2. I/O device (communications) subsystems
3. Data-entry and validation subsystems
4. Process-control subsystems
5. Transaction-support subsystems
6. Information-retrieval subsystems
7. Decision-support subsystems
8. Restart and recovery subsystems
9. Audit/diagnostic subsystems

With the preceding components, we can define a *basic information-system architecture* (see Fig. 12.6) that uses existing software-support produces as well as *custom-built* software to work together and use the system's data bank.

BASIC INFORMATION-SYSTEM ARCHITECTURE

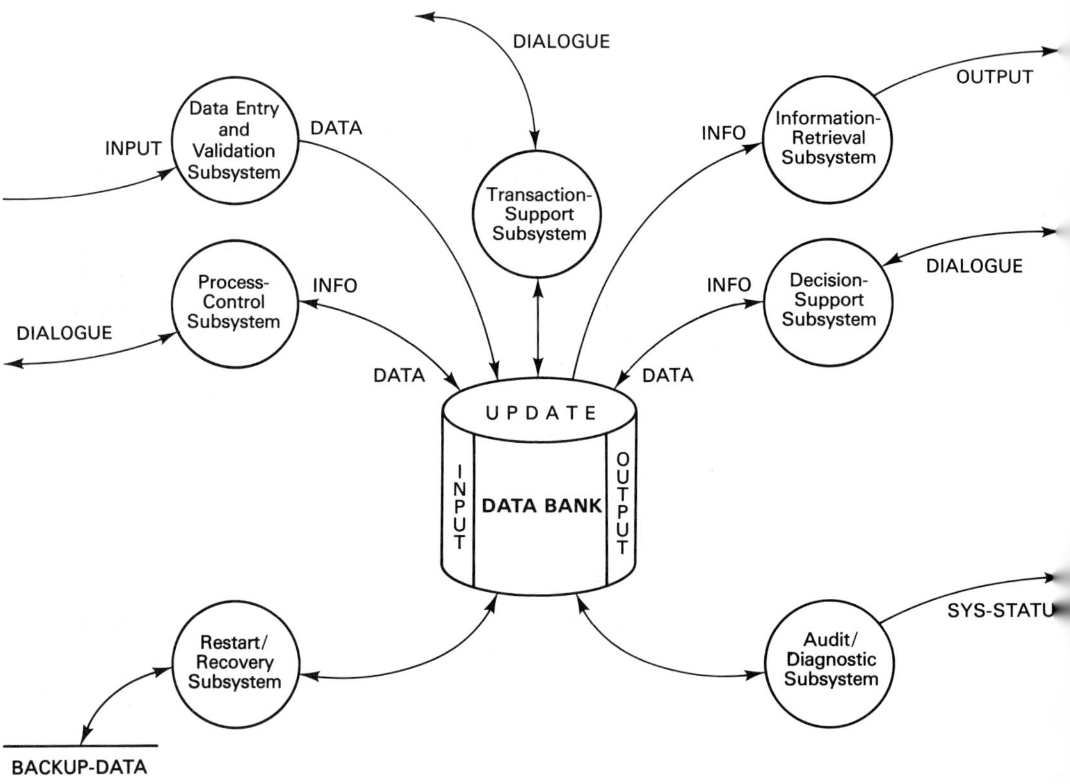

Figure 12.6

Data-entry and update subsystems, information-retrieval subsystems, decision-support subsystems, and restart and recovery subsystems, as well as data base management systems, are functions that often can make use of existing software. We do not have to reinvent this software to be able to include these functions in a business system. We must be judicious in the selection of such software. The data bank must not be restricted by any defined structure relating to a specific application. Software packages that use a relational data base and have a fourth-generation nonprocedural query language are among the most flexible and interface easily with *custom processing* software. We have to analyze the data stored in a relational data base management system, beginning with an overall business purpose of the enterprise. The information retrieval system can best provide information to the unanticipated query without (or with very little) information-system development work.

With the basic information-system architecture, we operate on and control the data of the enterprise in an environment that is responsive to change. Without individual subsystems providing the necessary functions and services, the system cannot respond to changes in technology without causing trauma within the system. We can incorporate new high-tech input/output devices as they become available by including the independent device-support software module in the system that translates the source data from the new device into machine-readable form. The I/O device-support software can even be provided by the hardware vendors of the I/O devices.

By using software packages to support decision making, as well as the user-generated report facilities of a fourth-generation nonprocedural language, the information-system development organization is relieved (for the most part) of developing reports. As a result, more time is available to develop *custom-built* data-transformation subsystems, process-control, and other data-processing subsystems. Even in the category of data-processing subsystems for accounting, process control, inventory management, etc., the make/buy decision to develop *custom-built* software or incorporate purchased software packages into the information system is applicable.

As technology advances and the industry pushes the processors to their physical limits, we will turn more and more toward engineering *systems of computers* to provide the functions and services needed to run the business enterprise. The basic information-system architecture lends itself to one where there are many concurrently operating independent processors. Taken to a logical conclusion, we could allocate a processor with the capability of sharing data with other processors in a non-interfering manner for each *event* to which the system must respond. Likewise, the activities of the data bank can also be managed and performed by a system of independent processors.

For the systems professional this means another round of new technology and another set of interesting and challenging problems. There will be new approaches and techniques for problem solving as well. Thus to remain a valuable problem-solver we must be committed to a life-long process of learning.

SUMMARY

. . . "one can argue indeed that economics has had certain negative effects, in that it has improved the efficiency of doing things that probably should not be done at all." W-2

- Kenneth Boulding

Now that you have been armed with the tools and methods to solve a large variety of problems, and before you go off to solve the problems of the world, I would like to relate one final story. It was told to me by the late Dr. Maurice Halstead while he was a professor of computer science at Purdue University.

During the mid-1960s there was much unrest on our college campuses. Frequently, riots would break out and the students often prepared bombs and attempted to bomb buildings on campus. The school's computer center was frequently the most popular target. For instance, rioting students bombed the data-processing facilities of the University of Wisconsin and elsewhere.

During that same time, there was similar unrest on the campuses in Mexico. However, a strange thing often occurred. It seems that when the rioting students would attempt to bomb buildings on campus, the students who did not participate in the rioting would form *human chains* by placing their arms over each other's shoulders. The students would completely surround the building that contained the school's computer. The only students allowed through the human barrier were those who had a deck of punched cards and said they had come to make a computer-run. This unusual behavior brought many American psychologists to the campuses of the Mexican universities in search of an explanation for this behavior.

It is believed that the explanation lies in the manner in which data-processing computers were introduced into the respective societies. If you were not employed by an American corporation when they purchased their first general-purpose business computer, find a senior member of the programming staff and ask, "What was the first application the company ran on its computer?" Eight out of ten will answer, "PAYROLL." Why? Didn't these companies get their payrolls out on time? Weren't these companies able to calculate accurately the amounts they had to pay their employees? No, the payroll job was getting done. However, by using this marvelous machine, companies were able to get the job done with fewer people. Now think about the effect that had on the employees.

I am well aware of the effect, because I was teaching in the data-processing department of a community college in the late 1960s and early 1970s. We had many people who came to learn about computers in fear of losing their jobs to a machine, a machine they did not understand. These people were not young enough to retire and old enough to have no formal education or experience with a computer. So they came to their local community college to learn about this machine to avoid being overcome by a technology that was encroaching on their livelihoods.

During the 1960s, computers were also available to businesses in Mexico. Likewise, in Mexico there were the same young aspiring professionals who sought approval from their senior management to invest in these marvelous but expensive machines. However, due to the low wages paid to the clerical help in Mexico, there was no way companies could cost-justify the purchase of a multimillion dollar computer by relieving the work of a few dozen payroll clerks. As a result, these aggressive young people came to their management and said, "Do you remember when we tried to solve this serious problem that plagues our company a few years ago? No matter how many people we hired, and how we organized them, there was just no way to overcome the obstacles. However, if we had one of these marvelous machines with the ability to calculate thousands of instructions per second and a tremendously large and accurate memory, perhaps we will be able to find the solution we are seeking." And many times they did. Many problems that were unsolved up until that time were solved by using the computer. Now, how do you think the society in Mexico viewed the computer?

Go forth and do likewise, and remember a quote from Gerald Weinberg: "A systems analyst should study people more than machines."[W-3]

QUESTIONS

1. What is a model? Why is it used in systems development?
2. What is systems analysis?
3. Who is a user?
4. What are the steps involved in creating a new system?
5. What is the uncertainty principle of data processing?
6. What is the spotlight effect?
7. What is an unanticipated query?
8. What are the characteristics of a fourth-generation language?
9. What elements define the basic information-system architecture? How does it differ from the conventional architecture of information systems?
10. What are the advantages of building an information system in accordance with the basic information-system architecture?

PROBLEMS

1. Consider Problem 2 in Chapter 2, discuss the system architecture you would recommend for the Purchasing System.

2. Using the proposed system architecture in Problem 1, give three examples of changes and/or enhancements to the system that your architecture could implement easily.

3. Using the proposed system architecture in Problem 1, give an example of a change or enhancement that would not easily be implemented with your architecture.

GLOSSARY

ACTION DIAGRAM is a diagramming notation and specification tool that uses nested layers of brackets that represent the structure of the logic used to transform input data into output data.

ADAPTIVE METHODOLOGY is one that chooses and/or modifies a development life cycle to fit the characteristics and needs of each particular project.

ALIAS is a synonym for a previously defined data item.

ALTERNATE KEY is a candidate key that is not chosen to be the primary key.

ANOMALOUS EVENT (see INTERNAL EVENT).

ATTRIBUTE is an elementary data element that describes an object or is a fact about an object.

BALANCED DATA FLOW DIAGRAM is one in which all of the inputs and outputs of the "child" diagram are exactly the same inputs and outputs of the corresponding "parent" process.

BASIC SYSTEM ARCHITECTURE consists of DATA BANKS whose structures are not application-dependent, that can independently organize data around the major objects, that can associate data elements of one object to those of another without software modifications, and where the format of the reports is defined separately from the structure used to store the data. Examples are DATA-ENTRY and UPDATE SUBSYSTEM, INFORMATION-RETRIEVAL SUBSYSTEM, "CUSTOM PROCESSING" SUBSYSTEM, RESTART AND RECOVERY SUBSYSTEM, and AUDIT/DIAGNOSTIC FUNCTIONS SUBSYSTEM.

BATCH SYSTEM is one whose data is collected and fed into the system only at the specific times when it is activated.

BEHAVIOR is a set of data flows, control flows, and processes that performs a specific function or accomplishes a result.

BEHAVIOR DIAGRAM is a level-1 data flow diagram that illustrates the data flows, control flows, and processes that perform a specific behavior within the system.

BIDIRECTIONAL DATA (or CONTROL) FLOW is a pipeline that sends and receives data (or control instructions). (See also DIALOGUE DATA FLOW and DIALOGUE CONTROL FLOW.)

BLACK BOX is a component of a system where only the input data and output data are known. Nothing is known explicitly about the internal process of transforming the input data into the output data.

BOTTOM-UP IMPLEMENTATION produces the modules containing the lowest level of detailed logic first, tests them separately, and then integrates them to form the complete system.

BOYCE/CODD NORMAL FORM (BCNF) of an object data store exists if and only if every determinant is a candidate key.

CANDIDATE KEY is a data element (or set of data elements) in a data store that has an unique value for each entry.

CARDINALITY is the number of unique entries in a data store.

CASE (COMPUTER-AIDED SOFTWARE ENGINEERING), automated tools that aid in the definition and implementation of software systems.

CLASSICAL DEVELOPMENT LIFE CYCLE is a sequential progression from the survey of user requirements, analysis, design, code, module test, subsystem test, integration or final system test to documentation and installation.

CLOSED SYSTEM is one that is influenced by its own past behavior.[F-2]

COHERENCY TEST is an examination of the input data flows and access to data stores that ensures that all the necessary inputs are available to produce the required outputs.

COMMON PROCESSING ROUTINE is procedural logic that is used as a subroutine or a function call that is invoked by the logic of a functional primitive.

CONCATENATION is combining or grouping a data item with another data item.

CONSERVATION OF DATA, PRINCIPLE OF, states that all of the inputs to a process must be sufficient to produce the outputs of the process.[W-2]

CONTEXT DIAGRAM is the highest-level data flow diagram that portrays all the net inputs and outputs of a system, but shows no decomposition.[D-2]

CONTINUOUS CONTROL FLOW is a command signal that is received continuously at every instant of time over a specific time interval and permits the continual operation of a process.

CONTINUOUS DATA FLOW represents a constant movement of data, whereby the flow of data exists at every instant of time over a specific time interval.

CONTROL is the action that initiates, terminates, or alters process behavior in carrying out an action (usually transforming input data into output data).

CONTROL FLOW is a pipeline along which a command signal is passed to control another process in the system. (See CONTINUOUS CONTROL FLOW and DISCRETE CONTROL FLOW.)

CONTROL PROCESS initiates, terminates, or alters process behavior in carrying out the transformation of input into output.

CONVERGING BRANCH is an arrow drawn on the bubble form of a state-transition diagram that points toward a state.

CRITERIA is the standard of measurement used to illustrate preference to a given alternative.[O-1]

DATA BANK refers to any collection of data, whether in the form of files, databases, or an information retrieval system.[M-4]

DATA BASE is two or more data stores with interrelating data elements that can be accessed in more than one way.[F-1] (Data base is a more precise term than data bank. Data base implies the formal techniques of data-base management.)

DATA DICTIONARY is an organized list of the composition of each data element, data flow, and data store used in the data flow diagrams, and the detailed logic specifications of commonly used processing routines and functions in a system.

DATA FLOW is pipeline along which information of known composition passes to other components of the system. (See CONTINUOUS DATA FLOW and DISCRETE DATA FLOW.)

DATA LITERAL is the exact characters, numbers, or symbols that make up the data element.

DATA STORE is a time-delayed repository of data.

DATA VALUE is a particular value of an attribute.

DECISION RULE describes a set of condition alternatives and a series of actions to be performed.[C-5]

DECISION-SUPPORT SYSTEM is an interactive computer-based system that provides managers with information to quantify and/or reduce the risks of decision making.

DECISION TABLE is a tabular expression of procedural decision situations which is characterized by one or more sets of actions that are executed based upon the state of the conditions.[C-5]

DECISION TREE is a graphic tool for portraying a hierarchy of independent conditions and the activities resulting from each valid combination of conditions.[P-1]

DEGREE is the number of key and nonkey attributes in a data store.

DEPENDENCY RELATIONSHIP exists between two object data stores if an entry from one object data store is dependent on the existence of one (or more) associated entries from another object data store.

DETERMINANT is any attribute on which some other attribute is (fully functionally) dependent.

DIALOGUE DATA (or CONTROL) FLOW is an interactive data flow (or control flow) that inputs and outputs data (or control) to a source/sink. *The time interval for the source/sink to respond to the dialogue is not a factor. The source/sink's ability to respond in a timely manner during a dialogue is a function of the technology used by the source/sink.*

DISCRETE CONTROL FLOW signals the target process at discrete moments in time. It is assumed that at all other times, the control flow is nonexistent.

DISCRETE DATA FLOW represents the movement of individual data elements at specific moments in time. It is assumed that at all other times, the data flow has no value (null value) or is undefined.

DIVERGING BRANCH is an arrow drawn on the bubble form of a state-transition diagram that points toward another state.

DOMAIN is the set of all possible values for a given attribute.

DOUBLE-HEADED DATA (or CONTROL) FLOW is used to represent a continuous data (or control) flow. (See also CONTINUOUS DATA-FLOW and CONTINUOUS CONTROL-FLOW.)

ELEMENTARY DATA FLOW ELEMENT is a data element that for the purposes of the system does not need to be further decomposed. *All elementary data elements can be defined by comments that explain the data element or by a data literal.*

ENTITY (see OBJECT).

ENTITY INTEGRITY (see INTEGRITY RULE #1).

ENTITY-RELATIONSHIP (E-R) DIAGRAM is a graphical illustration of the objects and attributes, and the relationships that exist between them.

ESSENTIAL COMPONENTS of a system are those essential activities and essential memories that must be present no matter what type of technology is used to implement the system.

ESSENTIAL CUSTODIAL ACTIVITY (ECA) is a process that establishes and maintains the essential memory, such that it is sufficiently correct to support the essential fundamental activities.

ESSENTIAL FUNDAMENTAL ACTIVITY (EFA) is a process that carries out the responses that justify the existence of the system.

ESSENTIAL MEMORY (EM) consists of the information that must be remembered to support the essential fundamental activities.

EVENT is an independent occurrence outside the system, an occurrence of time, or (in a real-time system) an occurrence detected inside the system that causes activity in, or results from, the system.

EVENT DIAGRAM is the highest-level, logical data flow diagram illustrating the association of the object data stores to each of the events to which the system has a planned response.

EVENT-DRIVEN SYSTEM is one that produces results upon the arrival of a stimulus, an occurrence of time, or the recognition of an abnormal condition within the system.

EVENT MODEL is a set of documentation consisting of the system-based diagrams (CONTEXT DIAGRAM, DATA DICTIONARY, E-R DIAGRAM, and EVENT DIAGRAM), the scenario-based documentation (SCENARIO DIAGRAMS and SCENARIO SPECS), and the behavior-based documentation (BEHAVIOR DIAGRAMS and MINI-SPECS).

EVENT SEGMENT is composed of all the processes, data flows, and data stores that are activated when a particular event occurs.

EVENT TABLE is a list of every event to which the system responds, together with the related set of essential processes and, optionally, the set of infrastructure processes associated to each event.

EXPANDED DATA FLOW DIAGRAM (DFD) consists of all of the functional primitives linked with the normal data flows shown on the various levels of data flow diagrams found in the physical model.

EXTERNAL CONTROL is enacted by a supervisory process that creates, activates, reactivates, deactivates, or suspends the operation of another process in the system.

EXTERNAL EVENTS is an independent occurrence that takes place outside the system and causes activity to take place inside the system when a stimulus is presented to the system.

FEEDBACK is output delivered as input back into the process from which the output was derived. The input can be a data flow or a control flow directive.

FEEDBACK CONTROL is a system function that compares output with a criteria.[O-1]

FEEDBACK CONTROL PROCESS is one influenced by the results of either its own past behavior or by the behavior of another process.[F-2]

FEEDBACK SYSTEM has a closed data flow structure that brings results from past actions of the system as input used to control future actions of the system.[F-2]

FINITE-STATE MACHINE (see FINITE-STATE PROCESS).

FINITE-STATE PROCESS is one whose internal logic controls the output delivered as a function of the input received and the time it was received. The process consists of a finite set of states, a finite set of input data flows, a finite set of output data flows, a mapping of the input and present state into next state, and a mapping of the input and present state into the output data flow(s).[M-3]

FIRST NORMAL FORM (1NF) occurs when all entries of the data store contain only elementary data elements and each element has precisely one value.

FOREIGN KEY is a data element that is a nonkey attribute in one data store and is used to identify a related entry in another data store where the attribute is the primary key.

FOURTH AND FIFTH NORMAL FORM: defined in an article by William Kent.[K-1]

FOURTH-GENERATION LANGUAGE is a computer programming language that is a very high-level and "user-friendly" way of specifying the results to be obtained from data usually stored in a data base management system. Elements of the language do not necessarily have to be specified in any logical order and the results can be obtained in an order-of-magnitude less time than with conventional programming languages.[M-6A]

FUNCTIONAL DECOMPOSITION is breaking a function of a system into a hierarchy of subfunctions. Each subfunction consists of a related set of activities and/or tasks that can be further factored into subfunctions until the functional primitives are reached.

FUNCTIONAL PRIMITIVE is a process within a system, consisting of a related set of tasks that perform a basic function and cannot be further decomposed.

GROUP DATA ELEMENT is a composite of elementary and/or other group data elements.

INFORMATION ENGINEERING is an integrated set of methodologies that organizes the data of an enterprise so that they can be easily accessed by end users. Information engineering entails (1) strategic requirements planning, (2) information analysis, (3) data modeling, (4) procedure formation, (5) data-use analysis, (6) distribution analysis, (7) physical data-base design, and (8) program specification synthesis.[M-6A]

INFORMATION ENGINEERING DEVELOPMENT LIFE CYCLE is a sequence of steps originating with strategic planning, which includes the top executives of the enterprise and culminates in a set of user-driven and data-processing-driven methods to obtain the desired information to meet the needs of the business. The cycle is characterized by a data-oriented approach that focuses on the data needs of the enterprise across all functional areas and emphasizes the use of data models, data normalization, and data structures that support user-written data-retrieval processes.

INFORMATION LEVEL is some measurable quantity about some aspect of the system, such as the quantity of the material being regulated or produced by a process.

INFORMATION RETRIEVAL SYSTEM is a collection or network of data and data relationships under the control of a data-management system that maintains the data and data relationships as well as monitoring and controlling access to the data.

INFRASTRUCTURE refers to a network of components that transports people, goods, and information from one location to another in a society.[M-1]

INPUT/OUTPUT PROCESSING CHART is a cross-reference table of all the system's inputs, outputs, data stores, and functional-primitive processes. This chart can be expanded to include the events to which the system responds.

INTEGRITY RULE #1 (Object Integrity) permits every data store entry in the system to have a unique identifier composed of the data store name and the primary key. No part of the primary key is permitted to have a null value. Integrity Rule #1 guarantees that every data element in the system specification has a unique address, namely,

< DATA-STORE-NAME, primary key value, ATTRIBUTE-NAME >

INTEGRITY RULE #2 (Referential Integrity) states given a data store that has a primary key made up of a single attribute taken from a domain D and if there is another data store that has an attribute A that is also defined on the domain D, then, at any given time, each value of A in the data store must be null or equal to some value V, where V is the primary-key value of some entry in a data store in the system.

INTERNAL ADMINISTRATIVE ACTIVITY (IAA) is a process that maintains orderly flow of data, manages the affairs of editing, auditing, approving, or measuring the performance of the system's activities, detects and possibly corrects operational errors, monitors and regulates the use of resources by the other processes, or keeps a historical record.

INTERNAL ADMINISTRATIVE MEMORY (IAM) is a data store that holds information about the system for either some current use of maintaining orderly flow of data, managing the system's resources, or for some future use that is not part of the normal activities or purpose of the system.

INTERNAL COMMUNICATIONS ACTIVITY (ICA) is a process that communicates the results of one activity to another or links another process to a data store. The main purpose of the activity is to *transport* data, and, in some cases, it may even *translate* data from one form of representation into another or *organize it by merging, sorting*, or *collating*.

INTERNAL COMMUNICATIONS MEMORY (ICM) is a data store that acts as a temporary storage area to synchronize communications by holding data until the receiving process is available to receive it.

INTERNAL CONTROL is a predefined plan that permits a process to perform differently based on input, time, or the variation in output as compared to the criteria.

INTERNAL EVENT is an occurrence of an anomaly that is recognized by the system, and that requires action to preserve or restore the system to a normal condition. (Most often applies to real-time and operating-system software.)

INTERSECTING DATA is an object data store or data element(s) that describes or contains fact(s) about the relationship that exists between two other object data stores.

ISOLATED STATE is a "parent" state that cannot lead into any other "parent" state and cannot be reached by any other "parent" state.

ITERATIVE PROCESS is one that is invoked externally and continuously repeats processing until some criteria is satisfied.

KEY or KEY ATTRIBUTE (see PRIMARY KEY).

LEGEND TABLE is a listing of the abbreviations used on a diagram. *Abbreviations are used for the input data flows or conditions, and output data flows or actions specified on data flow and state-transition diagrams.*

LEVEL is an elementary data element that represents the condition or state of the system.[F-2]

LOGICAL ANALYSIS is the process of separating the physical characteristics of a system (physical model) into the supporting infrastructure and the essential activities and memories that are required for the system to achieve its intended purpose.

MANY-TO-MANY RELATIONSHIP exists when more than one entry in a data store is associated with more than one entry in another data store.

MANY-TO-ONE RELATIONSHIP exists when more than one entry in a data store is associated with exactly one entry in another data store.

MECHANIZED DATA DICTIONARY is a vendor software product that permits entering, storing, and updating the characteristics of data.

METHODOLOGY is a documented set of practices and procedures that defines the development life cycle and specifies how systems are to be developed.

MINI-SPECIFICATION (mini-spec) provides a detailed explanation of the internal processing policy of a functional primitive that transforms input data into output data or simply provides output data.

MODEL is a mathematical (or logical) representation of a process, device, or concept.[S-1] It is an inexpensive replica and is used to study the behavior characteristics of a system without having to expend large amounts of time or money.

NEGATIVE-FEEDBACK SYSTEM is one that *seeks a goal* and responds as a consequence of failing to achieve the goal.[F-2]

NON-KEY ATTRIBUTE is a data element of a data store that is not the primary key (see KEY or KEY ATTRIBUTE).

NON-TRIVIAL REJECT (see TRIVIAL REJECT).

NORMALIZATION is the decomposition of complex data structures into the normal forms of data, based on the rules of data dependency.

NULL VALUE means that *no data value* is assigned to the attribute.

OBJECT is some distinct identifiable item for which the system must store data in order to perform the fundamental activities in the system.

OBJECT ANALYSIS is the process that reorganizes the essential memories of a system into the objects, attributes, and relationships between the objects and attributes of a system.

OBJECT INTEGRITY (see INTEGRITY RULE #1).

OBJECT SUBTYPE is a distinguishable category or class of an object that has its own attributes, characteristics, and/or relationships.

OBJECT SUPERTYPE is an object or object subtype that contains distinguishable categories or classes, each having its own attributes, characteristics, and/or relationships.

ONE-TO-MANY RELATIONSHIP exists when one entry in a data store is associated with more than one entry in another data store.

ONE-TO-ONE RELATIONSHIP exists when one entry in a data store is associated with only one entry in another data store.

ON-LINE SYSTEM is one that carries on direct data flows that collect input data and/or provide output information while the system is processing data.

OPEN SYSTEM is one that is not aware of its own performance.[F-2]

ORDER of a feedback system designates the number of independent levels controlled within the system.

OWNER is the manager or person funding the development and continued operation of the system. (See USER.)

PAGING is a memory management technique in which the memory address space is split into fixed length blocks called *page frames* and the software is also split into the same-size blocks called *pages*.[T-1]

PARTITIONING is a two-dimensional factoring without any hierarchical or inherent structure.

PERSISTENT STATE is a "parent" state that contains a finite-state process that can be reached from at least one other "parent" state, but cannot be left once it is reached.

POSITIVE-FEEDBACK SYSTEM generates activity whose action builds a result that generates still greater action.[F-2]

PRIMARY DOMAIN exists if there is at least one data store in the system whose primary key is a single attribute defined on the domain.

PRIMARY KEY is the particular candidate key chosen to identify a single entry from all others in a data store.

PROCESS is a defined activity or task that can be accomplished completely in a finite period of time. It is used to transform input data flows into output data flows.

PRODUCT PROPOSAL is a document that specifies the intended objective of the new product, the targeted customers, the planned operating environment (type and size of computer, memory, and peripheral options), and the structured specifications, including the events that the product will support.

PROJECT DEVELOPMENT LIFE CYCLE is a well-defined consistent set of activities that is used to translate the customer's requirements into a working system, and contains checkpoints for management control.

QUERY is a request for data made of a data base.

REAL TIME is defined as the condition to operate concurrently with a regular or predictable time relationship with independent real-world events.

REAL-TIME SYSTEM is one that contains processes that operate concurrently with a regular or predictable time relationship with external independent, or temporal events.

RECURSIVE PROCESS is one that, during an execution cycle, invokes itself and initiates another execution cycle.

REFLECTING BRANCH is an arrow drawn on the bubble form of a state-transition diagram that originates and terminates in the same state.

REFLECTIVE RELATIONSHIP is one defined on a single object data store.

REJECT is a special data element name that is not defined in the data dictionary and is used everywhere data is rejected from a process. (See also TRIVIAL REJECT.)

RELATED EVENTS are two or more events that use the same behavior, process, or data to produce their response. (Sharing the same common processing routine or function is not sufficient to qualify two events as related.)

REPETITIVE is the multiple occurrences of data item(s).

RESPONSE is an output data flow from the system to a sink, or some internal activity in the system, and is the consequence of the system recognizing that an event occurred.

RESPONSIBLE USER is the professional responsible for specifying the operating procedures of the system from a system user's perspective. (See USER.)

RESTRICTIONS are those specifications that stipulate the internally or externally imposed guidelines that bound the problem or dictate operating constraints.[O-1]

RIGOR is defined as the exactness implicit in a precise and complete statement of the systems requirements.[O-1]

ROOT NODE is the single, highest-level node in a decision tree from which the entire tree is generated.

SCENARIO is a set of input and output data (or control) flows, and behaviors of a system brought about by an instance of an event.

SCENARIO DIAGRAM is a data flow diagram that illustrates a single scenario of an event.

SECOND NORMAL FORM (2NF) occurs when all nonkey attributes provide a fact about the whole key and not just part of the key.

SELECTION is a choice of one of several data elements or literal values for a data item.

SEMAPHORE is a technique modeled as a data store on the data flow diagram and indicates whether a process is active or inactive. The data store is *not* included in the data dictionary; it contains only an ON/OFF designator for each associated process.

SERVICE DESCRIPTION is a document that specifies in narrative text the functions, features, and services provided by the product.

SET (CODASYL Data-Base Task-Group definition) is a relationship of a two-level hierarchy of records made up of a "parent" or owner record and a "child" or member record. Each record contains one or more data elements.

SOFTWARE consists of the computer programs and associated documentation required to develop, operate, and maintain the programs.[B-1]

SOFTWARE ENGINEERING is the practical application of scientific knowledge in the design and construction of computer programs and the associated documentation required to develop, operate, and maintain the programs.[B-1]

SOFTWARE ENGINEERING DEVELOPMENT LIFE CYCLE is a sequence of steps with certain identifiable deliverables that signal the end of one step and the beginning of another. The cycle is characterized by a process-oriented approach that focuses on the system being developed, emphasizing reusable code, peer reviews, checkpoints, and metrics for management control of productivity, quality, and costs.

SPONSOR is a person or organization that provides the funds to develop a new product, feature, or enhancement.

SPOTLIGHT EFFECT focuses management's attention on a specific aspect of the organization until the results or expectations are improved.

STACK is a last in, first out (LIFO) queue of data elements.

STARVING THE BUBBLES is the process that replaces the group data elements and data store with the minimum elementary data elements that are absolutely necessary for each functional primitive.

STATE is an operating circumstance that has a predefined result that is triggered by an input at a particular moment in time.

STATE-TRANSITION TABLE lists all possible input data flows, process states, and the respective response made by the process when a particular input is received in each state.

STIMULUS is an input data flow from a source and is the consequence of the occurrence of an external event.

STRUCTURED DEVELOPMENT LIFE CYCLE consists of a sequential progression from survey, structured analysis, structured design, structured programming, structured testing, and installation.[D-2]

STUB MODULE is a program module that contains simple code that responds with some fixed constant values each time the module is invoked.

SUBRELATION is a relationship that is used as an object in another relationship.

SUBTYPE (see OBJECT SUBTYPE).

SUCCESSIVE DECOMPOSITION is a hierarchical factoring where each "parent" node contains processes that are either functional primitives (i.e., cannot be further decomposed) or can be decomposed into a network of "child" processes, data flows, and data stores that are illustrated on the next lower-level diagram.

SUPERTYPE (see OBJECT SUPERTYPE).

SUPERVISORY PROCESS initiates, terminates, or otherwise controls the activities or operations of a target process.

SYSTEM is an association of interdependent devices, people, rules, and/or procedures organized to form an integral whole to achieve a common purpose.[W-4]

SYSTEM USER is the person(s) that interacts with the system. (See USER.)

SYSTEMS ANALYSIS is the process of examination, identification, and evaluation of the components and interrelationships involved in systems development effort; definition of problems, objectives, requirements, priorities, and constraints of systems, plus identification of cost, benefit, and schedule estimates for potential solutions.[W-4]

TABLE is a collection of data items organized by a data element(s) known as the key that is held somewhere in the system for reference during processing.

TARGET PROCESS is a subordinate process that is activated, deactivated, or otherwise controlled by one or more independent supervisory processes.

TEMPORAL EVENT is an occurrence of time that causes activity within the system and usually results in an output data flow.

THIRD NORMAL FORM (3NF) occurs when all nonkey attributes are functionally dependent on the primary key and are independent of each other. (See also BOYCE/CODD NORMAL FORM.)

TIME-DEPENDENCY CHART is a graphical illustration of the sequence, concurrency and duration requirements, and/or restrictions among processes in a system. *The chart is used to show the relationships among processes with respect to the amount of time a process is required to be active as well as specifying any time interval whereby a process is required to be inactive or specifying overall system behavior with respect to time.*

TIMEOUT is a temporal event that initiates action if no anticipated action or response from a process or source is received within a specific time interval.

TOP-DOWN IMPLEMENTATION produces higher-level modules first. The higher-level modules call upon modules that are planned to perform tasks at a lower-level of detail. Initially "stubs" are substituted for the lower-level modules during the initial testing and while the lower-level modules are being developed.

TRANSACTION- SUPPORT SYSTEM is an interactive computer-based system that provides the services of collecting, validating, editing, and displaying data upon a request from a user.

TRANSIENT STATE is a "parent" state that contains a finite-state process that can lead into at least one other "parent" state, but cannot be reached once it is left.

TRIVIAL REJECT is a data flow that fails to meet the criteria for processing in a functional primitive and is discarded without any further processing. NONTRIVIAL REJECTS are shown as normal data flows with unique data flow names.

UNANTICIPATED QUERY is a request for information by the user that was not expected to be provided by the system when it was developed.

UNCERTAINTY PRINCIPLE IN DATA PROCESSING states that the act of providing what an end user says he needs, changes his perception of those needs.[M-6A]

USER is a person, persons, or organizational entity expected to derive benefits from the development of a system; a client of systems development effort.[W-4] (See also SYSTEM USER, RESPONSIBLE USER, and OWNER.)

REFERENCES

B-1. Boehm, B.W. "Software Engineering," *IEEE Transactions on Computers,* **C-25**, No. 12 (December 1976), pp. 1226-1241. Reprinted in E.N. Yourdon, ed., *Classics in Software Engineering.* Englewood Cliffs, NJ: Prentice-Hall, Yourdon Press Series, 1979, pp. 325-361.

B-2. Boehm, B.W. *Software Engineering Economics.* Englewoods Cliffs, NJ: Prentice-Hall, 1981.

B-3. Bohm, C., and G. Jacopini. "Flow Diagrams, Turing Machines and Languages with Only Two Formation Rules," *Communications of the ACM,* Vol. **9**, No. 5 (1966), pp. 366-371. Reprinted in E.N. Yourdon, ed., *Classics in Software Engineering.* Englewood Cliffs, NJ: Prentice-Hall, Yourdon Press Series, 1979, pp. 13-25.

C-1. Caine, S., and E.K. Gordon. "PDL - A Tool for Software Design," *AFIPS Proceedings of the 1975 National Computer Conference,* Vol. **44** (Montvale, NJ: AFIPS Press, 1975), pp 271-76. Copyright (c) 1975 by the American Federation of Information Processing Societies, Inc. Reprinted in E.N. Yourdon, ed., *Writings of the Revolution, Selected Readings in Software Engineering.* Englewood Cliffs, NJ: Prentice-Hall, Yourdon Press Series, 1979, pp. 13-25.

C-2. Chen, P.P. " The Entity-Relationship Model -- Towards a Unified View of Data," *ACM Transactions on Database Systems*, Vol. **1**, No. 1 (March 1976), pp. 9-36.

C-3. Chen, P.P. *The Entity Relationship Approach to Logical Data Base Design.* Wellesley, MA: Q.E.D. Information Science Inc., 1977.

C-4. Chen, P.P., ed. *E-R Approach to Systems Analysis & Design.* New York: North Holland, 1979.

C-5. Codasyl Task Group. *A Modern Appraisal of Decision Tables*, A Codasyl Report. New York: ACM, 1982.

C-6. Codd, E.J. "Further Normalization of the Data Base Relational Model," in *Data Base Systems,* Courant Computer Science Symposia Series, Vol. **6**. Englewood Cliffs, NJ: Prentice-Hall, 1975.

C-7. Codd, E.J. "Recent Investigations into Relational Data Base Systems," *Proceedings of IFIP Congress* (1974).

D-1. Date, C.J. *An Introduction to Database Systems,* fourth ed. Reading, MA: Addison-Wesley, 1984.

D-2. DeMarco, T. *Structured Analysis and System Specification.* Englewood Cliffs, NJ: Prentice-Hall, Yourdon Press Series, 1978.

D-3. Dijkstra, E.W. "Cooperating Sequential Processes," in F. Genuys, ed., *Programming Languages*. New York: Academic Press, 1968.

D-4. Dijkstra, E.W. "Go To Statements Considered Harmful," *Communications of the ACM*, Vol. **11**, No. 3 (May 1966), pp. 147-148. Reprinted in E.N. Yourdon, ed., *Classics in Software Engineering.* Englewood Cliffs, NJ: Prentice-Hall, Yourdon Press Series, 1979, pp. 29-33.

F-1. Flavin, M. *Fundamental Concepts of Information Modeling.* Englewood Cliffs, NJ: Prentice-Hall, Yourdon Press Series, 1981.

F-2. Forrester, J.W. *Principles of Systems*. Cambridge, MA: Wright-Allen Press, 1968.

G-1. Gane, C., and T. Sarson. *Structured Systems Analysis: Tools and Techniques.* Englewood Cliffs, NJ: Prentice Hall, 1979.

G-2. Gill, A. *Introduction to the Theory of Finite State Machines*. New York: McGraw-Hill, 1962.

H-1. Hopper, Grace M. "Information: Don't Take It at Face Value," *Viewpoint*, Vol. **24**, Number 34, Springfield, NJ: AT&T Technologies Inc., August 13, 1984.

H-2. Huffman, D.A. "The Synthesis of Sequential Switching Networking," *Journal of Franklin Institute*, Vol. **257** (1954), pp. 161-190.

K-1. Kent, W. "A Simple Guide to Five Normal Forms in Relational Database Theory," *Communications of the ACM*, Vol. **26**, No. 2 (February 1983), pp. 120-125.

K-2. Kleene, S.C. "Representation of Events in Nerve Nets and Finite Automata," in *Automata Studies.* Princeton, NJ: Princeton University Press, 1956, pp. 3-41.

M-1. McMenamin, S.M., and J.F. Palmer. *Advanced Structured Analysis Workshop,* 4.1. ed. Course offered by Yourdon, Inc., 1981.

M-2. McMenamin, S.M., and J.F. Palmer. *Essential Systems Analysis.* Englewood Cliffs, NJ: Prentice-Hall, Yourdon Press Series, 1984.

M-3. McMenamin, S.M., and J.F. Palmer. "The Transition Between Analysis and Design." Paper presented at Guide 54, May 1982.

M-4. Martin, J., *Managing the Data-Base Environment*. Englewood Cliffs, NJ: Prentice-Hall, 1983.

M-5. Martin, J. *Recommended Diagramming Standards for Analysts and Programmers*, second ed. Carnforth, Lancashire, England: Savant Institute, 1985.

M-6A. Martin, J., and C. Finkelstein. *Information Engineering, Volumes I and II.* Carnforth, Lancashire, England: Savant Institute, 1981.

M-6B. Martin, J. *Information Engineering, Volumes 1 to 4.* Carnforth, Lancashire, England: Savant Institute, 1987.

M-7. Martin, J., and C. McClure. *Diagramming Techniques for Analysts and Programmers*. Englewood Cliffs, NJ: Prentice-Hall, 1985.

M-8. Miller, G.A. "The Magical Number Seven, Plus or Minus Two: Some Limits on Our Capacity for Processing Information," *The Psychological Review,* Vol. **63**, No. 2 (1956), pp. 81-97. Reprinted in E.N. Yourdon, ed., *Writings of the Revolution, Selected Readings in Software Engineering*. Englewood Cliffs, NJ: Prentice-Hall, Yourdon Press Series, 1982, pp. 443-460.

M-9. Moore, E.F. "Gedanken-Experiments on Sequential Machines," *Automata Studies.* Princeton, NJ: Princeton University Press, 1956, pp. 129-153.

M-10. Moret, B. "Decision Trees and Diagrams," *Computer Surveys*, Vol. **14**, No. 4 (December 1982), pp. 593-623.

M-11. Myers, G.J. *The Art of Software Testing.* New York: John Wiley & Sons, 1979.

O-1. Optner, S.L. *Systems Analysis.* Englewood Cliffs, NJ: Prentice-Hall, 1960, 1968.

P-1. Page-Jones, M. *The Practical Guide to Structured Systems Design.* Englewood Cliffs, NJ: Prentice-Hall, Yourdon Press Series, 1980.

R-1. Ross, D.T., and K.E. Schoman, Jr. "Structured Analysis for Requirements Definition," *IEEE Transactions on Software Engineering*, Vol. **SE-3**, No. 5. (Jan., 1977), pp. 6-15. Reprinted in E.N. Yourdon, ed., *Classics in Software Engineering*. Englewood Cliffs, NJ: Prentice-Hall, Yourdon Press Series, 1979, pp. 365-386.

S-1. Shannon, C.E. "A Mathematical Theory of Communications," *Bell System Technical Journal*, Vol. **27** (1948), pp. 379-420, 623-656.

S-2. Sippl, C.J. *Computer Dictionary and Handbook*. New York: Howard W. Sams & Co., 1966.

S-3. Snider, M.L., *"Software Engineering"*. Seminar offered by Battelle Memorial Institute, Seattle, Washington, September 1981.

S-4. Stevens, W., G. Myers, and L. Constantine. "Structured Design," *IBM Systems Journal*, Vol. **13**, No. 2 (May 1974), pp. 115-139. Reprinted in E.N. Yourdon, ed., *Classics in Software Engineering*. Englewood Cliffs, NJ: Prentice-Hall, Yourdon Press Series, 1979, pp. 207-232.

T-1. Tsichritzis, D.C., and P.A. Bernstein. *Operating Systems*. New York: Academic Press, 1974.

T-2. Turing, A.M. "On Computable Numbers with an Application to the Entscheidungsproblem," *Proceedings of the London Math. Society*, Ser. **2**, Vol. **42** (1936-1937), pp. 230-265.

W-1. Ward, P.T. *Systems Development Without Pain: A User's Guide to Modeling Organizational Patterns.* Englewood Cliffs, NJ: Prentice-Hall, Yourdon Press Series, 1984.

W-2. Ward, P.T., and S.J. Mellor. *Structured Development for Real-Time Systems, Volumes 1, 2, and 3.* Englewood Cliffs, NJ: Prentice-Hall, Yourdon Press Series, 1986.

W-3. Weinberg, G.M. *Rethinking Systems Analysis and Design*. Boston: Little, Brown & Co., 1982.

W-4. Weinberg, V. *Structured Analysis*. Englewood Cliffs, NJ: Prentice-Hall, Yourdon Press Series, 1978.

Y-1. Yourdon, E. *Managing the System Life Cycle.* Englewood Cliffs, NJ: Prentice-Hall, Yourdon Press Series, 1982.

Y-2. Yourdon, E.N., ed. *Classics in Software Engineering*. Englewood Cliffs, NJ: Prentice-Hall, Yourdon Press Series, 1979.

Y-3. Yourdon, E.N., ed. *Writings of the Revolution, Selected Readings in Software Engineering*. Englewood Cliffs, NJ: Prentice-Hall, Yourdon Press Series, 1982.

Y-4. Yourdon, E., and L.L. Constantine. *Structured Design: Fundamentals of a Discipline of Computer Program and System Design.* Englewood Cliffs, NJ: Prentice-Hall, Yourdon Press Series, 1978.

Z-1. Zloof, M.M. "Query-by-Example: A Data Base Language," *IBM Systems Journal,* Vol. **16**, No. 4 (1977).

Z-2. Zimmermann, A.J. "Physical Objects in Logical Data Flow Diagrams," *The Yourdon Report; The Journal of Structured Systems Development.* New York: Yourdon Inc.

INDEX

Bold page numbers reference pages where a formal definition is given.

D

E

K

L

M

N

O

S